NUCLEIC ACIDS

IN

INNATE IMMUNITY

NUCLEIC ACIDS
IN
INNATE
IMMUNITY

EDITED BY
KEN J. ISHII
SHIZUO AKIRA

CRC Press
Taylor & Francis Group
Boca Raton London New York

CRC Press is an imprint of the
Taylor & Francis Group, an **informa** business

CRC Press
Taylor & Francis Group
6000 Broken Sound Parkway NW, Suite 300
Boca Raton, FL 33487-2742

First issued in paperback 2019

© 2008 by Taylor & Francis Group, LLC
CRC Press is an imprint of Taylor & Francis Group, an Informa business

No claim to original U.S. Government works

ISBN-13: 978-1-4200-6825-2 (hbk)
ISBN-13: 978-0-367-38729-7 (pbk)

Library of Congress Cataloging-in-Publication Data

Nucleic acids in innate immunity / [edited by] Ken J. Ishii and Shizuo Akira.
    p. ; cm.
  Includes bibliographical references and index.
  ISBN 978-1-4200-6825-2 (alk. paper)
    1. Natural immunity. 2. Nucleic acids. I. Ishii, Ken J. II. Akira, S. III. Title.
  [DNLM: 1. Immunity, Natural--physiology. 2. Autoimmunity. 3.
DNA--immunology. 4. RNA--immunology. 5. Receptors, Pattern
Recognition--immunology. 6. Self Tolerance. QW 541 N964 2008]

  QR185.2.N82 2008
  616.07'9--dc22
                                      2007047405

Visit the Taylor & Francis Web site at
http://www.taylorandfrancis.com

and the CRC Press Web site at
http://www.crcpress.com

Contents

Section I
Roles of Nucleic Acids in Immunity

Section II
Mechanisms and Therapeutic Applications of Immunomodulatory DNA

Preface

All living organisms are continuously exposed to foreign entities including food, microorganisms, and unnecessary self-metabolites, thus creating a need to discriminate dangerous non-self from safe self entities, particularly life-threatening microorganisms that invade the body. The vertebrate immune system has evolved two arms of defense against invading pathogens: innate (natural) immunity and adaptive (acquired) immunity. A great deal of immunology research revealed that adaptive immunity has two sophisticated systems designated self- and non-self-discrimination by which T and B cells, both of which express highly diverse antigen receptors generated through DNA rearrangement, are thereby able to respond to a wide range of potential antigens.

In contrast, innate immunity had been regarded as a relatively non-specific system whose two main roles were engulfing and destroying pathogens. The innate system also triggers pro-inflammatory responses and is involved in antigen presentations to prime adaptive immune responses.

Recent studies have shown that the innate immune system has a greater degree of specificity than was previously thought. The system has a highly developed ability to discriminate between self and foreign entities, including microorganisms and unnecessary self molecules including proteins, and lipids, as well as nucleic acids that constitute the main topic of this book. Why and how the innate immune system discriminates self and non-self nucleic acids will be discussed in particular detail by the groups of Wagner and Stacey (DNA) and Kariko, Diebold, Hornung, Bauer, and Colonna (RNA).

This discrimination relies, to a great extent, on pattern-recognition receptors (PRRs) including Toll-like receptors (TLRs), Nod-like receptors (NLRs) and the recently described RIG-I-like receptors (RLRs) that play a crucial role in early host defenses against invading pathogens, as described in the chapters contributed by the groups of Wilson, Kawai, Kato, Kaisho and Krug. These germ line-encoded PRRs are expressed constitutively on both immune and non-immune cells, and recognize conserved microbial components known as pathogen-associated molecular patterns (PAMPs).[1] After recognition, each PRR activates specific signaling pathways, leading to robust but highly defined innate immune responses, followed by protective adaptive (antigen-specific) immune responses to pathogens.

PAMPs have triggered considerable interest in nucleic acids in the field of immunology. While nucleic acids such as DNA and RNA are essential components of all living organisms, accumulating evidence over the last several decades suggests that nucleic acids function as essential ultimate units of life and also stimulate the immune system when they are released from pathogens.[2,3] Their connection to pathogens attracted little attention in the past, but the link is in the limelight after the recent discovery of TLRs.[4,5]

Structure- and sequence-dependent immune recognitions of nucleic acids by TLRs were shown to play an important role in both innate and adaptive immune responses to infectious organisms, including bacteria, viruses, and parasites.[6,7] Novel therapeutics include nucleic acid-based agonists and antagonists via TLR-mediated immunomodulation, and including the use of CpG DNA as a potent TLR9 agonist are under development for multiple applications to prevent or treat infectious diseases, allergic disorders, and cancers (detailed by Krieg, Verthelyi, Klinman and Broide).

On the other hand, the innate immune system that fights infection also seems to have an important role in clearing unnecessary or abnormal host molecules including nucleic acids. In fact, the system possesses specialized sets of genes including TLRs that facilitate clearance in cases of trauma, tumor, and autoimmune diseases,[8,9] as summarized in the chapter by Rothstein. This role of the innate immune system is important. Initial dogma dictating that the system including TLRs discriminates infectious non-self from non-infectious self nucleic acids has been challenged by findings that host (self) nucleic acids are no longer inert in the immune system under certain conditions. Thus, the one or more elements within DNA and RNA (sequence, modification, structure) recognized by the innate immune system constitute an important issue that must be clarified to further explain this system. This book reviews recent advances in our understanding of the innate immune recognition of nucleic acids, and describes the resulting immune modulation through TLR-dependent or -independent pathways.

We would like to acknowledge some of the pioneering works that appeared long before the studies in this book were published. In 1963, two independent groups, including one led by Alick Isaacs who discovered interferon, reported that DNA and RNA derived from pathogens or host cells activated chicken and mouse fibroblasts to produce interferon (IFN).[11] The other pioneering work in 1984 by Tokunaga and colleagues showed that a DNA fraction isolated from BCG activated both human and mouse non-B, non-T cells to produce type I IFNs.[12] They also demonstrated that bacterial, and not mammalian, DNAs are immunostimulatory and can be reproduced by short, single-stranded (ss) oligodeoxyribonuleotides (ODNs) containing palindromic GC-rich sequences.[13]

Finally, we hope that our book will provide insight into the new areas of immunology, nucleic acid recognition, and regulation by innate immune systems.

Ken J. Ishii and Shizuo Akira
Research Institute for Microbial Diseases
Osaka University

REFERENCES

1. C. A. Janeway, Jr. and R. Medzhitov, *Annu .Rev. Immunol.* 20, 197 (2002).
2. A. Isaacs, R. A. Cox, and Z. Rotem, *Lancet* 2, 113 (1963).
3. T. Tokunaga, T. Yamamoto, and S. Yamamoto, *Jpn. J. Infect. Dis.* 52, 1 (1999).
4. R. Medzhitov and C. A. Janeway, Jr., *Science* 296, 298 (2002).
5. S. Akira and K. Takeda, *Nat. Rev. Immunol.* 4, 499 (2004).
6. H. Wagner, *Trends Immunol.* 25, 381-386 (2004).
7. K. J. Ishii and S. Akira, *Trends Immunol.* 27, 525 (2006).
8. P. Matzinger, *Science* 296, 301 (2002).

9. B. Beutler, *Nature* 430, 257 (2004).
10. Z. Rotem, R. A. Cox, and A. Isaacs, *Nature* 197, 564 (1963).
11. K. E. Jensen, A. L. Neal, R. E. Owens, and J. Warren, *Nature* 200, 433 (1963).
12. T. Tokunaga et al., *J. Natl. Cancer Inst.* 72, 955 (1984).
13. S. Yamamoto et al., *J. Immunol.* 148, 4072 (1992).

Editors

Dr. Ken J. Ishii is currently an associate professor at the Research Institute for Microbial Diseases of Osaka University in Japan. He is also affiliated with the Japan Science and Technology Agency where he serves as a group leader of the Akira Innate Immunity Project. Dr. Ishii graduated from the medical school of Yokohama City University, also in Japan. After years of clinical work, he joined the United States Food and Drug Administration where he trained as an immunologist and reviewer for vaccine clinical trials. Since then, his research has focused on how the immune system can recognize nucleic acids, including RNA and DNA, and the physiological relevance of RNA and DNA to infectious diseases, allergies, cancers, and autoimmune diseases.

Dr. Ishii is also interested in developing nucleic acid-based immunotherapies including vaccines against infectious, allergic, neoplastic, and autoimmune diseases. He was singled out in September 2007 as a Rising Star by Thomson Scientific, Inc. (Institute for Scientific Information) after achieving the highest percentage increase in total citations in the immunology field. He was recently named an expert and member of the Independent Peer Review Panel of the National Institutes of Health in 2007.

Dr. Shizuo Akira is currently a professor at the Research Institute for Microbial Diseases of Osaka University, and a group leader of the Akira Innate Immunity Project of the Japan Science and Technology Agency. He was recently appointed a director of the Frontier Immunology Research Center at Osaka University.

Dr. Akira is a leading immunologist and has made many contributions establishing the importance of innate immunity. He has received many awards in Japan and abroad, including the Robert Koch Prize. His work on Toll-like receptors is highly regarded worldwide, and he has lectured at a number of international conferences including Keystone Symposia, Gordon Conferences, Nobel Forum, and meetings dealing with immunology and infectious diseases. He has authored over 600 papers, and is one of the most cited immunologists. He was recognized in 2006 and 2007 by the Institute for Scientific Information as the scientist who published the greatest number of "hot papers" during the preceding two years.

Contributors

Andrea Ablasser
Division of Clinical Pharmacology
Department of Internal Medicine
University of Munich
Munich, Germany

Stefan Bauer
Institute for Immunology
Philipps University Marburg
Marburg, Germany

David Broide
Department of Medicine
University of California
San Diego, California, USA

Jungwoo Choe
Department of Life Sciences
University of Seoul
Seoul, Korea

Francis Clark
Institute for Molecular Bioscience and
Cooperative Research Centre for
  Chronic Inflammatory Diseases
University of Queensland
Brisbane, Australia

Marco Colonna
Department of Pathology and
  Immunology
School of Medicine
Washington University
St. Louis, Missouri, USA

Debbie Currie
Section of Retroviral Research
Center for Biologics Evaluation and
  Research
United States Food and Drug
  Administration
Bethesda, Maryland, USA

Sandra Diebold
Peter Gorer Department of
  Immunobiology
King's College and Guy's Hospital
London, UK

Chiaki Fujimoto
Laboratory of Immunology
National Eye Institute
National Institutes of Health
Bethesda, Maryland, USA

Igal Gery
Laboratory of Immunology
National Eye Institute
National Institutes of Health
Bethesda, Maryland, USA

Leonid Gitlin
Department of Pathology and
  Immunology
School of Medicine
Washington University
St. Louis, Missouri, USA

Tobias Haas
Institute for Medicine, Microbiology,
  Immunology, and Hygiene
Technical University of Munich
Munich, Germany

Svetlana Hamm
Institute for Medicine, Microbiology,
  Immunology, and Hygiene
Technical University of Munich
Munich, Germany

Gunther Hartmann
Division of Clinical Pharmacology
University Hospital
University of Bonn
Bonn, Germany

Antje Heit
Institute for Medicine, Microbiology,
  Immunology, and Hygiene
Technical University of Munich
Munich, Germany

Julie L. Himes
Coley Pharmaceutical Group, Inc.
Wellesley, Massachusetts, USA

Veit Hornung
Division of Clinical Pharmacology
Department of Internal Medicine
University of Munich
Munich, Germany

Katsuaki Hoshino
Laboratory for Host Defense
Riken Research Center for Allergy and
  Immunology
Yokohama, Japan

Tsuneyasu Kaisho
Laboratory for Host Defense
Riken Research Center for Allergy and
  Immunology
Yokohama, Japan

Katalin Karikó
School of Medicine
University of Pennsylvania
Philadelphia, Pennsylvania, USA

Hiroki Kato
Department of Host Defense
Research Institute for Microbial
  Diseases
Osaka University
Osaka, Japan

Taro Kawai
Department of Host Defense
Research Institute for Microbial
  Diseases
Osaka University
Osaka, Japan

Dennis M. Klinman
Section of Retroviral Research
Center for Biologics Evaluation and
  Research
United States Food and Drug
  Administration
Bethesda, Maryland, USA

Arthur M. Krieg
Coley Pharmaceutical Group, Inc.
Wellesley, Massachusetts, USA

Anne Krug
Department of Medicine
Technical University of Munich
Munich, Germany

Ann Marshak-Rothstein
Department of Microbiology
School of Medicine
Boston University
Boston, Massachusetts, USA

Montserrat Puig
Division of Therapeutic Proteins
Center for Drug Evaluation
United States Food and Drug
  Administration
Bethesda, Maryland, USA

Wolfgang Reindl
Department of Medicine
Technical University of Munich
Munich, Germany

Tara L. Roberts
Institute for Molecular Bioscience and
Cooperative Research Centre for
  Chronic Inflammatory Diseases
University of Queensland
Brisbane, Australia

Frank Schmitz
Institute for Medicine, Microbiology,
  Immunology, and Hygiene
Technical University of Munich
Munich, Germany

Hidekazu Shirota
Section of Retroviral Research
Center for Biologics Evaluation and
  Research
United States Food and Drug
  Administration
Bethesda, Maryland, USA

Mark J. Shlomchik
Department of Laboratory Medicine
School of Medicine
Yale University
New Haven, Connecticut, USA

Katryn J. Stacey
Institute for Molecular Bioscience and
Cooperative Research Centre for
  Chronic Inflammatory Diseases
University of Queensland
Brisbane, Australia

Takahiro Sugiyama
Laboratory for Host Defense
Riken Research Center for Allergy and
  Immunology
Yokohama, Japan

Osamu Takeuchi
Department of Host Defense
Research Institute for Microbial
  Diseases
Osaka University
Osaka, Japan

Daniela Verthelyi
Division of Therapeutic Proteins
Center for Drug Evaluation
United States Food and Drug
  Administration
Bethesda, Maryland, USA

Hermann Wagner
Institute for Medicine, Microbiology,
  Immunology, and Hygiene
Technical University of Munich
Munich, Germany

Drew Weissman
School of Medicine
University of Pennsylvania
Philadelphia, Pennsylvania, USA

Ian A. Wilson
Department of Molecular Biology
Skaggs Institute for Chemical Biology
Scripps Research Institute
La Jolla, California,USA

Greg R. Young
Institute for Molecular Bioscience and
Cooperative Research Centre for
  Chronic Inflammatory Diseases
University of Queensland
Brisbane, Australia

I

---

*Roles of Nucleic Acids
in Immunity*

1 Structural Analysis of Toll-Like Receptors

Jungwoo Choe and Ian A. Wilson

ABSTRACT

Toll-like receptors (TLRs) represent a primary line of defense against invading pathogens, including bacteria, viruses, fungi, and parasites.[1,2] Recognition of conserved microbial components by these receptors triggers innate immune responses that result in inflammation, antiviral responses, and maturation of dendritic cells, and ultimately leads to the clearance of the infectious agents. TLRs contain extracellular domains with leucine-rich repeats for ligand binding and intracellular Toll/interleukin-1 receptor (TIR) domains for signaling. The ligand binding domain of a TLR recognizes various pathogen-associated molecules, including lipopeptide (ligand of TLR2), double-stranded RNA (TLR3), lipopolysaccharide (TLR4), flagellin (TLR5), single-stranded RNA (TLR7), and unmethylated CpG DNA (TLR9).[3,4] The intracellular TIR domain interacts with TIR domains in adaptor molecules such as MyD88, TIRAP, TRIF, and TRAM to initiate pathogen-specific immune responses.[5,6]

INTRODUCTION

All living organisms face the challenge of defending themselves against microorganisms in the environment. Although the adaptive immune system has been subject to considerable study, the contribution of the innate immune system to defense against microbial pathogens has been less well appreciated. Innate immunity is often regarded as relatively non-specific. However, recent studies have shown that the innate immune system has a much greater specificity than previously thought, and can indeed respond to specific antigens. The first step in innate immunity is the recognition of microorganisms by receptors that recognize specific molecules that are present in the pathogen but not in self tissues. In mammals, a family of Toll-like receptors (TLRs) plays a central role in this discrimination, and ten human TLRs have been identified to date (Figure 1.1).

These pathogen-associated molecules are likely to be essential for the survival of microbes and therefore cannot change rapidly under innate immunity selection pressure. A TLR contains a number of leucine-rich repeats (LRRs) in its ectodomain, and a Toll/interleukin-1 receptor (TIR) domain in the cytoplasmic region. The TLR ectodomain is responsible for ligand binding, and the TIR domain recruits cytoplasmic adapter proteins that also contain TIR domains such as MyD88, TIRAP, TRIF, and TRAM to carry signals into the cytoplasm. Most TLRs activate the MyD88-dependent pathway to induce a core response such as inflammation. However, individual TLRs

3

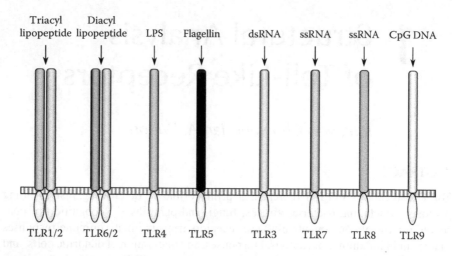

FIGURE 1.1   Human Toll-like receptors and their representative ligands.

can also induce immune responses tailored to a specific microbial infection.[7,8] In addition, TLRs control adaptive immune responses via multiple dendritic cell functions.[9]

Knowledge of the innate immune system, particularly related to TLRs as key receptor molecules and downstream adaptor molecules involved in the signaling pathway, has provided the research community with novel targets for drug design.[10,11] Potential therapeutic agents include: (1) small molecule antagonists or neutralizing antibodies to TLR ectodomains to block ligand binding; (2) small molecule agonists such as imiquimod that can exert adjuvant effects for immunostimulation; and (3) small molecules that can interfere with the downstream signaling cascade, such as the TIR–TIR domain interaction between TLR and adaptor molecules.[12] These small molecules can be used to interrupt specific TLR signaling pathways while leaving other beneficial TLR signaling intact because of the different adaptor usage and signaling specificity of TLRs.

STRUCTURE OF HUMAN TOLL-LIKE RECEPTOR 3 (TLR3)

OVERALL STRUCTURE

The human TLR3 ectodomain structure at 2.1 Å revealed a large horseshoe-shaped solenoid structure assembled from 23 LRRs forming a right-handed solenoid[13] (Figure 1.2). The inner and outer diameters of the horseshoe are about 50 and 80 Å, respectively, thus representing the largest structure of an LRR-containing protein known to date. The inner concave surface forms a continuous β sheet composed of 25 parallel β strands. Twenty-three of the 25 parallel β strands come from the LRRs in the TLR3 ectodomain, and two come from the N- and C-terminal regions, respectively.

The concave surface contains mostly β strands, while the outer convex surface contains more diverse secondary structures such as loops, α helices, 3_{10} helices, and β strands. The outer surface of TLR3 also contains TLR3-specific insertions, with segments LRR12 and LRR20 containing insertions longer than 10 residues that may

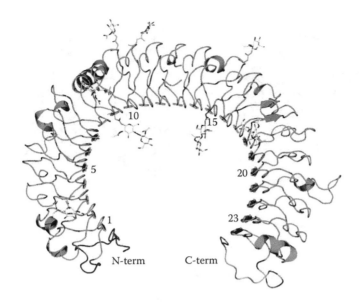

FIGURE 1.2 Structure of human TLR3. Leucine-rich repeats (LRRs) are numbered 1 through 23, and observed glycosylation sites are shown in detail.

be involved in TLR3 function. Superposition of the 23 LRRs revealed that the residues that form the inner concave surface are structurally similar, with RMSDs ranging from 0.3 to 0.8 Å. The outer surface adopts more diverse structures, with RMSDs ranging from 1.0 to 2.5 Å when equivalent Cα atoms are used for the superposition.

LEUCINE-RICH REPEATS

As do other TLRs, the LRRs of the TLR3 ectodomain contain the typical 24-residue LRR motif consisting of xLxxLxLxxNxLxxLxxxxFxxLx, where L represents hydrophobic residues (mostly leucines), F a conserved phenylalanine, and N a conserved asparagine. Fourteen of the 23 LRRs in TLR3 follow the 24-residue motif, and only LRR12 and LRR20 have insertions longer than five residues. The lengths and positions of insertions in LRR motifs vary from TLR to TLR and can be important in determining the ligand specificity of each TLR.[14] Hence, the location and structure of LRR12 and LRR20 in TLR3 may be important for its function and will be discussed later.

Within each LRR, seven conserved hydrophobic residues at LRR motif positions 2, 5, 7, 12, 15, 20, and 23 point in toward the solenoid and form a tight hydrophobic core that contributes to the structural stability of the elongated solenoid structure (Figure 1.3A). The side chains of the asparagine residues at LRR motif position 10 form extensive hydrogen bonding networks linking neighboring LRR motifs, and further contribute to the stability of the overall structure (Figure 1.3B). The inner concave surface is formed by a continuous standard parallel β sheet, except that at both ends of the β strand the backbone carbonyl group forms bifurcated hydrogen bonds with two main chain amide groups (Figure 1.3C).

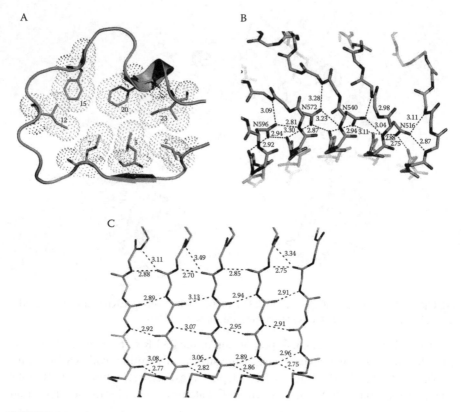

FIGURE 1.3 Contribution of leucine-rich repeats to the structural stability of the TLR ect-odomain (A) Seven conserved hydrophobic residues form a tight hydrophobic core within each leucine-rich repeat as shown for LRR7. (B) The conserved asparagines at position 10 form three hydrogen bonds with the previous LRR motif and a single hydrogen bond within their own motif. (C) A concave surface is formed by 25 continuous β strands connected by numerous hydrogen bonds including bifurcated hydrogen bonds at both ends of the β strands.

GLYCOSYLATION PATTERN

In the crystal structure, electron density is observed for carbohydrates at 8 of the 15 possible N-linked glycosylation sites. If all the 15 N-linked glycosylation sites were post-translationally modified, the TLR3 surface would be largely masked by carbohydrate. The inner concave surface also contains three predicted glycosylation sites in LRR1, LRR9, and LRR15, and glycosylation was observed in LRR9 and LRR15. The carbohydrates from these sites can fill the space inside the concave surface and may hinder the binding of large ligand molecules such as double-stranded RNA. Only one face is glycosylation-free [Figure 1.4(B) and (C)], suggesting its potential role in ligand binding or oligomerization.

Investigation of predicted glycosylation sites in other human TLRs shows that all human TLRs contain at least two predicted N-linked glycosylation sites in the concave surface (residues 2 through 9 in the 24-residue motif). The exact *in vivo* compositions and structures of carbohydrates on different TLRs are not well characterized and may depend on the localization of each TLR. For example, TLRs that reside in the endosome, including TLR7, TLR8, and TLR9, may have different glycosylation patterns from the TLR4 targeted to the cell surface. Results indicate that some glycosylation sites may play critical roles in protein stability and targeting.[15–17] Other results have shown that glycosylation may be important in ligand recognition.[18]

SURFACE ELECTROSTATIC POTENTIAL

Analysis of the surface electrostatic potential of the TLR3 ectodomain using the GRASP program revealed that the inner concave surface is predominantly negatively charged [Figure 1.5(B)]. This finding was surprising because the inner concave surface was thought to be the binding site for double-stranded RNA, which is also highly negatively charged. The largest cluster of positively charged residues was found on the glycosylation-free face of TLR3 [Figure 1.5(C)].

A recent paper showed that the ligand binding of TLR3 is pH-dependent (optimum pH around 5), and that ligand binding is abolished at pH values above 7.[19] The

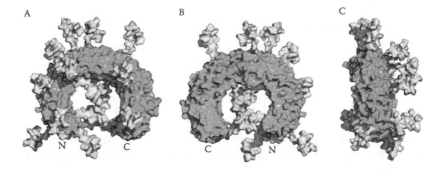

FIGURE 1.4 Predicted glycosylation of TLR3 ectodomain. (A) When oligomannose-type glycosylation is modeled for all the 15 predicted N-linked glycosylation sites, the carbohydrates cover a large portion of the TLR3 ectodomain including the concave surface. (B) View rotated 180° from (A) showing the glycosylation-free face. (C) View rotated 90° from (A) showing dramatic differences in the extent of glycosylation on the two side faces.

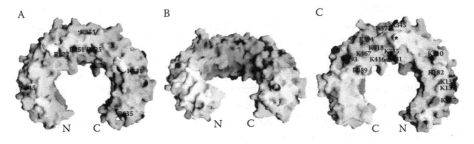

FIGURE 1.5 Surface electrostatic potential of TLR3 ectodomain calculated with GRASP. (A) same orientation as Figure 1.4(A). (B) Inside concave surface shows predominantly negative electrostatic potential. (C) Glycosylation-free face contains the largest number of positively charged residues that may be involved in double-stranded RNA recognition. The locations of four histidine residues near the N-terminus are indicated by dots.

authors suggested that double-stranded RNA may undergo pH-dependent conformational changes, thus affecting its binding to TLR3. Alternately, the pH-dependent change in histidine protonation may affect the ability of TLR3 to bind ligand. Interestingly, a patch of four histidine residues is located near the N-terminal region on the glycosylation-free face, as indicated by dots in Figure 1.5(C).

CONSERVED SURFACE RESIDUES AND POSSIBLE DIMERIZATION

In the crystal, two molecules related by two-fold crystallographic symmetry form a putative homodimer. The dimer interface is near the C-terminus on the glycosylation-free face. The dimer interaction is largely hydrophilic, including ionic interactions (Glu442–Lys467 and Lys547–Asp575), hydrogen bonds between side chains, and water-mediated hydrogen bonds. The buried surface area is relatively small at 640 Å², and the dimer interface largely overlaps with the most highly conserved surface residues on TLR3. (Figure 1.6 A–E). The dimer interface also includes a TLR3-specific insertion from LRR20 and three residues (Lys[547], Ala[549], and Pro[551]) from the LRR20 motif.

The dimerization of TLR3 *in vivo* has not yet been experimentally confirmed, but some evidence indicates multimerization of TLR3.[19] However, other TLRs are known to form functional homodimers or heterodimers *in vivo*. For example, TLR2 forms heterodimers with TLR1 and TLR6.[20] TLR4 is known to form a homodimer for signaling.[21]

TLR3-SPECIFIC INSERTIONS

The TLR-specific insertions are likely to be involved in the specific functions (ligand recognition and signaling) of individual TLRs. In the case of TLR3, two large insertions are found in LRR12 (10 residues) and LRR20 (11 residues). Although the insertion in LRR12 is disordered in the crystal structure, it was observed in a second human TLR3 structure.[22] The insertion protrudes about 13 Å from the horseshoe and contributes to the proposed dimer formation. Although the LRR12 insertions within the proposed homodimer are spatially close, interactions between them are limited.

The role of the LRR12 insertion is not clear and its deletion does not have an effect on the ligand binding ability of TLR3.[19,23] It is interesting that TLR3 contains a structure that protrudes so far from the horseshoe. The patch of positively charged residues on the glycosylation-free face is only about 15 Å from the LRR12 insertion; thus LRR12 may interact with the grooves of dsRNA. It is also possible that this LRR12 insertion may act as a spacer to separate the dimer to make space for the dsRNA interaction. The other TLR3-specific insertion at LRR20 plays a significant role in the dimer interface, thus supporting the idea that two TLR3s form a biologically relevant dimer. The LRR20 insertion may play an important role in reinforcing ligand-induced dimerization and conformational changes.

Other TLRs contain varying numbers of insertions at different positions and can give rise to the structural differences of TLRs.[14] The positions of the insertions are usually on the sides or outside convex surfaces. Interestingly, two groups of TLRs (1, 6, and 10 in one group, and 7, 8, and 9 in the other) have similar patterns of insertions different from other TLRs. The 1–6–10 and the 7–8–9 groups share high sequence

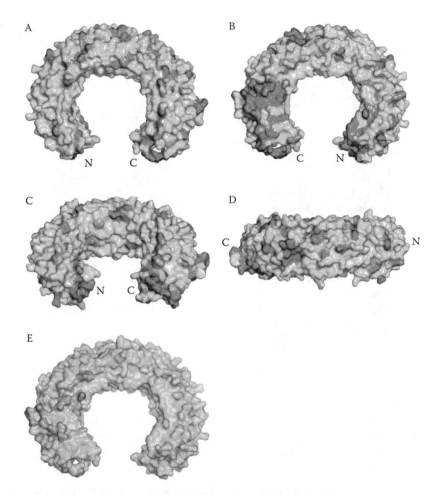

FIGURE 1.6 Conserved residues on TLR3 surface. (A) through (D) Completely conserved residues among five TLR3s (*Homo sapiens, Pan troglodytes, Bos taurus, Rattus norvegicus,* and *Takifugu rubripes*). (E) Residues involved in proposed dimerization. The dimer interface largely overlaps with the large patch of conserved residues in (B).

identities. Each group is known to recognize structurally- and chemically-related ligands (lipopeptides and nucleotides, respectively). TLRs 7, 8, and 9 also contain undefined regions with no sequence similarity to one of the known LRR motifs. These trends imply that the locations and structures of insertions in different TLRs are likely to be important in the functioning of TLRs by shaping ligand binding sites and determining the ligand binding specificity of each TLR.

SITE-DIRECTED MUTAGENESIS AND EFFECTS ON FUNCTION

To identify amino acid residues that are important in ligand recognition, over 50 residues in TLR3 have been mutated using site-directed mutagenesis.[23] The most dramatic change in ligand binding ability was observed for H539E and N541A mutations (Figure 1.7). H539 coordinates one of the two sulfate molecules observed in the

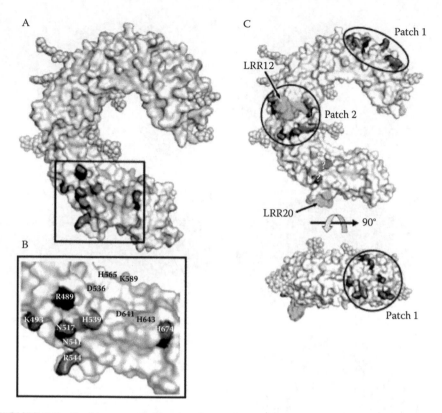

FIGURE 1.7 Residues mutated by site-directed mutagenesis. (A) and (B) Residues mutated on concave and lateral surfaces. Bell et al.[23] showed that H539 and N541 are important for functioning. (C) LRR12 and LRR20 insertions and positively charged residues in patches 1 and 2.

crystal structure that may mimic the phosphate groups of dsRNA. However, other residues that coordinate the sulfate ion with H539, including R488, N515, Q538, and E570, had no effect on ligand binding when mutated.

Residues that interact with the second sulfate, such as Y326, H359, and N361, also had no effect on RNA binding when mutated, thus indicating that the two sulfate binding sites located in the concave surface are not directly involved in ligand binding. Mutation of residues proximal to H539 revealed that the N541A mutation almost completely abrogates activity. N541, located on the glycosylation-free face, is also essential for RNA recognition (Figure 1.7). H539E and N541A mutants showed negligible change in their size-exclusion chromatography elution patterns in contrast to wild-type TLR3-ECD, whose elution pattern changes in the presence of its ligand, poly I:C. The authors suggested that these two residues, H539 and N541, are directly involved in dsRNA binding by interacting with the phosphate group and 2′ hydroxyl group of the RNA ribose, respectively. It is, however, noteworthy that these two residues are intimately involved in the proposed dimer interface, and the lack of RNA binding affinity may also reflect disruption of dimerization necessary for ligand recognition.

Deletion mutants of TLR3-specific insertions (LRR12 and LRR20) showed that the loss of LRR12 had no effect on TLR3 stimulation, but the LRR20 deletion resulted in significant, but not total, loss of activity.[23] The mutation of individual positively charged residues in the two basic patches on the glycosylation-free face also had no significant effect on ligand recognition. Only when a number of basic residues were mutated together (eight residues in patch 1 and seven in patch 2) was a significant loss of activity observed [Figure 1.7(C)]. Because such a large number of mutations were required for this loss of activity, whether these positively charged residues are directly involved in dsRNA binding, or whether loss of activity is due to alteration of the TLR3 structure or stability, is inconclusive. Mutation of each of the 15 asparagine residues in consensus N-linked glycosylation motif (Asn-X-Ser/Thr) to aspartate also had no significant effect on TLR3 expression or activity.

PROPOSED DOUBLE-STRANDED RNA BINDING SITE

Contrary to the original idea that the inner concave surface would act as the ligand binding site of TLRs,[24] evidence from the TLR3 structure indicates that the inner concave surface is not suitable for double-stranded RNA binding. First, the inner concave surface contains three predicted glycosylation sites that fill most of the space. Although the exact nature and extent of the glycosylation are still unknown, it is likely that any degree of glycosylation may hinder the binding of dsRNA, considering the dimensions of the inner space (diameter ~50 Å) and dsRNA (diameter ~26 Å).

It is possible that the carbohydrate groups in the inner surface may interact with the ligand, but examples of carbohydrates involved in ligand recognition are limited. The results of extensive site-directed mutagenesis experiments also indicated that residues located in the inner concave surface are not critical for dsRNA binding.[19] The electrostatic potential of the inner concave surface is predominantly negatively charged. This negative potential would make interaction with highly negatively charged dsRNA unfavorable.

Interestingly, two patches of positively charged residues and a TLR3-specific insertion in LRR12 are located in proximity to each other on the glycosylation-free side surface and may provide the binding site for double-stranded RNA, although mutagenesis experiments to confirm this theory were inconclusive.[23] This putative RNA binding site is on the same face as the dimer interface seen in the crystal, and dimerization of TLR3 could then create a deep binding cleft between two TLR3 molecules for dsRNA binding (Figure 1.8).

Based on site-directed mutagenesis results, Bell et al.[23] proposed that the dsRNA ligand binding site is located on the glycosylation-free face near the C-terminus. Specifically, H539 can interact with the phosphate group in the minor groove of RNA and N541 can form a hydrogen bond with the 2′ OH group of the ribose in the minor groove. These interactions can provide a model in which two TLR3-ECDs related by a 180° rotation can bind both sides of one dsRNA molecule (Figure 1.9).

TLRs including TLR3 should discriminate self from non-self molecules to specifically respond to invasions of pathogens. A subgroup of TLRs has become specialized to detect non-self nucleic acids derived primarily from viruses. TLR3 recognizes dsRNA; TLR7 and TLR8 bind single-stranded RNA and small immunomodulatory

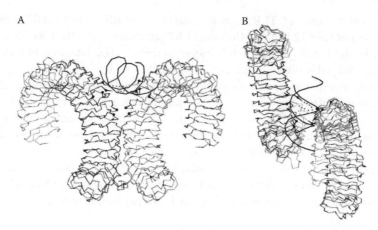

FIGURE 1.8 Proposed double-stranded RNA binding site of TLR3. (A) Dimerization of TLR3 ectodomain brings the positively charged residues (dots) and TLR3-specific insertion at LRR12 close together to form a potential binding site for a dsRNA molecule. Cα positions of the disordered LRR12 region are indicated as dashed lines. (B) Top view of (A).

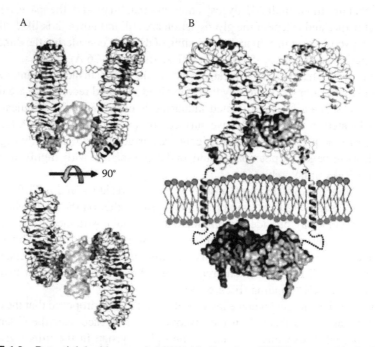

FIGURE 1.9 Potential double-stranded RNA binding site of TLR3 proposed by Bell et al.[23] (A) H539 and N541 of TLR3-ECD interact with minor grooves of dsRNA to bring the two TLR3 molecules close together for intracellular signaling (B). TLRs with TIR domains (Protein Data Bank ID code 1FYV). Linker regions between TLR3-ECD–TM (7 residues) and TM–TIR (19 residues) domains are denoted by black dotted lines. The homologous position of the *Lpsd* mutation in the BB loop of the TLR4 TIR domain postulated to interact with MyD88[25] is indicated. The separation of TLR3-ECD in the proposed oligomer easily accommodates the formation of a TIR dimer.

compounds such as imidazoquinolines; and TLR9 interacts with unmethylated CpG DNA. It is unclear how these TLRs achieve their specificities. The key to this question may be found in the intracellular localization of these subtypes of TLRs.

A recent paper indicates that subcellular localization is very important for specificity and, in particular, for discrimination of viral nucleic acids from self nucleic acids[26] by TLRs. The intracellular localization of TLR9 is specified by the transmembrane domain. When a chimeric receptor composed of the TLR9 ectodomain and the TLR4 transmembrane and cytoplasmic domain is made, it can be redirected to the cell surface. These hybrids are then able to signal in response to exogenous self nucleic acids, whereas TLR9 (present in the endoplasmic reticulum, endosomes, and lysosomes) responds only to viral or bacterial CpG-containing DNA. Thus, the discrimination of viral DNA from self DNA is due to delivery by the virus of its nucleic acid to an intracellular compartment where self DNA is not normally found.[26] Although the exact localization of TLR3 is still unclear, it is very likely that the intracellular compartmentalization of TLR3 plays a critical role in discriminating self and non-self dsRNA. It is interesting that CD14 is involved in enhancing dsRNA-mediated TLR3 activation by physical interaction and internalization of dsRNA.[27]

POSSIBLE MODES OF DIMERIZATION AND SIGNAL TRANSDUCTION

Although the exact binding sites for dsRNA recognition are different in the two proposed models,[13,23] the proposed RNA binding sites still reside within the putative dimer interface. This suggests that binding of dsRNA may strengthen dimerization or induce conformational changes of the dimer that then bring the intracellular TIR domains closer together for signal transduction (Figure 1.10). This type of signal transduction mechanism is often used by cell surface receptors.[28,29]

The basic mechanisms of signaling by Toll and TLRs seem to involve ligand-induced dimerization and oligomerization or conformational changes.[30,31] Evidence from the interaction of Spätzle with *Drosophila* Toll showed, both *in vivo* and *in vitro*, that the binding of Spätzle crosslinks two *Drosophila* Toll ectodomains. This crosslinking is necessary and sufficient for signal transduction by Toll.[32] Recent evidence illustrates that chimeric TLR4 molecules in which the TLR4 ectodomain is replaced by *Drosophila* Toll are activated by Spätzle and have very similar biochemical characteristics to Toll. This finding also indicates that the TIR domains can initiate downstream signaling if the receptor extracellular domain is crosslinked in a symmetrical manner.

Although receptor crosslinking is a crucial event in signal transduction, recent data suggest that the signaling process is also likely to involve a series of conformational changes.[33] Studies using analytical ultracentrifugation showed that truncation of the N-terminal region of Toll ECD allows the formation of stable dimeric complexes in solution and causes constitutive activation of the signaling cascade.[34] These results also indicate that the N-terminal domain in Toll provides structural hindrance, preventing the self-association of the receptors, and that binding of Spätzle can relieve this constraint by inducing conformational changes.

After TLR activation by possible ligand-induced dimerization and/or conformational changes, TLRs recruit a specific set of adaptor proteins that give rise to

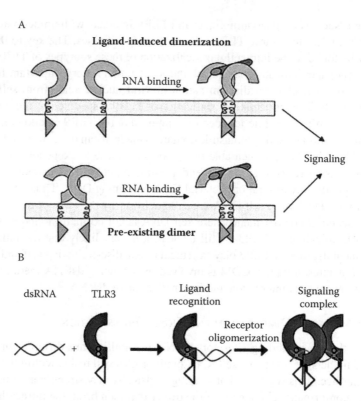

FIGURE 1.10 Hypothetical mechanism of ligand-induced signal transduction. (A) Double-stranded RNA binds to positively charged residues located on the glycosylation-free face and can induce either dimerization or conformational changes in pre-existing dimers to bring intracellular TIR domains closer together for signal transduction. (B) Residues near the C-terminal region of TLR3-ECD, such as H539 and N541, may interact with dsRNA and form a signaling complex by receptor oligomerization.

pathogen-specific immune responses.[7] These adaptors also contain TIR domains. Ligand binding to TLRs must create a new arrangement of the TIR domains formed by the TLR dimer that provides the binding site for the recruitment of appropriate adaptor molecules. Understanding this TLR–adaptor specificity requires detailed structural information of the TIR–TIR domain complexes. Although certain regions of the TIR domain, for example, the BB-loop region, are known to be involved in the TIR–TIR domain interaction, it is likely that other areas are also involved in the interaction with different adaptor molecules.

Recent data obtained from germ line mutagenesis experiments revealed that mutations in the TIR domain can block signaling with one adaptor molecule, but allow signaling via another.[35] Determining how such adaptor specificity is achieved awaits further biochemical and structural information. Understanding adaptor specificity can then provide opportunities for designing specific inhibitors to disrupt specific signaling pathways without disturbing other TLR-initiated signaling pathways that may be beneficial or essential to the host.

ACKNOWLEDGEMENT

IAW is supported by Grant AI-42266 of the National Institutes of Health..

REFERENCES

1. Akira, S. 2003, *Curr. Opin. Immunol.*, 15, 5.
2. Medzhitov, R. 2001, *Nat. Rev. Immunol.*, 1, 135.
3. Barton, G. M. and Medzhitov, R. 2002, *Curr. Top. Microbiol. Immunol.*, 270, 81.
4. Takeda, K. and Akira, S. 2005, *Int. Immunol.*, 17, 1.
5. Takeda, K. and Akira, S. 2004, *Semin. Immunol.*, 16, 3.
6. O'Neill, L. A., Fitzgerald, K. A., and Bowie, A. G. 2003, *Trends Immunol.*, 24, 286.
7. Yamamoto, M., Takeda, K., and Akira, S. 2004, *Mol. Immunol.*, 40, 861.
8. O'Neill, L. A. 2003, *Biochem. Soc. Trans.*, 31, 643.
9. Iwasaki, A. and Medzhitov, R. 2004, *Nat. Immunol.*, 5, 987.
10. Zuany-Amorim, C., Hastewell, J., and Walker, C. 2002, *Nat. Rev. Drug Discov.*, 1, 797.
11. Ulevitch, R. J. 2004, *Nat. Rev. Immunol.*, 4, 512.
12. Bartfai, T. et al. 2003, *Proc. Natl. Acad. Sci. USA*, 100, 7971.
13. Choe, J., Kelker, M. S., and Wilson, I. A. 2005, *Science*, 309, 581.
14. Bell, J. K. et al. 2003, *Trends Immunol.*, 24, 528.
15. Sun, J. et al. 2006, *J. Biol. Chem.*, 281, 11144.
16. Weber, A. N., Morse, M. A., and Gay, N. J. 2004, *J. Biol. Chem.*, 279, 34589.
17. da Silva Correia, J. and Ulevitch, R. J. 2002, *J. Biol. Chem.*, 277, 1845.
18. Kataoka, H. et al. 2006, *Cell Microbiol.*, 8, 1199.
19. de Bouteiller, O. et al. 2005, *J. Biol. Chem.*, 280, 38133.
20. Ozinsky, A. et al. 2000, *Proc. Natl. Acad. Sci. USA*, 97, 13766.
21. Lee, H. K., Dunzendorfer, S., and Tobias, P. S. 2004, *J. Biol. Chem.*, 279, 10564.
22. Bell, J. K. et al. 2005, *Proc. Natl. Acad. Sci. USA*, 102, 10976.
23. Bell, J. K. et al. 2006, *Proc. Natl. Acad. Sci. USA*, 103, 8792.
24. Kobe, B. and Kajava, A. V. 2001, *Curr. Opin. Struct. Biol.*, 11, 725.
25. Xu, Y. et al. 2000, *Nature*, 408, 111.
26. Barton, G. M., Kagan, J. C., and Medzhitov, R. 2006, *Nat. Immunol.*, 7, 49.
27. Lee, H. K. Et al. 2006, *Immunity*, 24, 153.
28. Livnah, O. et al. 1999, *Science*, 283, 987.
29. de Vos, A. M., Ultsch, M., and Kossiakoff, A. A. 1992, *Science*, 255, 306.
30. Gay, N. J., Gangloff, M., and Weber, A. N. 2006, *Nat. Rev. Immunol.*, 6, 693.
31. Gangloff, M., Weber, A. N., and Gay, N. J. 2005, *J. Endotoxin Res.*, 11, 294.
32. Weber, A. N. et al. 2003, *Nat. Immunol.*, 4, 794.
33. Weber, A. N. et al. 2005, *J. Biol. Chem.*, 280, 22793.
34. Winans, K. A. and Hashimoto, C. 1995, *Mol. Biol. Cell*, 6, 587.
35. Jiang, Z. et al. 2006, *Proc. Natl. Acad. Sci. USA*, 103, 10961.

2 Antiviral Signaling Through TLRs and RLHs

Taro Kawai

ABSTRACT

Innate immune cells, such as dendritic cells and macrophages, detect invading microorganisms including viruses through a limited number of receptors. Two pathways for the detection of viral nucleic acids have been identified. One is mediated by members of the Toll-like receptor (TLR) family that recognize viral double-stranded RNA, single-stranded RNA, and DNA. The other pathway is utilized by RIG-I-like RNA helicases (RLHs) that detect viral RNA in the cytoplasm. These receptors use specific intracellular adaptor proteins to activate the key IRF and NF-κB transcription factors that promote synthesis of various cytokines required to eliminate infected viruses. This review summarizes recent insights into the antiviral signaling pathways activated by TLRs and RLHs.

INTRODUCTION

Virus infection is sensed by the innate immune system. Within a host, innate immune cells such as dendritic cells (DCs) and macrophages express a variety of pattern recognition receptors (PRRs) that recognize specific pathogen-associated molecular patterns (PAMPs) within microbial structures such as viral nucleic acid.[1,2] Following recognition, PRRs initiate signaling pathways that induce the production of a variety of cytokines including inflammatory cytokines such as IL (interleukin)-6 and TNFα, type I interferons (IFNs; both α and β), chemokines, and IL-12, along with increased surface expression of co-stimulatory molecules to support the proliferation of T cells and their differentiation into Th cells.

Type I IFN in particular enhances DC maturation, natural killer (NK) cell cytotoxicity, and differentiation of virus-specific cytotoxic T lymphocytes. Type I IFN also upregulates transcription of many IFN-inducible genes that influence protein synthesis, growth arrest, and apoptosis to establish an antiviral state.[3]

PRRs that recognize viral PAMPs are classified into several families. The Toll-like receptor (TLR) family consists of more than ten members that respond to PAMPs derived from many pathogens. Within the TLR family, TLR3, TLR7, TLR8, and TLR9 represent a subfamily that recognizes viral nucleic acids. RLHs such as RIG-I and Mda5 participate in the detection of RNAs of infected viruses. Although TLRs and RLHs recognize viral nucleic acids to induce antiviral responses, they differ in cellular localization, ligand specificity, and downstream signal transduction pathways.[1-4]

TLRS 3, 7, 8, AND 9

Double-stranded (ds) RNA is synthesized during the course of replication of many viruses and serves as a potent activator of innate immune cells that induce type I IFN. A synthetic analog of viral dsRNA, polyinosinic acid:cytidylic acid (poly I:C), has been used extensively to mimic viral infection. Induction of type I IFN and pro-inflammatory cytokines in response to poly I:C or genomic RNA purified from a dsRNA virus such as reovirus was abrogated in macrophages derived from TLR3-deficient mice.[5] TLR3-deficient mice are consistently resistant to I:C-induced shock, indicating that TLR3 recognizes poly I:C and possibly senses viral dsRNA.

TLR3 is also implicated in recognizing dsRNA derived from ssRNA viruses such as the respiratory syncytial, encephalomyocarditis, and West Nile viruses.[6,7] TLR3 is expressed on a CD4$^-$ CD8$^+$ subset of cDC that has high phagocytic activity. The apoptotic bodies of virus-infected or dsRNA-loaded cells are taken up by CD8$^+$ DC, and TLR3 recognizes the dsRNA within these cells. This process triggers cross-presentation, a pathway important for the development of the CD8 cytotoxic T cell response against viruses that do not infect DC. Thus, the TLR3-dependent pathway is important for cross-presentation of CD8$^+$ DC.[6]

TLR7 was initially identified as a receptor that recognizes imidazoquinoline derivatives with antiviral activity, such as imiquimod and resiquimod (R-848), and guanine analogues such as loxoribine.[8] Subsequently, guanosine- or uridine-rich ssRNA derived from the human immunodeficiency virus (HIV) and the influenza virus was identified as a natural ligand for TLR7.[9,10] In addition, TLR7 recognizes synthetic poly U RNA and certain small interfering RNAs.[9,11] TLR8 is phylogenetically similar to TLR7. Human TLR8 preferentially mediates the recognition of HIV-derived ssRNA and R-848, although mice deficient for TLR8 respond normally to these molecules, suggesting that mouse TLR8 may not be functional.[9,10,12] TLR9 recognizes unmethylated 2′ deoxyribo(cytidine-phosphate-guanosine) (CpG) DNA motifs that are frequently present in pathogens such as viruses and bacteria but are rare in vertebrates.[13]

TLR7 and TLR9 are highly expressed on plasmacytoid DCs (pDCs; also known as IFN-producing cells)—a subset of DCs that have plasmacytoid morphology and primarily secrete vast amounts of type I IFN in response to viral infection.[14–16] In TLR7-deficient mice, IFNα production by pDCs is impaired after infection with influenza virus or vesicular stomatitis virus.[9,17] Moreover, IFNα production by pDCs in response to DNA viruses such as mouse cytomegalovirus, HSV-1, and HSV-2 is dependent on TLR9.[18–20] Inactivated HSV-2 or genomic DNA purified from this virus triggers IFNα secretion in pDCs in a TLR9-dependent manner.[20] pDCs rely on TLR7 and TLR9 to detect viral infection, but viral detection by pDCs does not seem to require viral replication within cells.

Unlike other TLRs, TLR 3, 7, 8, and 9 are not expressed at the plasma membrane and are exclusively localized to intracellular compartments like endosomes, suggesting that these intracellular TLRs recognize nucleic acids following the internalization and lysing of viruses.[1] Intracellular localization is also important to prevent contact with self DNA, because TLR9 can respond to self DNA if it is relocalized to plasma membranes.[21]

RIG-I AND MDA5

Because TLRs are localized to endosomes, they are unable to sense viruses that have entered the cytosol and initiated replication to produce dsRNA. Numerous studies implicated TLR-independent mechanisms in the detection of viral infection. For example, induction of IFNβ following transfection of poly I:C or infection with RNA viruses is normally observed in the absence of TLR3 or TRIF, suggesting that host cells have a mechanism to sense actively replicating viruses in the cytoplasm.[22-25]

RIG-I, a member of the RNA helicase family, was identified as a molecule that senses dsRNA and induces type I IFN responses.[25] RIG-I contains a DExD/H box RNA helicase and two caspase recruiting domain (CARD)-like domains. The helicase domain interacts with dsRNA, whereas the CARD-like domains are required for activating downstream signaling pathways. Furthermore, Mda5 and LGP2 were subsequently identified as members of the RLH's.[26,27] Mda5 contains two CARD-like domains and a helicase domain. LGP2 lacks the CARD-like domains and is thought to negatively regulate RIG-I and Mda5.

Studies of RIG-I- and Mda5-deficient mice revealed that RIG-I is essential for the recognition of a series of ssRNA viruses including flaviviruses, paramyxoviruses, orthomyxoviruses, and rhabdoviruses. Mda5 is required for the recognition of a different set of RNA viruses that includes picornaviruses.[28-30] Furthermore, Mda5 and RIG-I detect poly I:C and long dsRNA, respectively, indicating that these RNA helicases detect different RNA viruses.[29] RIG-I-mediated detection of RNA has recently been shown to depend on the 5' triphosphate end of RNA generated by viral polymerases.[31,32]

The production of type I IFN is still observed in pDCs derived from RIG-I- or Mda5-deficient mice, although cDCs, macrophages, and fibroblast cells derived from these mice showed impaired type I IFN induction after infection with their respective RNA viruses.[28,29] It is notable that the TLR system is required for pDC induction of the antiviral response. Collectively, these observations indicate that the TLR system plays a pivotal role in the detection of viruses by pDCs.

TLR7 AND TLR9 SIGNALING

TLRs contain extracellular LRRs that mediate ligand recognition, a transmembrane domain, and a cytosolic TIR domain required for downstream signaling pathways.[33] Upon recognition of nucleic acids, TLR7 and TLR9 recruit a TIR-containing MyD88 adaptor molecule that is universally utilized by all TLRs with the exception of TLR3 (Figure 2.1). The association of TLRs and MyD88 results in the recruitment of members of the IRAK family, including IRAK1, IRAK2, IRAK4, and IRAK-M. In particular, IRAK4 and IRAK1 are sequentially phosphorylated and involved in activation of the MyD88-dependent signaling pathway, while IRAK-M negatively regulates the MyD-88-dependent pathway. The function of IRAK2 remains unknown. Once phosphorylated, IRAK4 and IRAK1 dissociate from MyD88 and interact with TRAF6, an E3 ligase that forms a complex with Ubc13 and Uev1A and promotes the synthesis of lysine 63-linked polyubiquitin chains.

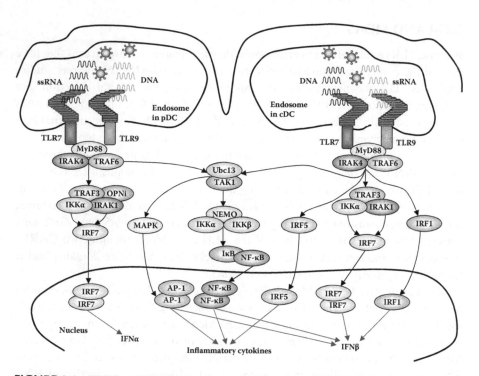

FIGURE 2.1 TLR7 and TLR9 signaling, which recruits MyD88, IRAK4, and TRAF6. TRAF6 then ubiquitin-dependently activates TAK1. The TAK1 complex activates the IKK complex consisting of IKKα, IKKβ, and NEMO/IKKγ to catalyze phosphorylation of IκB. IκBs are destroyed by the proteasome pathway, allowing NF-κB to translocate to the nucleus. TAK1 simultaneously activates the MAPK pathway, resulting in phosphorylation and activation of AP-1. NF-κB and AP-1 control inflammatory responses by inducing pro-inflammatory cytokines. MyD88 forms a signaling complex with IRAK1, IKKα, TRAF3, OPN-I, and IRF7 in pDC. In response to ligand stimulation, IRF7 is phosphorylated in a IRAK1- and IKKα-dependent manner, forms a dimer, and translocates to the nucleus to regulate expression of type I IFN genes, especially IFNα. IRF5 and IRF1 also interact with MyD88 and participate in induction of inflammatory cytokines and type I IFN, respectively, in cDCs.

TAK1, a member of MAPKKK, is activated by TRAF6-dependent ubiquitination, and in combination with TAB1, TAB2 and TAB3, activates two downstream pathways involving the IKK complex and the MAPK family. The IKK complex composed of the catalytic subunits IKKα, IKKβ, and a regulatory subunit known as NEMO/IKKγ, induce the phosphorylation and subsequent degradation of the IκB proteins that allow the NF-κB transcription factor to translocate into the nucleus. The MAPK family (JNK, p38, ERK) phosphorylates and activates the NF-κB transcription factor AP-1, a dimer of basic region leucine zipper proteins of the Jun, Fos and ATF subfamilies. NF-κB and AP-1 play central roles in the induction of genes encoding inflammatory cytokines[33] (Figure 2.1).

Although *in vitro* analyses have implicated Ubc13 in the control of both NF-κB and MAPK activation, studies of Ubc13-deficient mice have demonstrated that it is dispensable for NF-κB activation.[34] Macrophages deficient in Ubc13 show defective induction of inflammatory cytokines after treatment with TLR7 and TLR9 ligands. It is notable that they display impaired MAPK activation, but normal NF-κB activation. TAK1 activation and TRAF6 ubiquitination are normally observed in Ubc13-deficient cells, suggesting that Ubc13 may activate the MAPK pathway through a TAK1/TRAF6-independent pathway, or a pathway located downstream of TAK1/TRAF6. The finding that Ubc13 deficiency results in a loss of NEMO ubiquitination suggests a link between NEMO ubiquitination and MAPK activation.[34]

TLR7- and TLR9-mediated type I IFN induction by pDCs is dependent on MyD88. IRF7—structurally the most similar to IRF3—is present in the cytoplasm and translocates to the nucleus after phosphorylation by one or more virus-activated kinases. IRF7 potently activates the promoters of IFNα and IFNβ genes. The expression of the IRF7 gene is weak in unstimulated conditions, but rapidly upregulates in response to TLR ligands or viral infection in most cell types, suggesting a positive feedback regulation of type I IFN induction.

In pDCs, however, IRF7 is constitutively expressed and it (but not IRF3) binds to MyD88[35,36] (Figure 2.1). pDCs lacking IRF7 consistently fail to produce IFNα in response to CpG DNA, whereas IRF3 is dispensable in these pathways.[37] IRF7 also forms a complex with IRAK1, IRAK4, IKKα, and TRAF6 in addition to MyD88.[35,36,38,39] While mice deficient in MyD88, IRAK4, or TRAF6 exhibit defects in both IRF7 and NF-κB activation associated with impaired induction of type I IFN and inflammatory cytokines in response to CpG DNA, pDCs derived from IRAK1- or IKKα-deficient mice specifically show loss of IRF7 activation and type I IFN induction. Moreover, IRAK1 and IKKα (but not IRAK4) are capable of phosphorylating IRF7.[38,39] Together, IRAK1 and IKKα are most likely the kinases that catalyze the phosphorylation of IRF7 in pDCs. However, the functional relationship of IRAK1 and IKKα remains unclear. It is possible that they function as a heterodimer to potentiate IRF7 activation, or they may phosphorylate different residues of IRF7, both of which are required for the activation.

Several additional components of the MyD88–IRF7 complex have recently been identified (Figure 2.1). TRAF3 binds MyD88 and IRAK1, and is critical for type I IFN induction in TLR7 and TLR9 signaling.[40,41] TRAF3 is also necessary for the induction of the IL-10 anti-inflammatory cytokine, but not for the induction of pro-inflammatory cytokines, in response to ligands for TLR7 and TLR9.[40]

Osteopontin (OPN) is a secreted protein involved in diverse cellular functions such as bone resorption, vascularization, inflammation, and Th1 polarization. OPN expression is induced by a TLR9 ligand through T-bet, a master transcription factor for Th1 differentiation. pDCs derived from T-bet- and OPN-deficient mice show defects in the induction of type I IFN in response to a TLR9 ligand, whereas IL-6 induction and NF-κB activation are unaffected in these cells. An OPN precursor, designated OPN-i, is sequestered in the cytoplasm, suggesting that OPN-i functions as an intracellular signaling molecule. OPN-i interacts and colocalizes with MyD88, and nuclear translocation of IRF7 in response to a TLR9 ligand is impaired in pDCs derived from OPN-deficient mice.[42] Thus, OPN-i is a component of the MyD88-IRF7 complex in pDCs.

IRF8 is also implicated in TLR9-mediated responses in pDCs. pDCs derived from IRF8-deficient mice show a loss of TLR9-mediated induction of type I IFN and inflammatory cytokines linked to impaired NF-κB DNA binding activity, suggesting the possibility that IRF8 facilitates NF-κB DNA-binding.[43]

TLR9 activates different signaling pathways between pDCs and cDCs (Figure 2.1). While cDCs derived from IRF1-deficient mice display impaired induction of IFNβ, inducible nitric oxide synthase, and IL-12 p35 in response to a TLR9 ligand, pDCs derived from IRF1-deficient mice show normal induction of IFNβ and IFNα.[44] IRF1 also interacts with MyD88 and is released into nuclei in response to ligand stimulation.

Cytokine induction in response to TLR ligands is enhanced by pretreatment of cells with IFNγ. Consistent with the findings that IFNγ stimulation induces IRF1 expression, IFNγ-mediated enhancement is impaired in IRF1-deficient mice. Thus, IFNγ-induced IRF1 is recruited to MyD88 and translocated into nuclei in response to TLR stimulation to induce a set of genes including IFNβ in cDCs. IRF5 is also involved in TLR signaling. IRF5-deficient cDCs and macrophages exhibit impaired inflammatory cytokine production in response to multiple TLR ligands, but exhibit normal secretion of type I IFN by pDCs.[45] IRF5 binds MyD88 and TRAF6 and translocates to nuclei after phosphorylation. In the nucleus, IRF5 binds ISRE motifs found in the promoter regions of genes encoding inflammatory cytokines to cause their expression, presumably via collaborative activation with NF-κB. IRF5-mediated responses are negatively regulated by IRF4, which competes with IRF5 for interaction with MyD88.[46]

Studies of synthetic CpG ODN led to classification of into three groups, based on biological effects. D/A type ODN induces a secretion of type I IFN by pDCs but has a low ability to induce B cell activation and IL-12 production. In contrast, K/B type ODN stimulates B cell activation and IL-12 production, but poorly induces type I IFN. C type ODN has the ability to induce both type I IFN induction and B cell activation. A/D type CpG ODN colocalizes with TLR9, MyD88, and IRF7 in endosomes in pDCs and is rapidly transferred and degraded in lysosomes in cDCs. However, when A/D type CpG ODN relocalizes to the endosomes in cDCs using a cationic lipid, these cells can produce IFNα through activation of the MyD88-IRF7 pathway.[47] B/K type CpG ODN also induces secretion of IFNα if it is manipulated to remain in the endosomes of cDCs for longer periods. These findings suggest that retention of the CpG DNA–TLR9 complex in endosomes may cause the induction of robust IFNα production. Thus, the spatiotemporal regulation of TLR9 signaling is also important in controlling IFNα production.

TLR3 SIGNALING

Signaling through TLR3 activates IRF3 and NF-κB via a TIR domain-containing adapter molecule designated TRIF (also known as TICAM1), but not via MyD88.[33] (Figure 2.2). TRIF is also shared by TLR4, which recognizes LPS.[33] TRIF then recruits non-canonical IKKs, TBK1 (also known as NAK or T2K), and IKKi (also known as IKKε), rather than IRAK1 and IKKα, which phosphorylate IRF3.[48,49] Phosphorylated IRF3 forms a dimer, translocates into the nucleus, and binds to target sequences to induce the secretion of IFNβ and IFNα4. These subtypes of IFN subsequently activate a transcriptional complex known as ISGF3 consisting of STAT1, STAT2, and

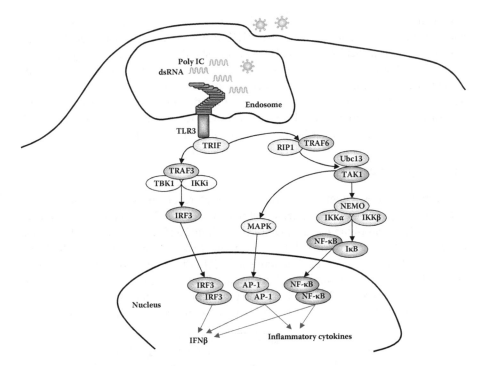

FIGURE 2.2 TLR3 signaling. TLR3 recruits the TRIF adapter that interacts with TRAF3, TBK1, and IKKi. TBK1 and IKKi mediate phosphorylation of IRF3. Phosphorylated IRF3 dimerizes and translocates to the nucleus where it binds DNA and induces expression of IFNβ. TRIF also interacts with TRAF6 and RIP1 to mediate NF-κB and AP-1 activation via TAK1.

IRF9 through the type I IFN receptor, stimulating the expression of IFN-inducible genes required for antiviral responses, including IRF7.[3] IRF7 is likely phosphorylated by TBK1 and IKKi in a manner similar to IRF3, inducing type I IFN.

TLR3-mediated inflammatory cytokine induction is controlled by TRIF-dependent NF-κB activation (Figure 2.2). The N-terminal and the C-terminal regions of TRIF have distinct functions with regard to the recruitment of downstream signaling molecules required for NF-κB activation. The N-terminal region recruits TRAF6 in addition to TBK1/IKKi.[50] Dominant-negative TRAF6 prevents TRIF-induced NF-κB activation, and mutations in the TRAF6 binding motifs of TRIF abrogate NF-κB activation. The C-terminal region of TRIF mediates its interaction with RIP1, a member of the RIP family involved in TNFR-mediated NF-κB activation, via its RIP homotypic interaction motif.[51] TLR3-mediated NF-B activation and the subsequent induction of target genes are impaired in the absence of RIP1. Furthermore, RIP1 polyubiquitination and complex formation with TRAF6 and TAK1 have been reported.[52] Thus, TRIF recruitment of RIP1 and TRAF6 may facilitate TAK1 activation, resulting in activation of NF-κB and MAPK. Whether Ubc13 is involved in RIP1 polyubiquitination is still unclear.

It has been suggested that TLR3-signaling is negatively regulated by several independent mechanisms. The TIR domain-containing protein SARM inhibits the TRIF-dependent pathway in human cell lines; however, the physiological function of

SARM in mice remains unknown.[53] While phosphorylation of the C-terminal serine and threonine clusters of IRF3 by TBK1/IKKi is essential for transcriptional activity, phosphorylation at Ser339 is linked to IRF3 destabilization. The Pin1 cytoplasmic peptidyl–prolyl–isomerase that catalyzes the cis–trans isomerization of peptide bonds located the N-terminus to proline residue to modulate substrate function binds IRF3 when phosphorylated at Ser339. This triggers ubiquitination and subsequent degradation of IRF3 by a proteasome-dependent pathway to terminate IFN responses.[54] A protein kinase responsible for Ser339 phosphorylation has not been identified.

RLH SIGNALING

Infection of viruses recognized by RIG-I or Mda5 results in the induction of type I IFN that is dependent on TBK1/IKKi but independent of TRIF and MyD88, indicating that RIG-I and Mda5 use an adaptor other than TRIF and MyD88.[22] IPS-1 (also known as MAVS, Cardif, or VISA) was identified as a potent activator of the IFNβ promoter[55–58] (Figure 2.3). IPS-1 contains a CARD-like domain that shows a homology to the CARD-like domains of RIG-I and Mda5. Overexpression of IPS-1 activates IFNα4, IFNα6, and NF-κB promoters in addition to the IFNβ promoter, and results in production of type I IFN sufficient for inhibition of viral replication. The

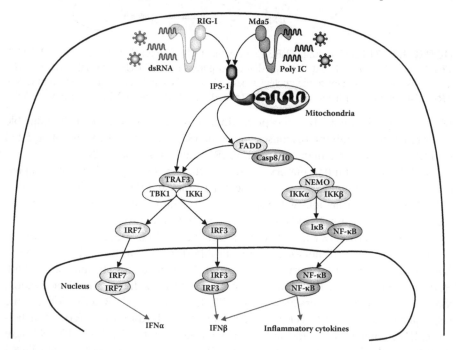

FIGURE 2.3   RLH signaling. RIG-I and Mda5 interact with adapter IPS-1 via CARD-like domains. IPS-1 is localized to mitochondria and initiates intracellular signaling pathways that lead to activation of IRF3, IRF7, and NF-κB via activation of TBK1/IKKi and IKKα/IKKβ, respectively. A complex consisting of FADD, caspase-8, and caspase-10 has been implicated in IPS-1-dependent NF-κB activation, while TRAF3 provides a physical link between IPS-1 and TBK1/IKKi.

activation of the IFNβ promoter after IPS-1 overexpression was ablated in TBK1-
and IKKi-null cells. The CARD-like domain of IPS-1 mediates interactions with
RIG-I and Mda5. IPS-1 is most likely an adaptor interacting with RIG-I and Mda5.

IPS-1-deficient mice have been recently developed.[59,60] Unlike RIG-I-deficient
mice that are embryonic lethal because of liver degeneration, IPS-1-deficient mice
are viable and develop normally. Macrophages, cDCs, and embryonic fibroblasts
derived from IPS-1-deficient mice failed to activate NF-κB and IRF3 with a concomitant loss of type I IFN and inflammatory cytokine induction after infection
with RNA viruses recognized by RIG-I or Mda5. Cytokine induction in response to
poly I:C or long dsRNA was also impaired in IPS-1-deficient mice. These mice are
consistently susceptible to infection with RNA viruses, indicating the importance
of IPS-1 in antiviral responses *in vivo*. However, IPS-1 is dispensable for type I IFN
induction after infection with a DNA virus such as modified Vaccinia Ankara, as
well as stimulation with TLR ligands. Collectively, IPS-1 is an essential adapter
utilized by both RIG-I and Mda5 that mediate antiviral innate immune responses. In
pDCs, however, TLR7 and TLR9 contribute more preferentially than RLHs to type
I IFN induction after virus infection, indicating a cell type-specific role of TLR and
RLH in the recognition of viruses and induction of antiviral immune responses.

IPS-1-mediated activation of IRF3 and IRF7 is dependent on TBK1/IKKi,
although IPS-1 does not directly interact with them. One report indicated that TRAF3
binds both IPS-1 and TBK1/IKKi, and TRAF3 deficiency results in impaired type I
IFN induction after virus infection, indicating that TRAF3 is a link between IPS-1
and TBK1/IKKi[61] (Figure 2.3). IPS-1 also interacts with FADD, a death domain containing adapter involved in death receptor signaling and RIP-1.[55] Cells deficient for
FADD display reduced induction of IFNβ and inflammatory cytokines in response
to poly I:C stimulation.[62,63] FADD forms a complex with caspase-10 and caspase-8.
These caspases are cleaved in response to poly I:C stimulation. The cleaved fragments of the caspases (encoding a death effecter domain) can activate NF-κB but not
the IFNβ promoter, suggesting that caspase-10 and caspase-8 selectively participate
in NF-κB activation downstream of FADD (Figure 2.3). Cells derived from cas-
pase-8-deficient mice accordingly show reduced NF-κB and inflammatory cytokine
induction in response to poly I:C.[63] How these cleaved fragments activate the IKK
complex remains unclear.

IPS-1 contains a transmembrane domain in the C-terminal tail that targets IPS-1
to the mitochondria[56] (Figure 2.3). Notably, mitochondrial retention of IPS-1 is
essential for IRF3 and NF-κB activation, suggesting that signaling from mitochondria plays an important role in antiviral immune responses. A report indicates that
the NS3/4A serine protease in the hepatitis C virus (HCV) targets IPS-1 for cleavage.[57,64–66] A putative cleavage site located upstream to the transmembrane domain
and HCV infection result in cleavage of IPS-1, as demonstrated in an *in vitro* cell culture infection system. A cleaved form of IPS-1 lacking a transmembrane region has
no ability to activate IRF3 and NF-κB. The NS3/4A combination is likely to change
the cellular localization of IPS-1 from mitochondria to the cytoplasm by cleavage,
thereby inhibiting type I IFN and the inflammatory response. NS3/4A also mediates
the cleavage of TRIF to inhibit the TLR3-dependent antiviral response.[67,68]

PERSPECTIVES

Host cells express multiple PRRs for the detection of viruses. These PRRs are expressed in different cellular compartments and recognize different types of nucleic acids. TLR3 recognizes dsRNA that is released into endosomes following phagocytosis of apoptotic bodies from virally infected cells, or detects dsRNA viruses that are internalized by endocytosis. In contrast, Mda5 and RIG-I are expressed in the cytoplasms of a variety of cells and recognize the dsRNA produced during viral replication. TLR7 and TLR9 are expressed by pDCs and act as sensors for viral ssRNA and DNA that trigger the production of large amounts of IFNα. Although these PRRs induce type I IFN, intracellular signaling pathways are different. TLR3 relies on the TRIF adapter to recruit TBK1/IKKi and induces IRF3 phosphorylation. TLR7 and TLR9 use MyD88 as an adapter to induce type I IFN via IRAK1/IKKα-dependent phosphorylation of IRF7. In RIG-I and Mda5 signaling, IPS-1 functions as a common adapter that mediates IRF3 and IRF7 activation via TBK1/IKKi.

Cells appear to express additional intracellular PRRs, such as a family of NOD proteins that serve to detect various PAMPs including RNA, and an unidentified receptor or receptors that recognize right-handed dsDNA released by DNA viruses, bacteria, and damaged host cells.[1,69,70] It will be important to understand how PPRs detect nucleic acid and induce antiviral innate immune responses. Such knowledge could improve therapeutic strategies for the treatment of infectious diseases and also autoimmune diseases associated with viral infection.

REFERENCES

1. Akira, S., S. Uematsu, and O. Takeuchi. 2006. Pathogen recognition and innate immunity. *Cell* 124:783.
2. Kawai, T. and S. Akira. 2006. Innate immune recognition of viral infection. *Nat Immunol.* 7:131.
3. Honda, K., A. Takaoka, and T. Taniguchi. 2006. Type I interferon gene induction by the interferon regulatory factor family of transcription factors. *Immunity* 25:349.
4. Meylan, E. and J. Tschopp. 2006. Toll-like receptors and RNA helicases: two parallel ways to trigger antiviral responses. *Mol. Cell* 22:561.
5. Alexopoulou, L. et al. Flavell. 2001. Recognition of double-stranded RNA and activation of NF-κB by Toll-like receptor 3. *Nature* 413:732.
6. Schulz, O. et al. 2005. Toll-like receptor 3 promotes cross-priming to virus-infected cells. *Nature* 433:887.
7. Wang, T. et al. 2004. Toll-like receptor 3 mediates West Nile virus entry into the brain causing lethal encephalitis. *Nat Med.* 10:1366.
8. Hemmi, H. et al. 2002. Small anti-viral compounds activate immune cells via the TLR7 MyD88-dependent signaling pathway. *Nat Immunol.* 3:196.
9. Diebold, S. S. et al. 2004. Innate antiviral responses by means of TLR7-mediated recognition of single-stranded RNA. *Science* 303:1529.
10. Heil, F. Et al. 2004. Species-specific recognition of single-stranded RNA via Toll-like receptor 7 and 8. *Science* 303:1526.
11. Hornung, V. et al. 2005. Sequence-specific potent induction of IFN-α by short interfering RNA in plasmacytoid dendritic cells through TLR7. *Nat Med.* 11:263.
12. Jurk, M. et al. 2002. Human TLR7 or TLR8 independently confer responsiveness to the antiviral compound R-848. *Nat Immunol.* 3:499.

13. Hemmi, H. et al. 2000. A Toll-like receptor recognizes bacterial DNA. *Nature* 408:740.

14. Liu, Y. J. 2005. IPC: professional type 1 interferon-producing cells and plasmacytoid dendritic cell precursors. *Annu Rev Immunol.* 23:275.

15. Colonna, M., G. Trinchieri, and Y. J. Liu. 2004. Plasmacytoid dendritic cells in immunity. *Nat Immunol.* 5:1219.

16. Kaisho, T. and S. Akira. 2006. Toll-like receptor function and signaling. *J Allergy Clin Immunol.* 117:979.

17. Lund, J. M. et al. 2004. Recognition of single-stranded RNA viruses by Toll-like receptor 7. *Proc Natl Acad Sci USA* 101:5598.

18. Krug, A. et al. 2004. TLR9-dependent recognition of MCMV by IPC and DC generates coordinated cytokine responses that activate antiviral NK cell function. *Immunity* 21:107.

19. Krug, A. et al. 2004. Herpes simplex virus type 1 activates murine natural interferon-producing cells through toll-like receptor 9. *Blood* 103:1433.

20. Lund, J. et al. 2003. Toll-like receptor 9-mediated recognition of Herpes simplex virus-2 by plasmacytoid dendritic cells. *J Exp Med.* 198:513.

21. Barton, G. M., J. C. Kagan, and R. Medzhitov. 2006. Intracellular localization of Toll-like receptor 9 prevents recognition of self DNA but facilitates access to viral DNA. *Nat Immunol.* 7:49.

22. Hemmi, H. et al. 2004. The roles of two IκB Kinase-related kinases in lipopolysaccharide and double stranded RNA signaling and viral infection. *J Exp Med.* 199:1641.

23. Yamamoto, M. et al. 2003. Role of adaptor TRIF in the MyD88-independent toll-like receptor signaling pathway. *Science* 301:640.

24. Hoebe, K. et al. 2003. Upregulation of costimulatory molecules induced by lipopolysaccharide and double-stranded RNA occurs by TRIF-dependent and TRIF-independent pathways. *Nat Immunol.* 4:1223.

25. Yoneyama, M. et al. 2004. The RNA helicase RIG-I has an essential function in double-stranded RNA-induced innate antiviral responses. *Nat Immunol.* 5:730.

26. Yoneyama, M. et al. 2005. Shared and unique fnctions of the DExD/H box helicases RIG-I, MDA5, and LGP2 in antiviral innate immunity. *J. Immunol.* 175:2851.

27. Rothenfusser, S. et al. 2005. The RNA helicase Lgp2 inhibits TLR-independent sensing of viral replication by retinoic acid-inducible gene-I. *J Immunol.* 175:5260.

28. Kato, H. et al. 2005. Cell type specific involvment of RIG-I in antiviral response. *Immunity* 23:19.

29. Kato, H. et al. 2006. Differential role of MDA5 and RIG-I in the recognition of RNA viruses. *Nature* 441:101.

30. Gitlin, L. et al. 2006. Essential role of mda-5 in type I IFN responses to polyriboinosinic:polyribocytidylic acid and encephalomyocarditis picornavirus. *Proc Natl Acad Sci USA* 103:8459.

31. Hornung, V. et al. 2006. 5′ triphosphate RNA is the ligand for RIG-I. *Science* 314:994.

32. Pichlmair, A. et al. 2006. RIG-I-mediated antiviral responses to single-stranded RNA bearing 5′ phosphates. *Science* 314:997.

33. Akira, S. and K. Takeda. 2004. Toll-like receptor signalling. *Nat Rev Immunol.* 4:499.

34. Yamamoto, M. et al. 2006. Key function for the Ubc13 E2 ubiquitin-conjugating enzyme in immune receptor signaling. *Nat Immunol.* 7:962.

35. Kawai, T. et al. 2004. Interferon-α induction through Toll-like receptors involves a direct interaction of IRF7 with MyD88 and TRAF6. *Nat Immunol.* 5:1061.

36. Honda, K. et al. 2004. Role of a transductional-transcriptional processor complex involving MyD88 and IRF-7 in Toll-like receptor signaling. *Proc Natl Acad Sci USA.* 101:15416.

37. Honda, K. et al. 2005. IRF-7 is the master regulator of type-I interferon-dependent immune responses. *Nature* 434:772.
38. Uematsu, S. et al. 2005. Interleukin-1 receptor-associated kinase-1 (IRAK-1) plays an essential role for TLR7- and TLR9-mediated interferon-α induction. *J Exp Med.* 201:915.
39. Hoshino, K. et al. 2006. IκB kinase-α is critical for interferon-α production induced by Toll-like receptors 7 and 9. *Nature* 440:949.
40. Hacker, H. et al. 2006. Specificity in Toll-like receptor signalling through distinct effector functions of TRAF3 and TRAF6. *Nature* 439:204.
41. Oganesyan, G. et al. 2006. Critical role of TRAF3 in the Toll-like receptor-dependent and -independent antiviral response. *Nature* 439:208.
42. Shinohara, M. L. et al. 2006. Osteopontin expression is essential for interferon-alpha production by plasmacytoid dendritic cells. *Nat Immunol.* 7:498.
43. Tsujimura, H. et al. 2004. Toll-like receptor 9 signaling activates NF-κB through IFN regulatory factor-8/IFN consensus sequence binding protein in dendritic cells. *J Immunol* 172:6820.
44. Negishi, H. et al. 2006. Evidence for licensing of IFN-γ-induced IFN regulatory factor 1 transcription factor by MyD88 in Toll-like receptor-dependent gene induction program. *Proc Natl Acad Sci USA* 103:15136.
45. Takaoka, A. et al. 2005. Integral role of IRF-5 in the gene induction programme activated by Toll-like receptors. *Nature* 434:243.
46. Negishi, H. et al. 2005. Negative regulation of Toll-like-receptor signaling by IRF-4. *Proc Natl Acad Sci USA* 102:15989.
47. Honda, K. et al. 2005. Spatiotemporal regulation of MyD88–IRF-7 signalling for robust type-I interferon induction. *Nature* 434:1035.
48. Sharma, S. et al. 2003. Triggering the interferon antiviral response through an IKK-related pathway. *Science* 300:1148.
49. Fitzgerald, K. A. et al. 2003. IKKε and TBK1 are essential components of the IRF3 signaling pathway. *Nat Immunol.* 4:491.
50. Sato, S. et al. 2003. Toll/IL-1 receptor domain-containing adaptor inducing IFN-β (TRIF) associates with TNF receptor-associated factor 6 and TANK-binding kinase 1, and activates two distinct transcription factors, NF-κB and IFN-regulatory factor-3, in the Toll-like receptor signaling. *J Immunol.* 171:4304.
51. Meylan, E. et al. 2004. RIP1 is an essential mediator of Toll-like receptor 3-induced NF-κB activation. *Nat Immunol* 5:503.
52. Cusson-Hermance, N. et al. 2005. RIP1 mediates the TRIF-dependent toll-like receptor 3 and 4-induced NF-κB activation but does not contribute to IRF-3 activation. *J Biol Chem.* 280:36560.
53. Carty, M. et al. 2006. The human adaptor SARM negatively regulates adaptor protein TRIF-dependent Toll-like receptor signaling. *Nat Immunol.* 7:1074.
54. Saitoh, T. et al. 2006. Negative regulation of interferon-regulatory factor 3-dependent innate antiviral response by the prolyl isomerase Pin1. *Nat Immunol.* 7:598.
55. Kawai, T. et al. 2005. IPS-1; an adaptor triggering RIG-I- and Mda5-mediated type I interferon induction. *Nat. Immunol.* 6:981.
56. Seth, R. et al. 2005. Identification and characterization of MAVS, a mitochondrial antiviral signaling protein that activates NF-κB and IRF3. *Cell* 122:669.
57. Meylan, E. et al. 2005. Cardif is an adaptor protein in the RIG-I antiviral pathway and is targeted by hepatitis C virus. *Nature* 437:1167.
58. Xu, L. G. et al. 2005. VISA is an adapter protein required for virus-triggered IFN-β signaling. *Mol. Cell* 19:727.
59. Kumar, H. et al. 2006. Essential role of IPS-1 in innate immune responses against RNA viruses. *J. Exp. Med.* 203:1795.

60. Sun, Q. et al. 2006. The specific and essential role of MAVS in antiviral innate immune responses. *Immunity* 24:633.
61. Saha, S. K. et al. 2006. Regulation of antiviral responses by a direct and specific interaction between TRAF3 and Cardif. *EMBO J.* 25:3257.
62. Balachandran, S., E. Thomas, and G. N. Barber. 2004. A FADD-dependent innate immune mechanism in mammalian cells. *Nature* 432:401.
63. Takahashi, K. et al. 2006. Roles of caspase-8 and caspase-10 in antiviral innate immune responses. *J. Immunol.* in press.
64. Li, X. et al. 2005. Hepatitis C virus protease NS3/4A cleaves mitochondrial antiviral signaling protein off the mitochondria to evade innate immunity. *Proc Natl Acad Sci USA.* 102:17717.
65. Lin, R. et al. 2006. Dissociation of a MAVS/IPS-1/VISA/Cardif-IKKepsilon molecular complex from the mitochondrial outer membrane by hepatitis C virus NS3-4A proteolytic cleavage. *J Virol.* 80:6072.
66. Loo, Y. M. et al. 2006. Viral and therapeutic control of IFN-β promoter stimulator 1 during hepatitis C virus infection. *Proc Natl Acad Sci USA.* 103:6001.
67. Ferreon, J. C. et al. 2005. Molecular determinants of TRIF proteolysis mediated by the hepatitis C virus NS3/4A protease. *J Biol Chem.* 280:20483.
68. Li, K. et al. 2005. Immune evasion by hepatitis C virus NS3/4A protease-mediated cleavage of the Toll-like receptor 3 adaptor protein TRIF. *Proc Natl Acad Sci USA.* 102:2992.
69. Ishii, K. J. et al. 2006. A Toll-like receptor-independent antiviral response induced by double-stranded B-form DNA. *Nat Immunol.* 7:40.
70. Stetson, D. B. and R. Medzhitov. 2006. Recognition of cytosolic DNA activates an IRF3-dependent innate immune response. *Immunity* 24:93.

3 Recognition of Virus Invasion by Toll-Like Receptors and RIG-I-Like Helicases

Hiroki Kato and Osamu Takeuchi

ABSTRACT

During viral infection, host cells sense the invasion by detecting viral components. Toll-like receptors (TLRs) comprise one of the systems that recognize viral nucleotides. TLR-mediated recognition of viruses leads to the production of type I interferons and pro-inflammatory cytokines. The TLR system plays an important role in plasmacytoid dendritic cells. In addition, the RIG-I and MDA5 RNA helicases located in the cytoplasm function as viral detectors by recognizing viral double-stranded (ds) RNA. Gene targeting has shown that RIG-I and MDA5 recognize different types of RNA viruses, as well as synthetic double-stranded RNAs. Moreover, 5′ triphosphate single-stranded (ss) RNA has also been shown to be a RIG-I ligand. The way these helicases sense distinct dsRNA structures, and the reason only RIG-I recognizes ssRNA structures, remain to be determined. Structural studies of the interactions of ligands with RIG-I and MDA5 are critical to the further understanding of antiviral immunity.

INTRODUCTION

When mammalian cells are infected with viruses, they produce type I interferons (IFNs) that interfere with viral replication. Many researchers have searched for the key molecules that sense viral invasion and trigger type I IFNs. One group of these molecules are called Toll-like receptors (TLRs). A TLR has a receptor with a transmembrane domain that recognizes components of pathogens at either cell surfaces or at lysosome or endosome membranes.[1–5]

Activation of intracellular signaling pathways triggered by TLRs induces the nuclear translocation of transcription factors, such as nuclear factor (NF)-κB and IFN regulatory factors (IRFs), that lead to the expression of IFN-inducible genes and genes encoding pro-inflammatory cytokines.[6] TLRs 3, 7, 8, and 9 are localized on the endosomal membrane. They recognize nucleotides derived from viruses and bacteria. TLR3 detects double-stranded (ds) RNA, while TLR7 and TLR9 recognize single-stranded (ss) RNA and unmethylated DNA with CpG motifs, respectively.[7–9]

TLR7 and TLR9 play important roles in the recognition of viruses in plasmacytoid dendritic cells (pDCs) that produce large amounts of type I IFNs in response to viral infection.[10] This system, however, may be ineffective for detecting pathogens that have invaded cytosols.

Accumulating evidence suggests that microorganisms invading a cell are recognized by cytoplasmic pattern recognition receptors (PRRs) such as RLHs (RIG-like helicases) and NLRs (NOD-like receptors).[11,12] RLHs are comprised of N-terminal caspase recruitment domains (CARDs) and C-terminal DExD/H box helicase domains. The RLH-mediated signaling triggered by RNA virus infection induces the expression of type I IFN genes by activating IRFs and NF-κB. This chapter will discuss the TLR and RLH system that recognizes viral nucleotides.

TLRS AND VIRAL RECOGNITION

Viruses contain nucleotides as their genomes and these are surrounded by structural proteins. Although TLR2 and TLR4 are also involved in the recognition of viral envelope proteins, this recognition mainly produces pro-inflammatory cytokines in a variety of cells.[13,14] By contrast, viral nucleotides are known to induce type I IFNs in immune cells.

TLRs 3, 7, 8 and 9 recognize viral nucleotides.[7-9,15] TLR3 recognizes dsRNA present as a viral genome or generated in virally-infected cells in the course of virus replication.[7] Upon stimulation with a synthetic dsRNA analog, polyinosine–polycytidylic acid (poly I:C), TLR3 triggers a signaling cascade via a cytoplasmic adaptor molecule known as TRIF (TIR domain-containing adaptor inducing interferon-β). TRIF associates with the downstream signaling molecules: TNF receptor-associated factor 3 (TRAF3), TRAF6, and receptor-interacting protein-1 (RIP1).[16-20] TRAF6 and RIP1

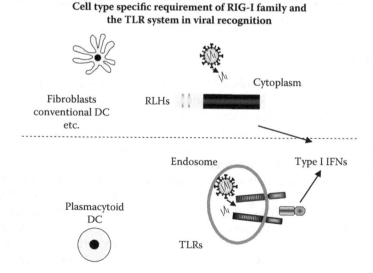

Cell type specific requirement of RIG-I family and the TLR system in viral recognition

Fibroblasts conventional DC etc.

RLHs

Cytoplasm

Endosome

Type I IFNs

Plasmacytoid DC

TLRs

FIGURE 3.1 Cell type-specific requirements of RLHs and the TLR system for viral recognition

activate NF-κB. TRAF3 is responsible for inducing type I IFNs. TRAF3 activates two IκB kinase-related kinases: TANK-binding Kinase 1 (TBK1) and inducible IκB Kinase (IKK-i), both of which are involved in the activation of IRF3 and/or IRF7.[21,22]

The role of the TLR3-dependent pathway is still controversial. It was shown that TLR3$^{-/-}$ mice were more resistant to infection with West Nile and influenza and other viruses, suggesting that these viruses take advantage of TLR3-mediated inflammatory responses for establishing infection.[23] Conversely, TLR3 has been reported to promote crosspresentation of virus-infected cells through the engagement of virus-derived RNAs.[24]

TLR7 and TLR8 have been shown to recognize ssRNA derived from RNA viruses. In addition, mouse TLR7 and human TLR8 recognize synthetic antiviral imidazoquinoline components including R-848 and imiquimods.[25] TLR7 and TLR8 are expressed within endosomal membranes. Although the role of mouse TLR8 is yet to be discovered, TLR7$^{-/-}$ mice are not responsive to challenge with R-848 or synthetic ssRNA. Furthermore, it was shown that pDCs from TLR7$^{-/-}$ mice failed to produce IFN-α in response to influenza virus infection. pDCs constitute a DC subpopulation characterized by the ability to secrete high levels of IFN-α in response to viral infection, indicating that TLR7-mediated recognition is critical for the recognition of viruses especially in pDCs.

DNAs from viruses and bacteria are known to be detected by TLR9 at the endosomes. Synthetic oligonucleotides with CpG motifs (CpG-DNAs) are known to be ligands for TLR9. They are subclassified into A/D- or B/K-type CpG-DNAs based on their sequences. A/D-type CpG-DNAs are characterized by phosphorothioate-modified poly G stretches at the 5′ and 3′ ends, and a phosphodiester CpG motif in the central portion; they are highly potent in inducing type I IFN in pDCs.[26,27] By contrast, B/K-type CpG-DNAs, conventional phosphorothioate-modified CpG-DNAs, are less potent in producing type I IFNs, although they can stimulate cells to produce pro-inflammatory cytokines. Recent studies suggested that A/D-type CpG-DNAs are retained in the endosomal vesicles for longer periods compared to B/K-type CpG-DNAs, facilitating TLR9-mediated recognition. TLR9 has been shown to recognize DNA viruses, including herpes simplex virus type 1 (HSV-1), HSV-2, and murine cytomegalovirus (MCMV), whose genomes are abundant in CpG-DNA motifs.[28-31]

In contrast to TLR3, TLR7 and TLR9 signaling is mediated by MyD88, an adaptor molecule with a TIR domain.[32] In various cell types, MyD88-dependent signaling leads to the activation of NF-κB and MAP kinases, and the ultimate expression of pro-inflammatory cytokine genes. By contrast, MyD88 can directly associate with IRF7 in pDCs, and activation of TLR7 or TLR9 signaling induces production of type I IFNs in this cell type.[33-35] Mice lacking MyD88 are susceptible to infection from viruses such as MCMV.[30,36] On the other hand, pDCs lacking both TBK1 and IKK-i can produce IFN-α in response to CpG-DNA stimulation or RNA virus infection, indicating that pDCs activate IRF7 independent of TBK1 and IKK-i, in contrast to other cell types.[37].

Although the involvement of TLRs in virus-induced cytokines and IFN production is well established, fibroblasts lacking both MyD88 and TRIF are still capable of expressing IFN-inducible genes in response to RNA virus infection. Therefore, the existence of virus detectors independent of TLRs had been predicted.

CYTOPLASMIC HELICASES FOR VIRAL DETECTION

An RNA helicase retinoic-acid-inducible gene I (RIG-I) that induces type I IFNs in a TLR-independent manner was identified as a cytosolic sensor for viral invasion.[38] Melanoma differentiation-associated gene 5 (MDA5) is homologous to RIG-I in possessing two N-terminal caspase recruitment domains (CARDs) followed by an RNA helicase domain. RIG-I and MDA5 recognize dsRNAs, unwind them with energy from ATPase activity, and undergo conformational changes for their signal transduction.[39,40] Their CARDs are responsible for this signal transduction via their adaptor molecule IFN-β promoter stimulator 1 (IPS-1), also called Cardif, MAVS, or VISA, leading to the activation of NF-κB, IRF3, and IRF7.[41–44]

TBK1 and IKK-*i* are also responsible for the activation of a TLR3-dependent pathway, indicating that the signaling pathways triggered by TLR stimulation and RIG-I converge at the level of TBK1/IKK-*i*.[21,22,45] The third member of the RLH group, LGP2, also contains a DExD box RNA helicase domain highly homologous to that of RIG-I, but lacks the CARD.[46,47] Overexpression of LGP2 inhibited RIG-I- and MDA5-mediated IFN responses, suggesting that LGP2 negatively regulates antiviral responses. LGP2 binds to dsRNA and interferes with the recognition of viral RNA by RIG-I or MDA5. It is also suggested that LGP2 competes with IKK-*i* for a common interaction site on IPS-1,[48] or shares a repressor domain with RIG-I.[49] However, the functional role of LGP2 *in vivo* remains to be determined.

Based on recent observations by gene targeting, we discuss how the RIG-I and MDA5 RNA helicases are involved in RNA viral recognition.

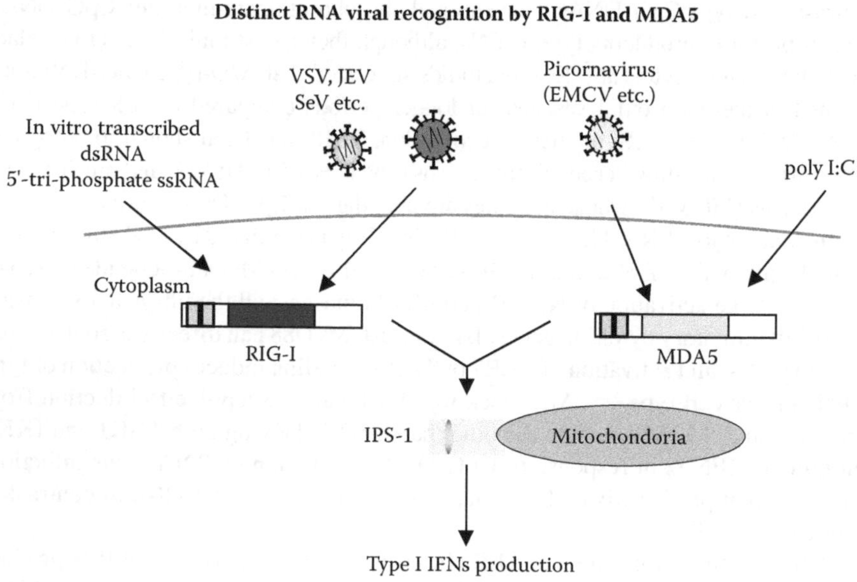

Distinct RNA viral recognition by RIG-I and MDA5

FIGURE 3.2   RIG-I and MDA5 sense distinct RNA viral infection.

REVERSE GENETIC APPROACHES FOR ASSESSING FUNCTIONS OF RIG-I AND MDA5 *IN VIVO*

To investigate the functional role of RIG-I and MDA5, mice lacking these helicases were generated by gene targeting. RIG-I$^{-/-}$ mice on a 129Sv/C57BL/6 background generally showed embryonic lethality due to liver degeneration at e13.5.[50] Although abrogation of TNF signaling failed to rescue liver apoptosis, adult RIG-I$^{-/-}$ mice were obtained by crossing with ICR mice.[51] RIG-I$^{-/-}$ mouse embryonic fibroblasts (MEFs) did not produce IFN-β, activate IFN-inducible genes, or induce pro-inflammatory cytokines in response to infection with RNA viruses, including Newcastle disease virus (NDV), Sendai virus, and vesicular stomatitis virus (VSV).[50,51] NDV-induced activation of NF-κB- and ISRE-containing genes was abrogated in RIG-I$^{-/-}$ cells. RIG-I overexpression in TBK1/IKK-i doubly deficient cells failed to activate the IFN-β promoter, indicating that RIG-I acts upstream of TBK1/IKK-i and governs both NF-κB and IRFs.

cDCs derived from the few RIG-I$^{-/-}$ mice born alive also failed to induce type I IFNs and cytokines upon exposure to NDV. Interestingly, upon NDV stimulation, pDCs from RIG-I$^{-/-}$ mice produced amounts of IFN-α comparable to production of pDCs from wild-type mice, indicating that cDCs and pDCs mainly utilize different mechanisms for the induction of IFNs (Figure 3.1). By contrast, MyD88$^{-/-}$ pDCs showed severely impaired IFN-α production in response to NDV infection, confirming that TLRs are essential for the recognition of viruses in pDCs.

In contrast, MDA5$^{-/-}$ mice showed no obvious developmental defects.[51,52] MDA5$^{-/-}$ MEFs produced IFN-β normally in response to infection with several RNA viruses, such as NDV, Sendai, and VSV, during which type I IFNs are produced in a RIG-I-dependent manner.[51] Production of type I IFNs in response to encephalomyocarditis virus (EMCV) was abrogated in MDA5$^{-/-}$ cells, but the response was not impaired in RIG-I$^{-/-}$ cells.[51] EMCV belongs to the picornavirus family whose members have positive sense ssRNA genomes. Members of the picornavirus family, including Theiler's virus and Mengo virus, were also recognized by MDA5, but not RIG-I, suggesting that MDA5 is responsible for the recognition of picornaviruses. These data indicate that RIG-I and MDA5 sense distinct RNA viral infections (Figure 3.2). During EMCV infection, cDCs derived from MDA5$^{-/-}$ mice failed to induce type I IFNs, while normal induction was seen in MDA5$^{-/-}$ pDCs, further supporting the cell type-specific involvement of the RIG-I family in RNA viral recognition (Figure 3.1). Also in EMCV infection, pDCs produced type I IFNs in a TLR-dependent manner, indicating that the TLR system is essential for viral detection in pDCs (Figure 3.2).

Compared with their littermate controls, MDA5$^{-/-}$ mice were highly susceptible to EMCV infection due to cardiomyopathy, while a deficiency in RIG-I did not affect the survival of mice infected with EMCV. These results are consistent with *in vitro* data showing that EMCV-induced IFNs are dependent on MDA5, but not on RIG-I. Furthermore, MyD88$^{-/-}$ mice were modestly susceptible to EMCV infection compared to wild-type mice, implying that pDC-mediated responses are not critical for eliminating infected EMCV.[34,51] In contrast, RIG-I$^{-/-}$ mice were susceptible to infection with VSV, while MDA5$^{-/-}$ mice were not, strongly suggesting that RIG-I and MDA5 sense distinct viral infections.

Some viruses, such as human hepatitis C (HCV), infect cells of human (but not mouse) origin. Analysis of a human hepatocyte cell line selected for permissiveness to HCV infection indicated that the cause of permissiveness to HCV infection was the mutational inactivation of RIG-I.[53,54] Also, HCV genomic RNA can bind RIG-I, but not MDA5. HCV is a member of the Flaviviridae that contain positive-sense ssRNA genomes. Therefore, this result suggests that RIG-I and MDA5 do not distinguish between positive and negative senses of the RNA virus genome.

One recent suggestion is that RNA viruses and also a DNA virus can be recognized by RIG-I. The Epstein Barr (EB) virus that produces dsRNAs with stem loops in infected cells induced IFNs in RIG-I-expressing cells.[55] However, the sensors for other DNA viruses such as herpes simplex are still unclear, although they are known to generate dsRNAs.[56]

It is well known that viruses evade host innate immune systems in various ways, and that the RIG-I/MDA5 pathway is one of the targets of viruses. Signaling molecules such as IPS1 and TBK1, were shown to be inactivated by proteins encoded in the viral genome. For example, IPS1 can be cleaved by HCV NS3/4A, disrupting its ability to induce type I IFNs.[42,57] Both the signaling molecules and receptors are inhibited by RNA viruses. The influenza virus NS1 protein has been shown to associate with RIG-I and inhibit RIG-I-mediated influenza virus recognition.[58–61] Furthermore, MDA5 is known to interact with V proteins from paramyxoviruses.[39,62] V proteins associate with MDA5 and inhibit its functions. However, generation of RIG-I$^{-/-}$ and MDA5$^{-/-}$ mice revealed that paramyxoviruses are recognized by RIG-I, but not MDA5.[51] Therefore, the functional role of the MDA5–V protein association remains unclear.

SYNTHETIC LIGANDS FOR RIG-I AND MDA5

Both RNA viruses and synthetic RNAs are differentially recognized by RIG-I and MDA5. For example, poly I:C induced the expression of type I IFNs in a RIG-I-independent and MDA5-dependent manner. By contrast, *in vitro* transcribed dsRNAs generated with T7 RNA polymerase induced the expression of type I IFNs normally in MDA5$^{-/-}$ MEFs, but not in RIG-I$^{-/-}$ MEFs. ssRNAs generated by *in vitro* transcription with T7 RNA polymerase possess triphosphate groups at their 5′ ends. It was previously shown that these 5′ triphosphate ssRNAs induced IFN production, whereas those without 5′ phosphate failed to activate cells.

An ssRNA with a 5′ triphosphate group has been shown to be a RIG-I ligand. RIG-I$^{-/-}$ MEFs failed to express type I IFNs in response to 5′ triphosphate ssRNA.[58,63] The 5′ end of the influenza genome harbors a triphosphate group. Genomic RNA purified from influenza virus has been shown to be sensed by RIG-I. Treatment of 5′ triphosphate RNA and influenza genomic RNA with a phosphatase greatly weakened the ability of these RNAs to stimulate cells, implicating the importance of the 5′ phosphates in activating innate immune cells.

Eukaryotic mRNA contains a 5′7-methylguanosine residue called the cap structure that suppresses nuclease-mediated mRNA degradation and facilitates translation. Since eukaryotic mRNA prevents recognition by host immune cells, it is possible that the 5′ cap structure also functions to prevent surveillance by the innate immune system.

However, it is not clear whether the 5′ triphosphate RNA recognition recapitulates the function of RIG-I in the recognition of RNA viruses. It is well known that genomic RNAs of various positive-sense ssRNA viruses contain 5′ cap structures for initiating the expression of viral proteins. Japanese encephalitis virus (JEV), one such RNA virus whose genome is 5′ capped, is also detected by RIG-I, leading to production of type I IFNs and cytokines.[51,64]

One possibility is that JEV genomic RNA transcribed in the cells is immediately recognized by RIG-I before the capping of the 5′ genomic end is completed. On the other hand, RIG-I may recognize RNA structures other than 5′ triphosphate ssRNA. dsRNA generated in the course of JEV replication is the candidate for the ligand. Because RIG-I exerts helicase activity to unwind short dsRNAs without 3′ overhangs, it is still possible that dsRNA in virus infected cells is the major target for RIG-I-mediated recognition.[65]

The precise structure of EMCV that is responsible for MDA5-mediated recognition is also unclear. Although EMCV genomic RNA does not have a 5′ cap structure, the 5′ end of this RNA is covalently linked to VPg, a peptide encoded by the EMCV genome. The EMCV appears to mask its 5′-triphosphate structure and block the detection by RIG-I through VPg. It has been reported that EMCV-infected cells are abundantly stained with an anti-dsRNA antibody, whereas dsRNA was hardly detected in cells infected with the influenza virus that is recognized by RIG-I. These findings suggest that dsRNAs generated during EMCV replication may be the predominant targets for MDA5-mediated recognition.[56] Indeed, poly I:C is recognized by MDA5, but not by RIG-I.

It is still possible that the sequence of the dsRNA determines the differential recognition by RIG-I and MDA5. Virulent picornavirus strains contain polycytidine stretches at the 5′ terminals of their genomes.[66] Although evidence is still insufficient, this poly C tract may be the target for MDA5-mediated recognition. Further studies are required to identify the precise structures recognized by RIG-I and MDA5.

PERSPECTIVES

This chapter describes the two types of RNA virus-recognizing receptors, TLRs and RLHs, and the specificity of their recognition. Distinct TLRs recognize different viral components, and TLR-mediated recognition plays a central role in the production of cytokines and type I IFNs in pDCs that are known to produce type I IFNs to combat viral infections. The recent expansion in our knowledge of cytoplasmic viral detectors has clarified their essential roles in the production of type I IFNs in immune cells such as cDCs, and non-immune cells including fibroblasts. These observations were made in gene targeting studies showing that RIG-I and MDA5 are effective for viral recognition in fibroblasts and cDCs, but not in pDCs in which the TLR system is required for sensing viral invasion.

Generation of IPS1$^{-/-}$ mice revealed that IPS1 is essential for the expression of type I IFNs and cytokines in response to viruses recognized by both RIG-I and

MDA5.[67] Nevertheless, IPS-1[−/−] mice and MDA5[−/−] mice are born in Mendelian ratios, grow healthily, and do not show gross developmental abnormalities.[51,52,67] In contrast, RIG-I[−/−] mice showed embryonic lethality due to liver degeneration at e13.5, depending on background. This observation implies that RIG-I is activated by unknown ligands during the development of embryonic hepatocytes, and that this activation may not require IPS-1 to induce responses. Understanding of the mechanisms by which RIG-I contributes to embryonic development will be interesting to explore.

RIG-I and MDA5 perform non-redundant functions in sensing different RNA viruses and distinct viral RNA products. The recognition of 5′ triphosphate ssRNA by RIG-I may explain the responses against infection with an influenza virus whose genome contains a 5′ triphosphate residue.[58] Whether triphosphate recognition is extended to the mechanisms used by RIG-I to recognize other RNA viruses is still unclear. We still do not know the precise structures of MDA5 ligands in picornaviruses. Further detailed studies of the functions of these two RNA helicases will help us understand how viral invasion is sensed.

ACKNOWLEDGMENTS

We thank all our laboratory colleagues for their help and thank M. Hashimoto for secretarial assistance. This work was in part supported by grants from the Japan Ministry of Education, Culture, Sports, Science, and Technology, the Twenty-First Century Center of Excellence Program of Japan, and the National Institutes of Health in the United States (Grant AI070167).

REFERENCES

1. Akira, S. and Takeda, K. 2004. Toll-like receptor signalling. *Nat Rev Immunol* 4: 499.
2. Akira, S., Uematsu, S., and Takeuchi, O. 2006. Pathogen recognition and innate immunity. *Cell* 124: 783.
3. Beutler, B. 2004. Inferences, questions and possibilities in Toll-like receptor signalling. *Nature* 430: 257.
4. Janeway, C. A., Jr. and Medzhitov, R. 2002. Innate immune recognition. *Annu Rev Immunol* 20: 197.
5. Hoffmann, J. A. 2003. The immune response of Drosophila. *Nature* 426: 33.
6. Taniguchi, T. et al. 2001. IRF family of transcription factors as regulators of host defense. *Annu Rev Immunol* 19: 623.
7. Alexopoulou, L. et al. 2001. Recognition of double-stranded RNA and activation of NF-κB by Toll-like receptor 3. *Nature* 413: 732.
8. Heil, F. et al. 2004. Species-specific recognition of single-stranded RNA via toll-like receptors 7 and 8. *Science* 303: 1526.
9. Diebold, S. S. et al. 2004. Innate antiviral responses by means of TLR7-mediated recognition of single-stranded RNA. *Science* 303: 1529.
10. Siegal, F. P. et al. 1999. The nature of the principal type 1 interferon-producing cells in human blood. *Science* 284: 1835.
11. Meylan, E., Tschopp, J., and Karin, M. 2006. Intracellular pattern recognition receptors in the host response. *Nature* 442: 39.
12. Stetson, D. B. and Medzhitov, R. 2006. Antiviral defense: interferons and beyond. *J Exp Med* 203: 1837.

13. Kurt-Jones, E. A. et al. 2004. Herpes simplex virus 1 interaction with Toll-like receptor 2 contributes to lethal encephalitis. *Proc Natl Acad Sci USA* 101: 1315.
14. Kurt-Jones, E. A. et al. 2000. Pattern recognition receptors TLR4 and CD14 mediate response to respiratory syncytial virus. *Nat Immunol* 1: 398.
15. Hemmi, H. et al. 2000. A Toll-like receptor recognizes bacterial DNA. *Nature* 408: 740.
16. Yamamoto, M. et al. 2003. Role of adaptor TRIF in the MyD88-independent toll-like receptor signaling pathway. *Science* 301: 640.
17. Hoebe, K. et al. 2003. Identification of Lps2 as a key transducer of MyD88-independent TIR signalling. *Nature* 424: 743.
18. Meylan, E. et al. 2004. RIP1 is an essential mediator of Toll-like receptor 3-induced NF-κB activation. *Nat Immunol* 5: 503.
19. Hacker, H. et al. 2006. Specificity in Toll-like receptor signaling through distinct effector functions of TRAF3 and TRAF6. *Nature* 439: 204.
20. Oganesyan, G. et al. 2006. Critical role of TRAF3 in the Toll-like receptor-dependent and -independent antiviral response. *Nature* 439: 208.
21. Fitzgerald, K. A. et al. 2003. IKK-ε and TBK1 are essential components of the IRF3 signaling pathway. *Nat Immunol* 4: 491.
22. Sharma, S. et al. 2003. Triggering the interferon antiviral response through an IKK-related pathway. *Science* 300: 1148.
23. Wang, T. et al. 2004. Toll-like receptor 3 mediates West Nile virus entry into the brain causing lethal encephalitis. *Nat Med* 10: 1366.
24. Schulz, O. et al. 2005. Toll-like receptor 3 promotes cross-priming to virus-infected cells. *Nature* 433: 887.
25. Hemmi, H. et al. 2002. Small anti-viral compounds activate immune cells via the TLR7 MyD88-dependent signaling pathway. *Nat Immunol* 3: 196.
26. Krug, A. et al. 2001. Identification of CpG oligonucleotide sequences with high induction of IFN-α/β in plasmacytoid dendritic cells. *Eur J Immunol* 31: 2154.
27. Verthelyi, D. et al. 2001. Human peripheral blood cells differentially recognize and respond to two distinct CPG motifs. *J Immunol* 166: 2372.
28. Lund, J. et al. 2003. Toll-like receptor 9-mediated recognition of herpes simplex virus-2 by plasmacytoid dendritic cells. *J Exp Med* 198: 513.
29. Hochrein, H. et al. 2004. Herpes simplex virus type-1 induces IFN-α production via Toll-like receptor 9-dependent and -independent pathways. *Proc Natl Acad Sci USA* 101: 11416.
30. Krug, A. et al. 2004. TLR9-dependent recognition of MCMV by IPC and DC generates coordinated cytokine responses that activate antiviral NK cell function. *Immunity* 21: 107.
31. Krug, A. et al. 2004. Herpes simplex virus type 1 activates murine natural interferon-producing cells through toll-like receptor 9. *Blood* 103: 1433.
32. Adachi, O. et al. 1998. Targeted disruption of the MyD88 gene results in loss of IL-1- and IL-18-mediated function. *Immunity* 9: 143.
33. Kawai, T. et al. 2004. Interferon-α induction through Toll-like receptors involves a direct interaction of IRF7 with MyD88 and TRAF6. *Nat Immunol* 5: 1061.
34. Honda, K. et al. 2005. IRF7 is the master regulator of type-I interferon-dependent immune responses. *Nature* 434: 772.
35. Honda, K. et al. 2004. Role of a transductional-transcriptional processor complex involving MyD88 and IRF7 in Toll-like receptor signaling. *Proc Natl Acad Sci USA* 101: 15416.
36. Delale, T. et al. 2005. MyD88-dependent and -independent murine cytomegalovirus sensing for IFN-α release and initiation of immune responses *in vivo*. *J Immunol* 175: 6723.

37. Matsui, K. et al. 2006. Cutting edge: Role of TANK-binding kinase 1 and inducible IκB kinase in IFN responses against viruses in innate immune cells. *J Immunol* 177: 5785.
38. Yoneyama, M. et al. 2004. The RNA helicase RIG-I has an essential function in double-stranded RNA-induced innate antiviral responses. *Nat Immunol* 5: 730.
39. Andrejeva, J. et al. 2004. The V proteins of paramyxoviruses bind the IFN-inducible RNA helicase, MDA5, and inhibit its activation of the IFN-β promoter. *Proc Natl Acad Sci USA* 101: 17264.
40. Kang, D. C. et al. 2002. MDA-5: An interferon-inducible putative RNA helicase with double-stranded RNA-dependent ATPase activity and melanoma growth-suppressive properties. *Proc Natl Acad Sci USA* 99: 637.
41. Kawai, T. et al. 2005. IPS-1, an adaptor triggering RIG-I- and Mda5-mediated type I interferon induction. *Nat Immunol*.
42. Meylan, E. et al. 2005. Cardif is an adaptor protein in the RIG-I antiviral pathway and is targeted by hepatitis C virus. *Nature* 437: 1167.
43. Seth, R. B. et al. 2005. Identification and characterization of MAVS, a mitochondrial antiviral signaling protein that activates NF-κB and IRF 3. *Cell* 122: 669.
44. Xu, L. G. et al. 2005. VISA is an adapter protein required for virus-triggered IFN-α signaling. *Mol Cell* 19: 727.
45. Hemmi, H. et al. 2004. Roles of two IκB kinase-related kinases in lipopolysaccharide and double stranded RNA signaling and viral infection. *J Exp Med* 199: 1641.
46. Rothenfusser, S. et al. 2005. The RNA helicase Lgp2 inhibits TLR-independent sensing of viral replication by retinoic acid-inducible gene-I. *J Immunol* 175: 5260.
47. Yoneyama, M. et al. 2005. Shared and unique functions of DExD/H box helicases RIG-I, MDA5, and LGP2 in antiviral innate immunity. *J Immunol* 175: 2851.
48. Komuro, A. and Horvath, C. M. 2006. RNA- and virus-independent inhibition of antiviral signaling by RNA helicase LGP2. *J Virol* 80: 12332.
49. Saito, T. et al. 2007. Regulation of innate antiviral defenses through a shared repressor domain in RIG-I and LGP2. *Proc Natl Acad Sci USA* 104: 582.
50. Kato, H. et al. 2005. Cell type-specific involvement of RIG-I in antiviral response. *Immunity* 23: 19.
51. Kato, H. et al. 2006. Differential roles of MDA5 and RIG-I helicases in the recognition of RNA viruses. *Nature* 441: 101.
52. Gitlin, L. et al. 2006. Essential role of MDA-5 in type I IFN responses to polyriboinosinic:polyribocytidylic acid and encephalomyocarditis picornavirus. *Proc Natl Acad Sci USA* 103: 8459.
53. Foy, E. et al. 2005. Control of antiviral defenses through hepatitis C virus disruption of retinoic acid-inducible gene-I signaling. *Proc Natl Acad Sci USA* 102: 2986.
54. Sumpter, R., Jr. et al. 2005. Regulating intracellular antiviral defense and permissiveness to hepatitis C virus RNA replication through a cellular RNA helicase, RIG-I. *J Virol* 79: 2689.
55. Samanta, M. et al. 2006. EB virus-encoded RNAs are recognized by RIG-I and activate signaling to induce type I IFN. *EMBO J* 25: 4207.
56. Weber, F. et al. 2006. Double-stranded RNA is produced by positive-strand RNA viruses and DNA viruses but not in detectable amounts by negative-strand RNA viruses. *J Virol* 80: 5059.
57. Lin, R. et al. 2006. Dissociation of a MAVS/IPS-1/VISA/Cardif-IKK-ε molecular complex from mitochondrial outer membrane by hepatitis C virus NS3-4A proteolytic cleavage. *J Virol* 80: 6072.
58. Pichlmair, A. et al. 2006. RIG-I-mediated antiviral responses to single-stranded RNA bearing 5′ phosphates. *Science* 314: 997.
59. Guo, Z. et al. 2006. NS1 Protein of influenza A virus inhibits the function of intracytoplasmic pathogen sensor RIG-I. *Am J Respir Cell Mol Biol*.

60. Mibayashi, M. et al. 2007. Inhibition of retinoic acid-inducible gene I-mediated induction of β interferon by NS1 protein of influenza A virus. *J Virol* 81: 514.
61. Opitz, B. et al. 2006. IFN-β induction by influenza A virus is mediated by RIG-I which is regulated by the viral NS1 protein. *Cell Microbiol.*
62. Childs, K. et al. 2006. MDA-5, but not RIG-I, is a common target for paramyxovirus V proteins. *Virology.*
63. Hornung, V. et al. 2006. 5' triphosphate RNA is the ligand for RIG-I. *Science* 314: 994.
64. Chang, T. H., Liao, C. L., and Lin, Y. L. 2006. Flavivirus induces interferon-β gene expression through a pathway involving RIG-I-dependent IRF3 and PI3K-dependent NF-κB activation. *Microbes Infect* 8: 157.
65. Marques, J. T. et al. 2006. A structural basis for discriminating between self and non-self double-stranded RNAs in mammalian cells. *Nat Biotechnol* 24: 559.
66. Martin, L. R. et al. 2000. Mengovirus and encephalomyocarditis virus poly(C) tract lengths can affect virus growth in murine cell culture. *J Virol* 74: 3074.
67. Kumar, H. et al. 2006. Essential role of IPS-1 in innate immune responses against RNA viruses. *J Exp Med* 203:1795.

4 Characteristics of Dendritic Cell Responses to Nucleic Acids

Tsuneyasu Kaisho, Takahiro Sugiyama, and Katsuaki Hoshino

ABSTRACT

Dendritic cells (DCs) are professional antigen-presenting cells linking innate and adaptive immunity. DCs recognize and are activated by nucleic acid adjuvants through the Toll-like receptor (TLRs) or non-TLR systems. Compared with lipid or protein adjuvants, nucleic acid adjuvants are distinguished by the ability to induce type I interferon (IFN), especially IFN-α. DCs are heterogeneous and respond to nucleic acids in a subset-specific manner. Plasmacytoid DCs (PDCs) constitute a unique DC subset specialized for nucleic acid sensing through TLRs that can secrete vast amounts of type I IFNs in response to TLR signaling. We review DC responses to nucleic acid adjuvants, focusing on PDCs. Clarifying the molecular mechanisms by which TLR-stimulated PDCs produce type I IFN should contribute to the establishment of effective therapeutic strategies to regulate antiviral immunity, autoimmunity, and allergies.

INTRODUCTION

Dendritic cells (DCs) sense pathogen infection through pattern recognition receptors (PRRs) consisting of internalizing or signaling PRRs. Internalizing PRRs such as lectins mainly function to incorporate invading microorganisms and are involved in Ag processing and presentation to Ag-specific T cells. Signaling PRRs can activate the signal transduction pathways that can induce production of pro-inflammatory cytokines or type I interferons (IFNs). Activated DCs can support T cell proliferation and direct helper T (Th) cell differentiation through a variety of costimulatory molecules or cytokines such as IL-12. Thus, with the help of PRRs, DCs function as innate immune cells and also contribute to the establishment of adaptive immunity.[1]

Toll-like receptor (TLR) family members are typical transmembrane signaling PRRs.[2,3] Cytosolic signaling PRRs include nucleotide binding oligomerization domain (NOD) family members and RNA helicases.[4,5] These PRRs can recognize a variety of molecular structures expressed mainly in microorganisms. These structures include lipids, proteins, and nucleic acids and can be referred to as microorganism-associated molecular patterns (MAMPs). They usually behave as non-self and possess common activities as immune adjuvants. However, nucleic acid adjuvants have several characteristics in comparison to other adjuvants. This chapter summarizes the ways DCs recognize and respond to nucleic acid adjuvants, focusing on a discrete DC subset, plasmacytoid DC (PDC).

NUCLEIC ACID RECOGNITION BY PRRs

Human TLR family members are numbered 1 through 10 and 12. Mouse TLRs are numbered 1 through 9 and 11 through 13. TLR3, TLR7, TLR8, and TLR9 are involved in recognizing nucleic acids and are localized in the endosomes. Lipid- or protein-recognizing TLRs are in the plasma membranes (Figure 4.1).

Bacterial and viral genomes contain DNAs with unmethylated CpG motifs (CpG DNAs); the motifs are rare in vertebrate genomes. Vertebrates possess DNA methyltransferase activity that can methylate cytosine residues. Because methylated cytosines can be read as thymidines during the replication phase, the enzyme activity decreases the frequency of unmethylated CpG motif in vertebrate DNA.

CpG DNA can exert potent immune-stimulatory activity and is recognized by TLR9 (Figure 4.1).[6] Synthetic oligonucleotides (ODNs) with such motifs also exhibit adjuvant activity. Screening a variety of CpG ODNs revealed functionally distinct types.[7,8] D/A-type CpG ODNs showed stronger effects in inducing type I IFNs, but less ability to induce B cell activation and IL-12 production than classical or K/B-CpG ODNs.

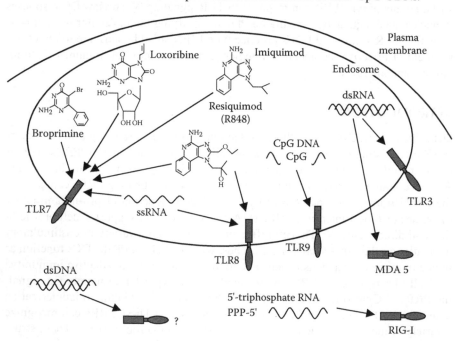

FIGURE 4.1 Nucleic acid recognition by signaling PRRs. TLRs and RNA helicases function in endosomes and cytosols, respectively. dsDNA is recognized in cytosols, although its receptor has not yet been defined.

Around the CpG motif, D/A-type CpG ODNs should have central palindromic base sequences that should be linked by phosphodiester bonds. D/A-type CpG ODNs should also carry phosphorothiolated polyG stretches at the 3' tails. Classical CpG ODNs do not necessarily require palindromic sequences or polyG tails, but their activity is stronger if bases are linked by phosphorothioated bonds. Vollmer et al.[9] reported C-type CpG ODNs that can both activate B cells and induce type I IFNs. All these CpG ODN effects are dependent on TLR9.[10]

D/A type CpG ODNs can form multimers through their polyG tails and palindromic sequences. The formation is facilitated by a flexible structure of phosphodiester-linked bases.[11] These multimers look like viral nanoparticles and are incorporated by DCs. Incorporated nanoparticles can be retained in the endosome longer than conventional CpG ODNs that are rapidly transported to lysosomes. This intracellular localization contributes to the type I IFN-inducing ability of D/A-type CpG ODNs.[12,13]

TLR7 ligands were first identified as antiviral chemical compounds [imidazoquinoline derivatives including imiquimod and resiquimod (R848)].[14] See Figure 4.1. R848 is about 100 times as potent as imiquimod. Several guanosine analogs and certain anti-tumor compounds, including 7-allyl-8-oxoguanosine (loxoribin) and 2-amin-5-bromo-6-phenyl-4(3)-pyrimidone (bropirimine), were also found to be TLR7 ligands.[15] These are synthetic compounds, but natural TLR7 ligands were later found to be single-stranded RNAs (ssRNAs) from certain viruses such as influenza and HIV-1.[16,17] Among synthesized homopolymeric ssRNAs, only polyuridylic RNAs can function as TLR7 ligands.

TLR8 closely resembles TLR7 in amino acid structure, and both genes are located about 50 kilobases apart on the X chromosome. R848 and other chemical compounds function as human TLR8 ligands[15,18] (Figure 4.1), but no evidence to date indicates that murine TLR8 functions in immune cells.

Upon viral infection, dsRNAs are produced in infected cells. Both dsRNAs and their synthetic mimic known as polyinosinic–polycytidylic acid [poly (I:C)] are recognized by TLR3 (Figure 4.1). Importantly, these dsRNAs are also recognized by cytosolic molecules such as retinoic-acid inducible gene-I (RIG-I) and melanoma differentiation associated gene 5 (MDA5).[19] Gene targeting experiments clarified their differential roles in pathogen recognition. RIG-I is essential for responses to paramyxoviruses and the influenza and Japanese encephalitis viruses. MDA5 is critical for detecting picornaviruses and poly I:C.[20] Extensive analysis clarified that RIG-I recognizes uncapped 5' triphosphate RNA rather than dsRNA.[21,22] Host-derived RNAs are modified at the 5' portion and not detected by RIG-I.

Another type of DNA-dependent IFN induction was also reported.[23–25] dsDNA functions as a ligand but its sequence specificity is not so strict and depends on the conformation.[23] The right-handed B-form (but not the left-handed Z-form) functions as a ligand. While the receptors have not been identified, the recognition is TLR-independent and mediated by cytosolic molecules (Figure 4.1). Thus, nucleic acids are recognized by TLRs and RNA helicases in endosomes and cytosols, respectively.

DC SUBSETS AND NUCLEIC ACID LIGANDS

PDC as Type I IFN-Producing Subset

Effects of nucleic acid adjuvants depend on DC subsets. PDC is one of a number of well characterized subsets.[26,27] For decades, a discrete type of cell similar to the plasma cell was known to exist in the T cell areas of human lymphoid tissues. It was designated the T-associated plasma cell or plasmacytoid T cell. Their identity and function were unclear, but CD4+CD3-CD11c- cells were found to be identical to plasmacytoid T cells and able to differentiate into mature DCs with IL-3 or IL-3+CD40L.[28] The cells acquired the ability to support Th2 cell differentiation after certain stimuli. Because of these characteristics, the cells were initially called type 2 DC precursors (pDC2s).

Although most cell types can produce type I IFNs upon viral infection, a subset of leucocytes can produce vigorous amounts of type I IFNs with viral infection.[29] The subset was found to be identical to pDC2.[30] These PDCs are now called IFN-producing cells (IPCs) based on morphology and function. PDCs were also identified in mice.[31-33]

PDCs originate from bone marrow, circulate in the blood, and reside in T cell areas in lymph nodes and spleen. PDCs express several lymphocyte-specific genes, including pre-T cell receptor-α or λ5, and carry DJ rearrangement in Ig heavy chain genes. Thus, the PDC is ontogenetically related to a lymphoid cell lineage.

PDCs express TLR7 and TLR9 exclusively, and therefore can be regarded as a nucleic acid-sensing DC. Whether PDCs utilize TLR-independent cytosolic recognition pathways is controversial,[34-36] but PDCs mainly utilize TLRs to sense nucleic acids. TLR7 and 9 signaling can induce both IFN-α and IFN-β. Type I IFNs consist of more than ten subtypes of IFN-α and a single IFN-β. All type I IFNs act through the same receptor complex, but expression of IFN-α and IFN-β is differentially regulated.

In mice, both PDCs and conventional DCs (cDCs) express TLR7 and TLR9. In response to TLR7 and TLR9 signaling, both PDCs and cDCs can secrete IFN-β and other pro-inflammatory cytokines, and upregulate surface expression of costimulatory molecules. However, only PDCs can produce IFN-α. Thus, PDCs constitute a unique DC subset that can produce IFN-α in response to TLR7 and TLR9 agonists.

In humans, PDCs and myeloid DCs (MDCs) can be defined as DC subsets in the peripheral blood. Human TLR9 is expressed on PDC and B cells, but not on MDCs. Human PDCs can produce type I IFN, but show poor ability to produce inflammatory cytokines. Thus, the DC systems of humans and mice are different. The DC system is conserved in that PDCs express TLR7 and TLR9 exclusively among TLRs, and PDCs possess a characteristic ability to produce type I IFNs. It is also notable that PDCs exist as IPCs in monkey, pigs, and rats.

How PDCs function except as IPCs remains unclear. PDCs show poor ability to induce T cell activation, likely due to low expression of major histocompatibility complex (MHC) class II and several co-stimulatory molecules including CD80 and CD86. Activated PDCs can still exhibit T cell stimulatory activity to an extent, and can activate CD8 T or NK cells through type I IFNs. Interestingly activated PDCs can induce interleukin (IL)-10 production from T cells and support regulatory T cell generation.[37] A tolerogenic function of PDCs was demonstrated in a murine asthma model. PDC depletion induced cardinal manifestations of atopic asthma, including Th2 responses and adoptive transfer of PDCs before sensitization prevented the disease.[38] PDCs are also involved in regulating graft rejection by presenting alloantigen.[39] Further studies are required to clarify whether PDC is tolerogenic in general.

CD8+ DC as Crosspresenting Subset

Murine cDCs can be further divided into three populations: CD4+CD8−, CD4−CD8+, and CD4−CD8− cells, according to the expression pattern.[40] A CD4−CD8+ DC is characterized by its ability to ingest apoptotic cells and involvement in crosspresentation.[41–44] Crosspresentation is an MHC-class I-restricted exogeneous includes crosspriming and crosstolerance, depending on whether they provoke or suppress immune responses.

For CD4 T cell activation, DCs incorporate pathogen products, process them into peptides to be associated with MHC class II, and present these Ags to T cells. The class II-restricted responses are directed at exogenous Ags. For CD8 T cell activation, Ags should be presented with MHC class I, which targets endogenous Ags expressed in DCs. If DCs are infected by a virus, antiviral CD8 T responses can be efficiently generated. However, not all viruses can infect DCs, and some infected DCs are functionally compromised by virus-derived proteins that can inhibit type I IFN expression and signaling. Therefore, crosspriming, which depends on the ability of non-infected DCs to capture virally-infected cells, is important for efficient antiviral immunity, especially against viruses that do not directly infect DCs.

TLR3 is expressed on CD4−CD8+ DCs but not on PDCs.[45] TLR3 is required for activation of CD4−CD8+ DCs by dsRNA from virally infected cells.[46] When immunized with virus-infected cells or dsRNA-containing cells, TLR3-dependent cytotoxic T cell responses are generated. Thus, TLR3 is critical for crosspresentation by CD4−CD8+ DCs.

LIAISON OF NUCLEIC ACIDS AND AUTOIMMUNITY THROUGH TYPE I IFN

Autoimmune Potentials of Nucleic Acid Adjuvants

Lipid or protein TLR ligands, such as LPS or flagellin, are apparently not expressed in vertebrates and therefore regarded definitely as non-self for a host. However, in the case of nucleic acids, structural distinctions between self and non-self are ambiguous. For example, CpG DNAs are observed in the host, albeit at a low frequency. Furthermore, host-derived mRNA and mRNA encoding green fluorescence protein can function as TLR7 ligands.[17] Host-derived nucleic acids can be released, for example, upon apoptosis in the body and function to stimulate the host immune system. In this context, nucleic acids have the potential to cause autoimmunity.

This potential danger is contained by several mechanisms. First, nucleic acids are prone to be degraded by nucleases and are unstable when released. Second, self nucleic acids and TLRs localize separately within cells. Self RNA is present in the cytoplasm, and self DNA is localized mainly in the nucleus. Thus, self nucleic acids normally fail to translocate into endosomal compartments where TLR7 and TLR9 are expressed. In contrast, virus-derived nucleic acids can be delivered into endosomes after viruses or virally-infected cells are endocytosed. The surfaces of plasma membranes are also unfavorable places for TLR9 expression because enforced expression on plasma membranes renders TLR9 responsive to self DNA.[47] Thus, endosomal expression of TLR7 and TLR9 is advantageous for the host to avoid self recognition and provoke antiviral immune responses. Furthermore, a regulated TLR expression pattern among DC subsets also contributes to prevention of self recognition. TLR7 expression is lowest in a CD4−CD8+ DC subset that has high phagocytic activity.[45] Thus, self RNA derived from incorporated cells has little chance to act through TLR7.

It is also notable that certain regions of the host DNA can inhibit TLR9 signaling. The frequency of the CpG motif in mammals is about 1/200 of the frequency in microorganisms. However, dose responses clarified that activity of mammalian DNA as CpG DNA was far less than anticipated.[48] Several inhibitory sequences have been identified in the telomere,[49] and TLR9 signaling can be inhibited by certain synthetic ODNs.[50] Interestingly, host-derived RNA is abundant in modified nucleosides and includes 5-methylcytidine, N6-methyladenosine, and pseudouridine. This modification significantly reduces immune-stimulatory activity.[51,52]

TYPE I IFN AND AUTOIMMUNITY

Type I IFNs can induce expression of a variety of anti-viral genes that encode factors to inhibit viral replication or induce apoptosis of infected cells. Type I IFNs also upregulate expression of MHC genes and induce DC maturation. They can also induce crosspriming during viral infection.[53] These effects are critical to anti-viral immunity.[54].

Clinical findings indicate potential pathogenic roles of type I IFN in autoimmune diseases including systemic lupus erythematosus (SLE).[55] SLE-like manifestations were observed in cancer patients during IFN-α therapy.[56,57] In addition, serum type I IFN levels are elevated in SLE patients and correlate with the severity of the disease.[58] DNA microarray analysis indicated that SLE patient-derived peripheral mononuclear cells show enhanced expression of type I IFN and IFN-inducible genes. This IFN signature can be diminished after treatment with steroids. Thus, type I IFN likely contributes to pathogenesis of autoimmunity.

TLR7 AND TLR9 AS TRIGGERS OF TYPE I IFN

In autoimmune diseases, auto antibodies (Abs) against DNAs and small nuclear ribonuclear proteins (RNPs) are generated. Such Abs form immune complexes (ICs) and prevent nucleic acids from degrading in serum (Figure 4.2). The ICs are rich in DNAs with unmethylated CpG motifs or uridine-rich RNAs that can function as TLR7 or TLR9 ligands. Thus, TLR7 and TLR9 signaling is a likely candidate as a type I IFN induction trigger in autoimmunity.

Several findings suggest that PDCs are responsible for enhanced production of type I IFN. PDC numbers in blood are decreased, but activated PDCs infiltrate the skin lesions of SLE or psoriasis,[59,60] likely due to mobilization of activated PDCs into damaged tissues. ICs can interact with fragment, crystallizable (Fc) receptors on PDCs, and these

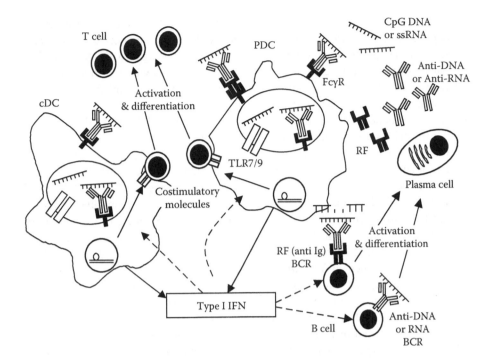

FIGURE 4.2 Autoimmune model of self nucleic acids. ICs consisting of Igs and nucleic acids are incorporated into the endosomes of DCs and stimulate DCs to produce type I IFN or inflammatory cytokines through TLR7 or TLR9. Type I IFN then activates DCs to mature or further produce type I IFN. Activated DCs can also produce inflammatory cytokines and further support T cell activation and differentiation. ICs or nucleic acids can also stimulate activation and differentiation of B cells bearing RF-BCR or anti-DNA/RNA-BCR through BCR and TLR7/9. B cell activation is also augmented by type I IFN or IL-6 from DCs.

interactions facilitate ICs to activate PDCs through internalization or dual engagement with TLR9[61,62] (Figure 4.2). It is still possible that cDCs also contribute to pathogenesis because cDCs can produce IFN-β in response to TLR7 and TLR9 signaling. Thus, type I IFN production through this signaling process contributes to both antiviral immune responses and pathogenesis of autoimmune diseases such as SLE (Figure 4.2).

TLR7- and TLR9-Induced B Cell Activation and Autoimmunity

TLR7/9 signaling can also activate both DCs and B cells. Rheumatoid factor (RF) is an immunoglobulin that reacts with certain types of IgG found in the sera of patients with autoimmune disorders. RF-positive B cells can recognize ICs through the interaction of their B cell receptors (BCRs) with Igs (Figure 4.2). Furthermore, B cells bearing anti-DNA or RNA BCR can recognize nucleic acids directly. These BCRs function in a similar manner to FcRs in PDCs by internalizing the nucleic acids into endosomes. TLR7 and TLR9 signaling then activates B cells to produce auto Abs, in cooperation with BCR.[63] Type I IFN can also activate B cells and upregulate TLR7 expression, thereby augmenting B cell activation through TLR7.[64]

TLR9 deficiency leads to decrease of anti-dsDNA and antichromatin Abs in lupus-prone mice.[65] Curiously TLR9 deficiency does not ameliorate the disease; on the contrary, the deficiency exacerbates it.[66] Auto Abs against RNA-containing Ags such as Smith (Sm) Ag are not decreased, and lymphocytes and PDCs are activated in TLR9-deficient lupus-prone mice. Importantly, these anti-RNA Ab titers are decreased and pathological findings are significantly improved in cases of TLR7 deficiency. This points to the interesting possibility that TLR7 and TLR9, although similar in various aspects, play opposite roles in the pathogenesis of autoimmunity.

Other murine lupus models also indicate that TLR7 signaling contributes to pathogenesis of autoimmunity. TLR7 deficiency led to decreases of auto Ab production in knock-in mice expressing Ig heavy and light chains that recognize RNA, ssDNA, and nucleosomes.[67] BXSB mice manifested higher incidences of autoimmune nephritis, with earlier onset in males.[68] The Y chromosome from BXSB mice is responsible for autoimmune-enhancing effects and called Yaa (Y-linked autoimmune accelerator). Yaa cells contain duplication of genomic DNA regions including the *TLR7* gene. TLR7 expression is enhanced in Yaa-carrying B cells.[69,70] Yaa B cells are biased toward nuclear Ags and hyperactive to TLR7 signaling likely due to enhanced expression of TLR7. This should account for the ability of Yaa to promote autoimmunity.

D/A-type CpG DNA has poor ability to activate B cells. However, B cells from lupus-prone mice can respond well to this type of CpG DNA.[71] The responsive B cells show phenotypes of marginal zone B cells that have immediate access to blood-borne particulate Ags and are involved in early Ig production. Thus, although the underlying mechanism is unknown in this lupus model, B cell hyper-responsiveness to TLR9 signaling is also suggested to contribute to lupus pathogenesis.

IKK FAMILY AND TLR SIGNALING

We recently clarified IκB kinase (IKKα as a critical molecule for type I IFN production from TLR-stimulated PDCs.[72] Before discussing the signaling mechanisms of PDCs, we will describe the functions of IKK family members.

IKKв

The IKK family consists of four members: IKKα, IKKβ, TANK-binding kinase 1 (TBK1), and inducible IκB kinase (IKKι), also known as IKKε and shown as IKKε/ι in this section. All four members are serine threonine kinases.[73,74] IKKα and IKKβ were originally identified as kinases that can phosphorylate IκB.[75] IκB retains a transcription factor, NF-κB, as an inactive form in the cytoplasm.

Upon various stimuli, IκB is phosphorylated and degraded. Then NF-κB is freed from IκB, translocates into the nucleus, and activates expression of target genes. Therefore, IκB phosphorylation by IKKs is a key step for NF-κB activation. All IKKs carry leucine zipper and helix-loop-helix domains, as well as a kinase domain, and IKKα and IKKβ show about 50% homology in amino acid structures. *In vitro* analysis did not reveal functional differences between them. Gene targeting experiments, however, clarified that IKKβ was more critically involved in NF-κB activation than IKKα.

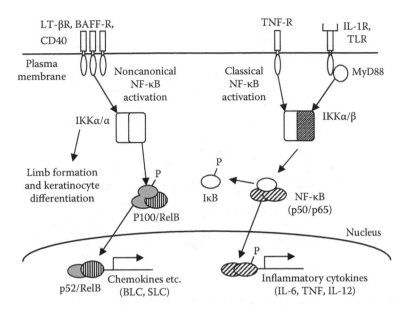

FIGURE 4.3   Distinct roles of IKKα and IKKβ in NF-κB activation. IKKα and IKKβ form heterodimers that are critically involved in the classical NF-κB pathway activated by TNF-R, IL-1R, and TLRs. The pathway induces expression of a variety of immune response genes including inflammatory cytokine genes. IKKα forms homodimers that mediate the non-canonical NF-κB pathway activated by certain TNF-R family members. The pathway leads to expression of certain chemokine genes. IKKα is also critical for limb formation and keratinocyte differentiation, although this pathway is not well characterized at present.

IKKβ-deficient mice are embryonic lethal due to liver apoptosis.[74] This phenotype is similar to that of mutant mice lacking the p65 NF-κB component. IKKβ-deficient cells showed severe impairment in NF-κB activation in response to TNF, IL-1, and TLR ligands. Impaired NF-κB activation leads to defective expression of various immune response genes, including pro-inflammatory cytokine genes. IKKβ forms heterodimers with IKKα and is critically involved in activation of a p50/p65 heterodimer. IKKα deficiency does not impair this pathway, implying that an IKKβ/IKKβ homodimer can function in the absence of IKKα. Thus, IKKβ plays a major role in this NF-κB pathway, also known as the classical or canonical NF-κB pathway (Figure 4.3).

TBK1 AND IKKε/ι

TBK1 and IKKε/ι are more similar to each other in amino acid structures than to IKKα or IKKβ. They fail to fully phosphorylate IκB, and IκB is not degraded by TBK1 or IKKε/ι. Furthermore, TBK1 and/or IKKε/ι deficiency do not lead to defective NF-κB activation. Thus, these kinases are not essential for NF-κB activation. They are, however, involved in phosphorylation and activation of IFN-regulatory factor (IRF-3).[76,77] See Figure 4.4.

IRF-3 is a transcription factor essential for IFN-β gene induction. IRF-3 is phosphorylated, homodimerized, and translocated into the nucleus by signaling through

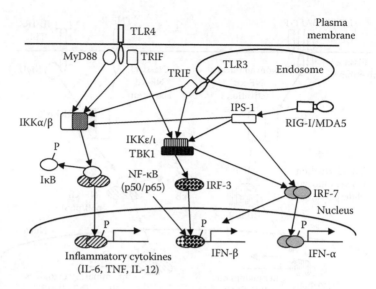

FIGURE 4.4 Roles of IKKs in TLR3/4 and RIG-I/MDA5 signaling. IKKα and IKKβ heterodimers are activated by TLR3/4 and RIG-I/MDA5. TLR3/4 and RIG-I/MDA5 require TRIF and IPS-1 as adapters, respectively. TBK1 and IKKε/ι form heterodimers that are critically involved in IRF-3 activation and subsequent IFN-β production. RIG-I/MDA5 can also activate IRF-7 and lead to production of IFN-α; TLR3/4 cannot do so.

TLR4 or TLR3. These TLRs activate IRF-3 through an adaptor, the TIR domain-containing adaptor-inducing IFN-β (TRIF).[78] This IRF-3 activation and subsequent IFN-β induction were impaired in TBK1-deficient (but not in IKKε/ι-deficient) cells, indicating that TBK1 plays a dominant role *in vivo*.[79,80] However, TBK1 can form heterodimers with IKKε/ι and analysis of mutant mice lacking both TBK1 and IKKε/ι demonstrated that IKKε/ι also functions cooperatively with TBK1.

In addition to TLR3 and TLR4, cytosolic viral recognition molecules such as RIG-I and MDA5 can also activate IRF-3 and induce IFN-β through IFN-β promoter stimulator-1 (IPS-1, also known as MAVS, VISA, and CARDIF).[81–84] This is also dependent on TBK1 and IKKε/ι. Thus, TBK1 and IKKε/ι are critical for IRF-3 activation and IFN-β induction in response to various PRRs (Figure 4.4).

IKKα

IKKα-deficient mice do not exhibit liver cell apoptosis. NF-κB activation was moderately impaired or almost normal in IKKα-deficient cells. The mutant mice, however, showed defects in skeletal morphogenesis and keratinocyte differentiation.[85–87] These phenotypes were not observed in IKKβ deficiency. IKKα deficiency also caused B cell maturation defect and lack of Peyer's patch formation.[88–90] B cell and Peyer's patch phenotype can be ascribed to a defect in B cell activation factor belonging to the TNF family receptor (BAFF-R) and lymphotoxin (LT)-βR signaling, respectively. These TNF-R family members require IKKα as a homodimer to lead to activation of NF-κB components such as p52 or RelB. This pathway is also known as the alternative or non-canonical NF-κB pathway[74] (Figure 4.3). Thus, IKKα plays a unique role in signaling pathways in a manner distinct from the other IKKs. However, the involvement of IKKα in innate immunity remained unclear until recently.

IKKα AS CRITICAL MOLECULE FOR PDC FUNCTION

TLR7/TLR9 signaling can activate PDCs to produce type I IFNs including IFN-α and IFN-β and inflammatory cytokines. All these effects require a cytoplasmic adapter molecule known as MyD88[10] and associated with intracytoplasmic domains of all TLRs except TLR3. All other MyD88-like adapters including TRIF are dispensable. As noted above, TLR3 and TLR4 can induce IFN-β, but this induction is dependent on TRIF—not on MyD88. TLR2 signaling that is totally dependent on MyD88 cannot induce any type I IFNs. Thus, TLR7/9 signaling is unique in that it can induce type I IFNs in a MyD88-dependent manner[2] (Figure 4.5).

TLR7/9-induced IFN-α production requires the IRF-7 transcription factor along with MyD88.[91] Interestingly, IRF-7-deficient PDCs retained the ability to produce inflammatory cytokines, indicating that IRF-7 is selectively involved in IFN-α-inducing pathways and functions independent of NF-κB.

IRF-5 also functions independently of (but cooperatively with) the NF-κB activating pathway, thereby leading to production of inflammatory cytokines.[92] MyD88 can directly interact with IRF-7,[93,94] but certain kinases should be involved because IRF-7 must be phosphorylated for its nuclear translocation and activation. In response to TLR7/9 signaling, PDCs can produce type I IFNs independently

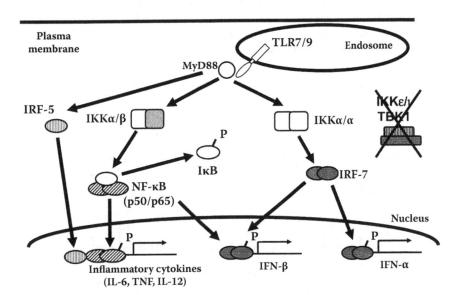

FIGURE 4.5 Roles of IKKs in TLR7/9-stimulated PDCs. TLR7/9 signaling can activate PDCs through MyD88. Inflammatory cytokines are induced by cooperative functions of IRF-5 and NF-κB. IKKα is critical for IRF-7 activation and subsequent induction of IFN-α and IFN-β. Type I IFN induction requires IRAK-1, osteopontin, and TRAF3 as well. TBK1 and IKKε/ι are dispensable in TLR7/9 signaling.

of TBK1 and IKKε/ι, indicating that TLR7/9 can induce type I IFNs in a distinct manner from TLR3/4.

IKKα-deficient mice die soon after birth because of abnormal keratinocyte differentiation. In order to investigate the function of IKKα in DCs, bone marrow chimeric mice were generated by transferring fetal liver cells into irradiated mice.[72] IKKα-deficient PDCs and cDCs were generated normally in terms of surface phenotype, indicating that DC development does not require IKKα. Analysis of purified PDCs and cDCs demonstrated that severe defects were observed in TLR7- or TLR9-induced IFN-α production from PDCs[72] (Figure 4.6). Production of inflammatory cytokines such as IL-12p40 and TNF-α was marginally or modestly impaired.

Vesicular stomatitis virus (VSV) can induce IFN-α production from PDCs in a TLR7-dependent manner. Production was also severely impaired in IKKα-deficient PDCs. Notably, the Newcastle disease virus can induce type I IFN production from IKKα-deficient embryonic fibroblasts. Because this type I IFN induction is dependent on RIG-I,[34] it can be assumed that the cytosolic viral recognition pathway does not require IKKα. Thus, IKKα plays a minor role in induction of inflammatory cytokines, but is critical for IFN-α production in TLR7/9-stimulated PDCs.[72]

IKK expression has been ubiquitously detected and is not specific to PDCs. This implies that IKKα is involved in type I IFN induction through certain PDC-specific molecules. IRF-7 is a critical molecule for type I IFN induction, as described above. Its protein level is constitutively high in PDCs.[95] Its phosphorylation must be activated, and it is thus a candidate target for IKKα. Several findings support the targeting of IRF-7 is targeted by IKKα.[72] First, overexpression experiments indicate a critical role of IKKα in IFN-α promoter activation. Enforced expression of IRF-7 can activate the IFN-α promoter, and this ability is enhanced by co-expression of MyD88. This synergistic effect of MyD88 and IRF-7 was inhibited by expression of kinase-deficient

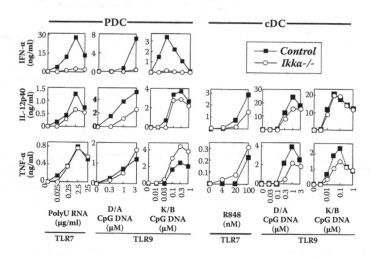

FIGURE 4.6 Cytokine production from TLR7/9-stimulated PDCs and cDCs. Bone marrow-derived DCs from control or IKKα-deficient mice was purified to PDCs or cDCs by FACS sorting and stimulated with various concentrations of TLR7/9 agonists. Amounts of cytokines were measured by ELISA.

IKKα. Second, IKKα associates with IRF7 in DCs. Molecular complexes containing IKKα and IRF-7 increase along with elevated expression levels of IRF-7 after TLR7/9 stimulation. Third, IRF-7 is as efficient an *in vitro* substrate as IκB for IKKα. Furthermore, the binding of IRF-7 to the 5′ regulatory region for IFN-α expression is prominently decreased, while NF-κB activity is retained in TLR-stimulated IKKα-deficient PDCs. Thus, IKKα is critically involved in IRF-7 activation in PDCs.

In the non-canonical NF-κB pathway, IKKα functions as a homodimer and requires another kinase, NF-κB-inducing kinase (NIK).[74] In TLR7/9-stimulated PDCs, IRF-7 is coprecipitated with IKKα and IKKβ, indicating the possibility that IKKβ is also involved in IRF-7 phosphorylation. Importantly, PDCs from *aly/aly* mice in which NIK is mutated showed normal IFN-α production in response to TLR7/9 signaling, suggesting that IKKα is involved in IRF-7 activation in a distinct manner from non-canonical NF-κB activation that depends on NIK.[72] Further studies are necessary to clarify the underlying molecular mechanism.

Several other molecules are also reported to be involved in this PDC-specific pathway. IRAK-1 is also a critical serine threonine kinase for IRF-7 activation.[96] Osteopontin, a phosphoprotein, was previously considered to be a cytokine to be secreted, but suggested to function as a MyD88-interacting intracellular molecule critical for type I IFN induction by TLR7/9.[97] Another adapter designated TRAF3 is a major regulator for type I induction by a variety of innate stimuli including TLR7/9.[98,99] These molecules may form a large molecular complex and contribute to the signaling pathway. Further analysis is required to clarify how these molecules cooperate.

PERSPECTIVES

We have clarified a novel function of IKKα, a well-known kinase, in TLR7/9-induced type I IFN induction from PDCs. The induction is involved in autoimmune pathogenesis, as well as antiviral immunity. Therefore, the signaling molecules including IKKα, should be potential targets for immune-regulatory manipulations for autoimmunity.

Notably, nucleic acids can be more easily manipulated than lipids or protein adjuvants because they are relatively small and can be synthesized *in vitro*. CpG ODNs can exhibit strong Th1-skewing effects and are expected to improve Th2-mediated allergic responses.[100] Successful treatments of human allergies were reported when CpG ODNs were conjugated with Ags.[101,102] Determining the molecular and cellular mechanisms by which nucleic acid adjuvants activate immune cells should also be useful in optimizing therapeutic regimes for allergies.

ACKNOWLEDGMENTS

We thank our laboratory colleagues for useful discussions and suggestions. This work was supported by the Special Coordination Fund for Promoting Science and Technology of the Japan Ministry of Education, Culture, Sports, Science, and Technology (MEXT), the Grant-in-Aid Program for Scientific Research of MEXT and JSPS, the Japan Science and Technology (JST) Corporation, the Uehara Memorial Foundation, the Naito Foundation, and the Novartis Foundation for the Promotion of

Science. The receptor for dsDNA (Figure 4.1) was found to be DNA-dependent activator of IFN-regulatory factors (DAI) (Takaoka, A. et al. 2007, *Nature*, 448, 502).

REFERENCES

1. Banchereau, J. and Steinman, R. M. 1998, *Nature*, 392, 245.
2. Takeda, K., Kaisho, T., and Akira, S. 2003, *Annu Rev Immunol*, 21, 335.
3. Medzhitov, R. 2001, *Nat Rev Immunol*, 1, 135.
4. Inohara, N. and Nunez, G. 2003, *Nat Rev Immunol*, 3, 371.
5. Akira, S., Uematsu, S., and Takeuchi, O. 2006, *Cell*, 124, 783.
6. Hemmi, H. et al. 2000, *Nature*, 408, 740.
7. Krug, A. et al. 2001, *Eur J Immunol*, 31, 2154.
8. Klinman, D. M. 2004, *Nat Rev Immunol*, 4, 249.
9. Vollmer, J. et al. 2004, *Eur J Immunol*, 34, 251.
10. Hemmi, H. et al. 2003, *J Immunol*, 170, 3059.
11. Kerkmann, M. et al. 2005, *J Biol Chem*, 280, 8086.
12. Honda, K. et al. 2005, *Nature*, 434, 1035.
13. Guiducci, C. et al. 2006, *J Exp Med*, 203, 1999.
14. Hemmi, H. et al. 2002, *Nat Immunol*, 3, 196.
15. Lee, J. et al. 2003, *Proc Natl Acad Sci USA*.
16. Heil, F. et al. 2003, *Eur J Immunol*, 33, 2987.
17. Diebold, S. S. et al. 2004, *Science*, 303, 1529.
18. Gorden, K. B. et al. 2005, *J Immunol*, 174, 1259.
19. Yoneyama, M. et al. 2004, *Nat Immunol*, 5, 730.
20. Kato, H. et al. 2006, *Nature*, 441, 101.
21. Hornung, V. et al. 2006, *Science*, online.
22. Pichlmair, A. et al. 2006, *Science*, online.
23. Ishii, K. J. et al. 2006, *Nat Immunol*, 7, 40.
24. Stetson, D. B. and Medzhitov, R. 2006, *Immunity*, 24, 93.
25. Okabe, Y. et al. 2005, *J Exp Med,* 202, 1333.
26. Liu, Y. J. 2005, *Annu Rev Immunol*, 23, 275.
27. Colonna, M., Trinchieri, G., and Liu, Y. J. 2004, *Nat Immunol*, 5, 1219.
28. Grouard, G. et al. 1997, *J Exp Med*, 185, 1101.
29. Perussia, B., Fanning, V. and Trinchieri, G. 1985, *Nat Immun Cell Growth Regul*, 4, 120.
30. Siegal, F. P. et al. 1999, *Science*, 284, 1835.
31. Nakano, H., Yanagita, M., and Gunn, M. D. 2001, *J Exp Med*, 194, 1171.
32. Asselin-Paturel, C. et al. 2001, *Nat Immunol*, 2, 1144.
33. Bjorck, P. 2001, *Blood*, 98, 3520.
34. Kato, H. et al. 2005, *Immunity*, 23, 19.
35. Hochrein, H. et al. 2004, *Proc Natl Acad Sci USA*, 101, 11416.
36. Krug, A. et al. 2004, *Blood*, 103, 1433.
37. Moseman, E. A. et al. 2004, *J Immunol*, 173, 4433.
38. de Heer, H. J. et al. 2004, *J Exp Med*, 200, 89.
39. Ochando, J. C. et al. 2006, *Nat Immunol*, 7, 652.
40. Liu, Y. J. 2001, *Cell,* 106, 259.
41. Iyoda, T. et al. 2002, *J Exp Med*, 195, 1289.
42. Heath, W. R. et al. 2004, *Immunol Rev*, 199, 9.
43. Bevan, M. J. 1976, *J Exp Med*, 143, 1283.
44. den Haan, J. M. et al. 2000, *J Exp Med*, 192, 1685.
45. Edwards, A. D. et al. 2003, *Eur J Immunol*, 33, 827.

46. Schulz, O. et al. 2005, *Nature*, 433, 887.
47. Barton, G. M., Kagan, J. C., and Medzhitov, R. 2006, *Nat Immunol*, 7, 49.
48. Sun, S. et al. 1997, *J Immunol*, 159, 3119.
49. Gursel, I. et al. 2003, *J Immunol*, 171, 1393.
50. Duramad, O. et al. 2005, *J Immunol*, 174, 5193.
51. Kariko, K. et al. 2005, *Immunity*, 23, 165.
52. Sioud, M. 2006, *Trends Mol Med*, 12, 167.
53. Le Bon, A. et al. 2003, *Nat Immunol*, 4, 1009.
54. Taniguchi, T. and Takaoka, A. 2002, *Curr Opin Immunol*, 14, 111.
55. Ronnblom, L., Eloranta, M. L., and Alm, G. V. 2006, *Arthritis Rheum*, 54, 408.
56. Gota, C. and Calabrese, L. 2003, *Autoimmunity*, 36, 511.
57. Stewart, T. A. 2003, *Cytokine Growth Factor Rev*, 14, 139.
58. Bennett, L. et al. 2003, *J Exp Med*, 197, 711.
59. Farkas, L. et al. 2001, *Am J Pathol*, 159, 237.
60. Nestle, F. O. et al. 2005, *J Exp Med*, 202, 135.
61. Bave, U. et al. 2003, *J Immunol*, 171, 3296.
62. Means, T. K. et al. 2005, *J Clin Invest*, 115, 407.
63. Leadbetter, E. A. et al. 2002, *Nature*, 416, 603.
64. Bekeredjian-Ding, I. B. et al. 2005, *J Immunol*, 174, 4043.
65. Christensen, S. R. et al. 2005, *J Exp Med*, 202, 321.
66. Christensen, S. R. et al. 2006, *Immunity,* 25, 417.
67. Berland, R. et al. 2006, *Immunity,* 25, 429.
68. Izui, S. et al. 1995, *Immunol Rev*, 144, 137.
69. Pisitkun, P. et al. 2006, *Science*, 312, 1669.
70. Subramanian, S. et al. 2006, *Proc Natl Acad Sci USA*, 103, 9970.
71. Brummel, R. and Lenert, P. 2005, *J Immunol*, 174, 2429.
72. Hoshino, K. et al. 2006, *Nature*, 440, 949.
73. Hayden, M. S. and Ghosh, S. 2004, *Genes Dev*, 18, 2195.
74. Bonizzi, G. and Karin, M. 2004, *Trends Immunol*, 25, 280.
75. Karin, M. 1999, *J Biol Chem*, 274, 27339.
76. Fitzgerald, K. A. et al. 2003, *Nat Immunol*, 4, 491.
77. Sharma, S. et al. 2003, *Science*, 300, 1148.
78. Yamamoto, M. et al. 2003, *Science*, 301, 640.
79. Hemmi, H. et al. 2004, *J Exp Med*, 199, 1641.
80. Perry, A. K. et al. 2004, *J Exp Med*, 199, 1651.
81. Kawai, T. et al. 2005, *Nat Immunol*.
82. Meylan, E. et al. 2005, *Nature*, 437, 1167.
83. Seth, R. B. et al. 2005, *Cell*, 122, 669.
84. Xu, L. G. et al. 2005, *Mol Cell*, 19, 727.
85. Takeda, K. et al. 1999, *Science*, 284, 313.
86. Hu, Y. et al. 1999, *Science*, 284, 316.
87. Li, Q. et al. 1999, *Genes Dev*, 13, 1322.
88. Kaisho, T. et al. 2001, *J Exp Med*, 193, 417.
89. Matsushima, A. et al. 2001, *J Exp Med*, 193, 631.
90. Senftleben, U. et al. 2001, *Science*, 293, 1495.
91. Honda, K. and Taniguchi, T. 2006, *Nat Rev Immunol*, 6, 644.
92. Takaoka, A. et al. 2005, *Nature*, 434, 243.
93. Kawai, T. et al. 2004, *Nat Immunol*, 5, 1061.
94. Honda, K. et al. 2004, *Proc Natl Acad Sci USA*, 101, 15416.
95. Izaguirre, A. et al. 2003, *J Leukoc Biol*, 74, 1125.
96. Uematsu, S. et al. 2005, *J Exp Med*, 201, 915.
97. Shinohara, M. L. et al. 2006, *Nat Immunol*, 7, 498.

 98. Oganesyan, G. et al. 2006, *Nature*, 439, 208.
 99. Hacker, H. et al. 2006, *Nature*, 439, 204.
100. Hessel, E. M. et al. 2005, *J Exp Med*, 202, 1563.
101. Tulic, M. K. et al. 2004, *J Allergy Clin Immunol*, 113, 235.
102. Simons, F. E. et al. 2004, *J Allergy Clin Immunol*, 113, 1144.

5 Dendritic Cells as Sensors for Foreign and Self Nucleic Acids

Anne Krug and Wolfgang Reindl

INTRODUCTION

In the search for inducers of anti-viral interferons (IFNs), several groups in the early 1960s found that viral and other nucleic acids (RNA as well as DNA) can trigger the production of interferons that protect host cells from the cytopathic effects of viruses.[1–3] Isaacs et al. noted that influenza virus particles containing nucleic acids were able to induce virus interference; virus particles lacking nucleic acids cannot do so.[3] In a series of papers by Hilleman's group published in 1967 in *PNAS*, it was shown that double-stranded (ds) RNA isolated from reovirus, virus-infected bacteria, and fungi, as well as dsRNA in the form of synthetic polynucleotides, were highly active IFN inducers.[4–7]

Field et al. demonstrated that poly I:C was one of the most potent polynucleotides to induce type I interferon responses in rabbits and mice.[5] Several decades after these observations were made, the molecular pathways that mediate immune responses to dsRNA were recognized. Toll-like receptor (TLR) 3 was identified as a receptor for dsRNA,[8] which we now know is mainly active intracellularly in specific dendritic cell (DC) subpopulations.[9] Two cytosolic receptors, the caspase recruitment domain (CARD) containing RNA-helicases RIG-I (retinoic acid-inducible gene I) and Mda5 (melanoma differentiation antigen 5), that sense RNAs of replicating viruses in all cell types, have been found only recently.[10–12]

In addition to these cytoplasmatic dsRNA recognition mechanisms, TLR7 and TLR8, which are localized in endosomal compartments, have been identified as receptors for GU-rich ssRNA derived from viruses (such as influenza) and autoimmune complexes containing self RNA.[13–16] The significance of these findings for anti-viral immune defense and autoimmunity will be discussed in detail below.

Tokunaga et al. recognized in 1984 that bacterial DNA is responsible for the immuno-stimulatory and anti-tumor activities of the *Mycobacterium bovis* BCG vaccine[17] currently used for immunotherapy of superficial bladder cancer.[18] Their studies showed for the first time that bacterial DNA was more immune-stimulatory than mammalian DNA, and that CG-rich oligodeoxyribonucleotides (ODNs) could mimic the effects of bacterial DNA.[19,20] Krieg et al. expanded this concept in 1995 by identifying unmethylated CpG dinucleotides with specific flanking sequences as stimulatory CpG motifs.[21]

The pattern recognition receptor for bacterial DNA and oligonucleotides containing CpG motifs was found to be TLR9[22] which, like TLR7, is localized in endosomes. TLR9, however, is not the only receptor capable of mediating immune responses to DNA. As indicated by the study of Aleck Isaacs' group in 1963, even mammalian DNA that lacks unmethylated CpG motifs can trigger immune responses.[3]

Subsequent studies demonstrated that dsDNA derived from dying mammalian cells, viruses, and bacteria, as well as synthetic dsDNA molecules, can activate dendritic cells or macrophages to upregulate costimulatory molecules and produce cytokines including type I IFN.[23,24] This activation was about 100-fold more potent when the dsDNA was transfected into the cells using a cationic liposome formulation. This novel intracellular DNA recognition pathway appears to be critically involved in type I IFN responses to intracellular bacteria,[25] and possibly DNA viruses in myeloid and non-immune cells. The receptors and signaling adaptors involved in recognition of intracellular non-CpG dsDNA have not been identified.

DENDRITIC CELLS SENSE NUCLEIC ACIDS WITHIN ENDOSOMES VIA TLRs

In contrast to the CARD-containing RNA helicases expressed ubiquitously in immune and non-immune cells throughout an entire organism, TLR7, TLR8, and TLR9 are active mainly in antigen-presenting cell types such as dendritic cells (DCs), monocytes, macrophages, and B cells. An exception to this rule was reported recently: regulatory T cells express TLR8 and respond to TLR8 ligands with reversal of their regulatory function.[26]

Occasional reports have also shown responsiveness of epithelial cells to TLR7 and TLR9 ligands. DC subpopulations differ remarkably in their TLR expression patterns. Human and murine plasmacytoid DCs express endosomally localized TLR7 and TLR9, but lack TLR2, TLR4, and TLR3 that are expressed in myeloid cells.[27-29] Human monocytes and myeloid DCs express mainly TLR8, whereas murine myeloid DCs express TLR7.[30,31] The role of TLR8 in the murine system remains to be elucidated.

In mice, a specialized DC subpopulation known as CD8α+ that is essential for the uptake of apoptotic cells and induction of tolerance has been identified.[32] Interestingly, this cell type lacks TLR7, possibly as a protective mechanism against recognition of self RNA from apoptotic material.[33] However, the CD8α+ DCs preferentially express TLR3 that is activated by dsRNA contained in internalized apoptotic virus-infected cells. TLR3-mediated activation of CD8α+ DCs is required for crosspriming of cytotoxic T lymphocytes against viruses that do not themselves infect DCs. In contrast to the human system, all murine DC subpopulations express functional TLR9.[33]

Due to their high endocytic capacity, DCs are especially well equipped to recognize nucleic acids contained in bacteria or viruses via endosomal TLRs. TLR engagement by particulate antigens (e.g., bacteria or viruses) within endosomes and phagosomes leads to specific maturation processes in the endosomes and phagosomes that prevent complete lysosomal degradation of the antigen and allow efficient antigen processing and presentation on major histocompatibility complex (MHC) class II. This mechanism of inducible phagosome maturation is a specific function of maturing DCs that contributes to their professional antigen-presenting capacity.[34,35]

CELL TYPE-SPECIFIC NUCLEIC ACID RECOGNITION PATHWAYS TRIGGER INNATE ANTI-VIRAL IMMUNITY

The hallmark of plasmacytoid DCs is their ability to produce large amounts of type I IFN in response to viruses or synthetic TLR7 and TLR9 ligands.[36–38] Human plasmacytoid DCs produce 10- to 100-fold more IFN-α than other cell types such as myeloid DCs or macrophages. One possible explanation for this extraordinary IFN production capacity is the high level of constitutive expression of IRF7 in plasmacytoid DCs.[39,40]

Transcription of IFN-α genes is controlled by IRF7. However, IRF7 is not expressed constitutively in most cell types; its expression is induced by a first wave of IFN-β and IFN-α4 released upon initial contact with the virus and signal through the type I IFN receptor, thus allowing further induction of IFN-α genes via IRF7 (type I IFN positive feedback loop).[41]

Plasmacytoid DCs are unique in that IFN-α expression is induced by ligands of TLR7 and TLR9 via a supramolecular complex formed by MyD88, TRAF6, IRAK1/4, and IRF7 that directly activates IRF7[42,43] (Figure 5.1). IRAK-1, IKK-α, and osteopontin have been implicated to participate in the phosphorylation of IRF7, specifically in plasmacytoid DCs.[44–46] Sustained production of type I IFN, however, depends on further IRF7 upregulation via the type I IFN feedback loop also in plasmacytoid DCs, confirming the role of IRF7 as the master regulator of the type I IFN response.[47]

Interestingly, the recruitment of signaling molecules necessary for direct type I IFN induction in plasmacytoid DCs by TLR9 ligands is dependent on retention of the ligand–receptor complexes within early endosomes at a pH value of 6.2 to 6.4.[48,49]

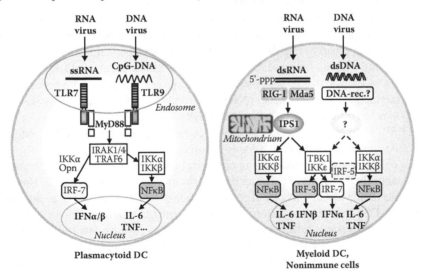

FIGURE 5.1 Cell-type specific nucleic acid recognition by the endosomal and cytosolic pathways.

Delivery of TLR9 ligands to late endosomes with lower pH values leads to lower type I IFN induction and higher expression of inflammatory cytokines and co-stimulatory molecules. Endosomal retention occurs spontaneously in plasmacytoid DCs when TLR9 ligands form aggregates as, for example, CpG ODN 2216,[38,50,51] or are delivered within cationic liposomes,[49,52] immune complexes,[53] and viral particles.

Honda et al. proposed that conventional DCs can also respond to TLR9 ligands with significant type I IFN production when endosomal retention of the ligands is achieved with cationic liposome delivery.[49] A similar mechanism has been proposed for type I IFN induction in plasmacytoid and conventional DCs by TLR7/8 ligands.

We and others have found that in murine plasmacytoid DCs isolated from spleen or cultured bone marrow cells, the production of type I IFN and pro-inflammatory cytokines in response to many enveloped DNA and RNA viruses is largely dependent on MyD88 signaling. *In vitro* studies identified TLR9 as the predominant receptor for herpes simplex virus (HSV) types 1 and 2,[54,55] murine cytomegalovirus (MCMV),[56,57] adenovirus,[58] and baculovirus[59] in plasmacytoid DCs. IL-12 production by conventional DCs isolated from murine spleen in response to HSV-1 and MCMV was also largely TLR9-mediated. However, several studies including our own data have shown that myeloid DCs and macrophages can produce type I IFN in response to DNA viruses in the absence of TLR9.[57,60-62]

The significant contribution of both TLR9–MyD88-dependent and -independent type I IFN induction pathways in the innate immune response against viral infection *in vivo* has been thoroughly dissected in the MCMV infection model.[56,57] We found significant reductions in IFN-α serum levels in TLR9- and MyD88-deficient mice 36 hours after infection when peak levels of IFN-α are observed in the sera of wild-type mice. Only 12 hours later, however, TLR9- and MyD88-deficient mice produced similar amounts of IFN-α as wild-type mice, suggesting that TLR9–MyD88-independent type I IFN production involves delayed kinetics.

Similarly, depletion of plasmacytoid DCs affected only the very early peak of type I IFN expression. In contrast, production of IL-12 (and other inflammatory mediators) was severely impaired in TLR9$^{-/-}$ and MyD88$^{-/-}$ mice as compared to wild-type mice at all time points, reflecting the requirement of TLR9 for IL-12 responses to MCMV in both plasmacytoid and conventional DCs. As a consequence, IFN-γ production by natural killer (NK) cells induced by IL-12 (as well as IL-18 and IL-15) was also reduced in TLR9$^{-/-}$ and MyD88$^{-/-}$ mice.

TLR9$^{-/-}$ and MyD88$^{-/-}$ mice are therefore more susceptible to MCMV infection than wild-type mice in the early phases of infection.

To date, the MCMV model is the only viral infection model that demonstrates a significant role for TLR9 in the innate antiviral immune response *in vivo*. However, the results obtained with this model clearly demonstrate the existence of an additional TLR9–MyD88-independent pattern recognition pathway that is critical for type I IFN production in response to herpes viruses *in vivo*. The putative intracellular dsDNA receptor postulated in several reports[24,25] may also be involved in this response.

Several studies convincingly demonstrated that TLR7 is responsible for recognition of influenza virus,[63,64] Newcastle disease virus (NDV),[65] vesicular stomatitis virus (VSV),[66] coronaviruses,[67] and probably the human immunodeficiency virus (HIV)[13,68] in plasmacytoid DCs. However we have observed that MyD88$^{-/-}$ mice are similarly susceptible as wild-type mice to intranasal infections with influenza virus. The type I IFN response to intravenously injected influenza virus was only partially reduced in the absence of MyD88.[64] The early IFN-α response to systemic infection with VSV is TLR7-dependent and mediated by plasmacytoid DCs only when high viral doses are used. At lower viral doses, TLR7-independent type I IFN production by other cell types is predominant.[66,69]

Interestingly, VSV can act directly on TLR7 in B cells to induce early antiviral IgM secretions.[69] In many cell types except plasmacytoid DCs (e.g., myeloid DCs, macrophages, and fibroblasts), the CARD-containing RIG-I RNA helicase is responsible for triggering the expression of type I IFN and NF-κB target genes in response to influenza virus,[70] VSV,[10] NDV,[12,65] Sendai virus (SV),[10,12,71] and Japanese encephalitis virus (JEV)[12] (Figure 5.1).

During transcription and replication, these viruses generate 5′ triphosphate ssRNA in cytosols. Two recent papers demonstrate that uncapped 5′ triphosphate ssRNA is specifically recognized by RIG-I.[72,73] In contrast, picornavirus family members such as encephalomyocarditis virus (EMCV) that generate cytosolic dsRNA lacking 5′ triphosphate ends are specifically recognized by the CARD-containing Mda5 RNA helicase,[12,74] but not by RIG-I.[72,73] These results demonstrate that protective innate immune responses to RNA viruses are generated by triggering endosomal TLRs and cytosolic RNA helicases in a cell type-specific manner.

Depending on route and kinetics of infection, cellular tropism, viral entry and replication, and immune evasion mechanisms of individual virus strains, one or the other nucleic acid recognition pathways may dominate innate anti-viral immune responses *in vivo*. For example, RIG-I[−/−] mice are clearly more susceptible to infection with VSV and JEV than wild-type or MyD88[−/−] mice.[12] This suggests that despite TLR7-dependent activation of plasmacytoid DCs by these viruses, triggering of the RIG-I pathway in other cell types is essential for immune defense against these infections.

In the case of coronavirus infection [mouse hepatitis virus (MHV) or SARS coronavirus], murine and human plasmacytoid DCs (but not conventional DCs) can produce protective amounts of type I IFN and this response is TLR7-mediated. Depletion of plasmacytoid DCs during MHV infection led to abrogation of the type I IFN response within the first 48 hours post-infection, and to increased disease severity, suggesting a critical role of the initial TLR7-mediated immune defense in infection by rapidly replicating coronaviruses.[67]

Most of the viral infection models described above involve intravenous or intraperitoneal infection leading to systemic spread of the virus. In contrast, Smit and colleagues achieved intratracheal infection of mice with the respiratory syncytial virus (RSV) that replicates in the lungs and causes severe pulmonary inflammation with airway hypersensitivity.[75] In the RSV model, plasmacytoid and conventional DCs were recruited to the lungs and depletion of plasmacytoid DCs led to abrogation of local type I IFN expression, an increase in RSV titers, and exacerbation of inflammatory responses in the lungs. These results support the notion that plasmacytoid DCs perform important immune-stimulatory and regulatory functions during organ-specific viral infection *in vivo*.

Results obtained with knockout mice indicate that RIG-I, Mda5, and IPS-1 (the central signaling adaptor downstream of RNA helicases) are not operative in plasmacytoid DCs.[65,76] However, transfection with *in vitro*-generated RNA or infection with replication-competent RSV triggers IFN-α production in isolated plasmacytoid DCs independently of endosomal TLRs.[77,78] We therefore propose the existence of an additional mechanism for TLR-independent detection of intracellular viral RNA in plasmacytoid DCs that may be relevant for anti-viral defenses. Most likely, TLR-dependent and TLR-independent nucleic acid recognition mechanisms are partially redundant and cooperate in generating protective anti-viral immune responses, especially with respect to type I IFNs that are critical for survival during viral infections before an effective adaptive immune response is generated. This principle of redundant viral pattern recognition mechanisms is underlined by the finding that patients with IRAK4-deficiency who cannot respond to MyD88-dependent TLR ligands are resistant to natural infection by most common viruses.[79] Peripheral blood mononuclear cells (PBMCs) from these patients can produce type I IFN in response to many DNA and RNA viruses.[80]

Most studies to date have concentrated on the role of nucleic acid recognition mechanisms during early phases of innate immune response to infection. Activation of DCs, B cells, and possibly subpopulations of T cells via pattern recognition receptors also shapes the adaptive immune response.[26,35,81,82] The addition of TLR7/8 or TLR9 agonists to vaccine preparations during the priming phase dramatically augments Th1 and CD8 T cell responses after boosting with the same antigen in mice and non-human primates.[83,84] Future studies in infection models will reveal the functions of the different nucleic acid recognition pathways for the adaptive immune response to infection.

PARTICIPATION OF NUCLEIC ACID RECOGNITION RECEPTORS IN PATHOGENESIS OF SYSTEMIC AUTOIMMUNE DISEASES

Humans with systemic lupus erythematosus (SLE) and mouse strains suffering from SLE-like disease often show excesses of apoptotic cells, in large part due to defects in the clearance of apoptotic materials.[85] For example, increased percentages of apoptotic neutrophils are found in the blood of SLE patients,[86,87] and apoptotic cells are found to accumulate in germinal centers in SLE patients and mice due to defects in removal by tingible body macrophages.[88]

Several genetic defects predisposing to SLE in humans and affecting the clearance of apoptotic cells (complement C1q, C2, C3, C4, and DNAse I)[89] have been identified. The nucleic acid recognition receptors, especially TLR9 and TLR7, have been implicated in the activation of autoreactive B cells, plasmacytoid DCs, and conventional DCs by DNA- and RNA-containing autoantigens.

It has been proposed that in specific situations (overloading with apoptotic and necrotic cells, defects in degradation, simultaneous stimulation by infection or exposure to ultraviolet light, defects in peripheral tolerance), tolerance can be broken and self nucleic acids can be recognized as autoadjuvants, directly by autoreactive B cells or by professional antigen-presenting cells, leading to the formation of antinuclear autoantibodies that are hallmarks of human SLE and murine SLE-like syndromes. The autoantibodies bind to the circulating nuclear material and form immune complexes that can then be internalized via the BCR into B cells or via Fc receptors into the endosomal compartments of dendritic cells.

Several groups including ours have found that these DNA- and RNA-containing immune complexes trigger the production of type I IFN and other inflammatory cytokines in plasmacytoid and conventional DCs[90,91] via TLR9[92,93] and TLR7,[15,16,51] respectively. In addition, direct activation of B cell proliferation and antibody secretion via TLR9 and TLR7 has been demonstrated in murine transgenic B cells.[94,95]

Type I interferons (IFN-α and IFN-β) are found at high levels in the sera of most patients with active SLE.[96] In PBMCs of SLE patients, the typical IFN "signature" of gene expression that correlates with disease activity is found.[97,98] Studies of the New Zealand black (NZB)/New Zealand white (NZW) mouse model of lupus,[99,100] and the recently appreciated association between trisomy of the type I IFN cluster on human chromosome 9p and lupus-like autoimmunity,[101] support a central role of IFN-α and -β in SLE pathogenesis. Type I IFN supports B cell differentiation and antibody production[102,103] along with DC differentiation and maturation,[104] and thus may contribute to autoimmunity in SLE. Plasmacytoid DCs identified as the major sources of IFN-α and IFN-β in human SLE[90] are thought to be important players in this systemic autoimmune disorder.[105]

Several lines of evidence support an important *in vivo* role for TLR7 in SLE. Two recent papers have shown that the y chromosome autoimmune accelerator (*yaa*) mutation is due to a duplication and y chromosome translocation of a cluster of x-linked genes containing TLR7. Mice carrying the *yaa* mutation showed increased expression of TLR7 and augmented responses to TLR7 ligands. Addition of the *yaa* mutation to B6.Sle1 or B6/FcγRIIb$^{-/-}$ mice leads to disease exacerbation in these mild SLE-like disease models.[106,107] Berland and colleagues generated knockin mice on the C57BL/6 background expressing autoreactive immunoglobulin that recognizes RNA, ssDNA, and nucleosomes. Backcrossing these mice with TLR7$^{-/-}$ mice showed that these knockin B cells were activated to produce autoantibodies *in vivo* in a TLR7-dependent manner, ultimately causing kidney pathology by immune complex deposition.[108] A paper by Christensen et al. shows that in the absence of TLR7, MRL/Mp$^{lpr/lpr}$ mice failed to produce antibodies against RNA autoantigens, but did produce anti-dsDNA antibodies.[109] TLR7-deficient mice showed lower activation of plasmacytoid DCs and lymphocytes and less kidney disease. The roles played by different cell types expressing TLR7—B cells, plasmacytoid DCs, and conventional DCs—in the pathogenesis of the disease are still unclear.

Results from mouse models focusing on the role of TLR9 in SLE are more controversial. Christensen et al. found impressive decreases in anti-dsDNA autoantibody production in TLR9-deficient MRL/Mp$^{lpr/lpr}$ mice (measured by Hep2 cells and *Crithidia luciliae* kinetoplast staining)[109,110] that were not confirmed in other studies measuring anti-DNA antibody titers by ELISA using the MRL/Mp$^{lpr/lpr}$, C57BL/6$^{lpr/lpr}$ or Ali5 model of SLE.[111–113] However, all models studied revealed significant increases in disease activity measured by glomerulonephritis severity and decreases in survival rate in TLR9-deficient mice. In some models, this exacerbation of disease in the absence of TLR9 was accompanied by an increase in the percentage of mice producing antibodies against RNA autoantigens and higher titers of these RNA-reactive autoantibodies.

The divergent results obtained with TLR7- and TLR9-deficient mice in murine lupus were unexpected, because both receptors share similar expression patterns and signaling pathways. The results can be interpreted in various ways. Triggering of TLR9 (but not TLR7) by nuclear autoantigens within immune complexes may serve a regulatory role in this disease, for example, by stimulating plasmacytoid DCs to promote regulatory T cell development, as demonstrated recently.[114] Differences in the signaling pathways downstream of TLR9 and TLR7 that have not been elucidated yet may confer different functions on these receptors in the same cell types. One report indicates that DNA-reactive antibodies may promote the clearance of apoptotic material and exert protective effects.[115]

Another important difference between TLR7 and TLR9 is the fact that TLR7 is efficiently upregulated in B cells and DCs upon stimulation by type I IFNs,[15,30,95] whereas TLR9 is not upregulated. This provides an enhancement mechanism for type I IFN induction by RNA autoantigens that is not available for TLR9 ligands. Therefore RNA-containing immune complexes that expand when DNA-containing immune complexes are reduced in the absence of TLR9 may be more pathogenic and lead to disease exacerbation in TLR9-deficient mice.[109]

It will be interesting in the future to investigate the specific contributions of the different cell types responding to autoantigens via TLR7 and TLR9 (mainly DC and B cells) in the course of lupus disease development. The roles TLR7 and TLR9 play in human SLE are also unclear. Several studies failed to find an association between polymorphisms in the TLR9 gene and occurrence of SLE.[116,117] Similar studies of TLR7 have not yet been performed. Other nucleic acid recognition receptors such as RIG-I, Mda5, and the putative cytosolic dsDNA receptor may be involved in the generation of autoimmune responses, but their functions in this regard remain to be investigated.

PREVENTION OF IMMUNE ACTIVATION BY SELF NUCLEIC ACIDS IN HEALTHY ORGANISMS

In the face of the constant emergence of a huge number of apoptotic cells in tissues with high cell turnover rates, the rapid clearance of apoptotic material and degradation of free nucleic acids by nucleases are critical to prevent immune stimulation by self nucleic acids leading to autoimmunity. The essential protective role of nucleases is revealed by the development of severe autoimmune disorders in humans and mice with genetic defects in these enzymes. DNAse I-deficient mice develop SLE-like diseases along with antinuclear antibodies and glomerulonephritis due to immune complex deposition in the kidneys.[118]

Interestingly, SLE patients have lower levels of DNAse I in their sera than healthy individuals,[118] and mutations in the DNAse I gene were found in two Japanese patients with SLE.[119] An association of DNAse I polymorphisms with anti-RNP and anti-DNA antibody production was found in Korean SLE patients.[120]

In addition to the presumed function of this extracellular nuclease in preventing autoimmunity leading to SLE, Nagata's group found that DNAse II localized in the phagosomal compartments of macrophages may also play an important role in preventing immune responses to self DNA. DNAse II-deficient embryos die *in utero* because of a strong activation of type I IFN and IFN-responsive genes that disturbs hematopoiesis and leads to lethal anemia.[121] DNAse II is reponsible for the cleavage of DNA from apoptotic cells and nuclei expelled from erythroid progenitors that are engulfed by macrophages.[121,122] In the absence of DNAse II, the macrophages in liver, thymus, spleen, and lymph nodes contain undigested DNA and strongly express IFN-β and IFN-stimulated genes.[121–123] Interestingly, however, TLR9 was not involved in the generation of this type I IFN response.[123]

Genetic association studies did not reveal an increased risk for SLE in general, but noted a higher risk of renal disease in SLE patients with single nucleotide polymorphisms in their DNAse II genes.[124] Mice lacking DNAse III (also called Trex1), a 3′-5′ DNA exonuclease localized in the nucleus, developed inflammatory myocarditis leading to heart failure and premature death.[125] The exact function of Trex1 and the mechanism of its protective role against inflammation are still unclear. However, four mutations were found recently in patients with an enigmatic autoimmune disease called Aicardi–Goutieres syndrome (AGS), one of which was reported in the Trex1 gene.[126] Three other mutations were found in genes encoding subunits of ribonuclease H2.[127] AGS is an inherited encephalopathy that in many ways resembles encephalitis caused by congenital viral infection not detected in this syndrome. Elevated IFN-α and increased numbers of lymphocytes in the cerebrospinal fluid are characteristic of AGS.[128,129]

A critical role in the pathogenesis of this disorder has been attributed to IFN-α because transgenic mice overexpressing IFN-α in astrocytes showed similar pathology.[130] Significant overlap between AGS and SLE was noted; both are associated with increased IFN-α levels.[131] Some AGS patients develop vasculitic skin lesions resembling those of SLE patients. Intracranial calcifications similar to those seen in AGS are found in some SLE patients with cerebral disease manifestations.[132] The nucleic acid recognition receptors leading to IFN-α induction in these syndromes and the cell types responsible for their production are still unknown. Plasmacytoid DCs may very well be involved because they are recruited to the brain during infection and inflammation.[133,134] These reports demonstrate that nucleases are critical for avoiding autoimmunity, and underscore the role of nucleic acid recognition pathways leading to type I IFN expression in different autoimmune disorders.

In addition to degradation of excess nucleic acids, other means of preventing immune activation by self nucleic acids have been suggested. For example, eukaryotic cells can distinguish intracellular self from viral RNA. The RNA transcribed in the nuclei of eukaryotic cells undergoes extensive post-transcriptional modification such as incorporation of 2′ O-methylated nucleotides, pseudouridines, and 2-thiouridines, as well as addition of 7-methyl-guanosine caps. These modifications can abrogate activation of TLR7/8 and triggering of RIG-I by RNA.[15,72,73,135] Thus inadvertent triggering of the anti-viral immune response program by self RNA is avoided. How DNA derived from pathogens can be distinguished from mammalian DNA on a molecular basis has not been determined conclusively; both DNAs can trigger immune responses when transfected into mammalian cells.

REFERENCES

1. Isaacs, A., Cox, R. A., and Rotem, Z. 1963. Foreign nucleic acids as the stimulus to make interferon. *Lancet* 2: 113.
2. Jensen, K. E. et al. 1963. Interferon responses of chick embryo fibroblasts to nucleic acids and related compounds. *Nature* 200: 433.
3. Rotem, Z., Cox, R. A., and Isaacs, A. 1963. Inhibition of virus multiplication by foreign nucleic acid. *Nature* 197: 564.
4. Field, A. K. et al. 1967. Inducers of interferon and host resistance IV. Double-stranded replicative form RNA (MS2-Ff-RNA) from *E. coli* infected with MS2 coliphage. *Proc Natl Acad Sci USA* 58: 2102.
5. Field, A. K. et al. 1967. Inducers of interferon and host resistance II. Multistranded synthetic polynucleotide complexes. *Proc Natl Acad Sci USA* 58: 1004.
6. Lampson, G. P. et al. 1967. Inducers of interferon and host resistance I. Double-stranded RNA from extracts of *Penicillium funiculosum*. *Proc Natl Acad Sci USA* 58: 782.
7. Tytell, A. A. et al. 1967. Inducers of interferon and host resistance III. Double-stranded RNA from reovirus type 3 virions (reo 3-RNA). *Proc Natl Acad Sci USA* 58: 1719.
8. Alexopoulou, L. et al. 2001. Recognition of double-stranded RNA and activation of NF-κB by Toll- like receptor 3. *Nature* 413: 732.
9. Schulz, O. et al. 2005. Toll-like receptor 3 promotes crosspriming to virus-infected cells. *Nature* 433: 887.
10. Yoneyama, M. et al. 2004. The RNA helicase RIG-I has an essential function in double-stranded RNA-induced innate antiviral responses. *Nat Immunol* 5: 730.
11. Yoneyama, M. et al. 2005. Shared and unique functions of the DExD/H box helicases RIG-I, MDA5, and LGP2 in antiviral innate immunity. *J Immunol* 175: 2851.
12. Kato, H. et al. 2006. Differential roles of MDA5 and RIG-I helicases in the recognition of RNA viruses. *Nature* 441: 101.
13. Heil, F. et al. 2004. Species-specific recognition of single-stranded RNA via toll-like receptors 7 and 8. *Science* 303: 1526.
14. Diebold, S. S. et al. 2003. Viral infection switches non-plasmacytoid dendritic cells into high interferon producers. *Nature* 424: 324.
15. Savarese, E. et al. 2006. U1 small nuclear ribonucleoprotein immune complexes induce type I interferon in plasmacytoid dendritic cells through TLR7. *Blood* 107: 3229.
16. Vollmer, J. et al. 2005. Immune stimulation mediated by autoantigen binding sites within small nuclear RNAs involves Toll-like receptors 7 and 8. *J Exp Med* 202: 1575.
17. Tokunaga, T. et al. 1984. Antitumor activity of deoxyribonucleic acid fraction from *Mycobacterium bovis* BCG I. Isolation, physicochemical characterization, and antitumor activity. *J Natl Cancer Inst* 72: 955.
18. Totterman, T. H., Loskog, A., and Essand, M. 2005. The immunotherapy of prostate and bladder cancer. *BJU Int* 96: 728.
19. Yamamoto, S. et al. 1992. Unique palindromic sequences in synthetic oligonucleotides are required to induce IFN [correction of INF] and augment IFN-mediated [correction of INF] natural killer activity. *J Immunol* 148: 4072.
20. Yamamoto, S. et al. 1992. DNA from bacteria, but not from vertebrates, induces interferons, activates natural killer cells and inhibits tumor growth. *Microbiol Immunol* 36: 983.
21. Krieg, A. M. et al. 1995. CpG motifs in bacterial DNA trigger direct B cell activation. *Nature* 374: 546.
22. Hemmi, H. et al. 2000. A Toll-like receptor recognizes bacterial DNA. *Nature* 408: 740.
23. Ishii, K. J. et al. 2001. Genomic DNA released by dying cells induces the maturation of APCs. *J Immunol* 167: 2602.

24. Ishii, K. J. et al. 2006. A Toll-like receptor-independent antiviral response induced by double-stranded B-form DNA. *Nat Immunol* 7: 40.
25. Stetson, D. B. and Medzhitov, R. 2006. Recognition of cytosolic DNA activates an IRF3-dependent innate immune response. *Immunity* 24: 93.
26. Peng, G. et al. 2005. Toll-like receptor 8-mediated reversal of CD4+ regulatory T cell function. *Science* 309: 1380.
27. Krug, A. et al. 2001. Toll-like receptor expression reveals CpG DNA as a unique microbial stimulus for plasmacytoid dendritic cells which synergizes with CD40 ligand to induce high amounts of IL-12. *Eur J Immunol* 31: 3026.
28. Kadowaki, N. et al. 2001. Subsets of human dendritic cell precursors express different Toll-like receptors and respond to different microbial antigens. *J Exp Med* 194: 863.
29. Jarrossay, D. et al. 2001. Specialization and complementarity in microbial molecule recognition by human myeloid and plasmacytoid dendritic cells. *Eur J Immunol* 31: 3388.
30. Bekeredjian-Ding, I. et al. 2006. T cell-independent, TLR-induced IL-12p70 production in primary human monocytes. *J Immunol* 176: 7438.
31. Gorden, K. B. et al. 2005. Synthetic TLR agonists reveal functional differences between human TLR7 and TLR8. *J Immunol* 174: 1259.
32. Liu, K. et al. 2002. Immune tolerance after delivery of dying cells to dendritic cells in situ. *J Exp Med* 196: 1091.
33. Edwards, A. D. et al. 2003. Toll-like receptor expression in murine DC subsets: lack of TLR7 expression by CD8α+ DCs correlates with unresponsiveness to imidazoquinolines. *Eur J Immunol* 33: 827.
34. Blander, J. M. and Medzhitov, R. 2006. Toll-dependent selection of microbial antigens for presentation by dendritic cells. *Nature* 440: 808.
35. Blander, J. M. and Medzhitov, R. 2006. On regulation of phagosome maturation and antigen presentation. *Nat Immunol* 7: 1029.
36. Cella, M. et al. 1999. Plasmacytoid monocytes migrate to inflamed lymph nodes and produce large amounts of type I interferon. *Nat Med* 5: 919.
37. Siegal, F. P. et al. 1999. The nature of the principal type 1 interferon-producing cells in human blood. *Science* 284: 1835.
38. Krug, A. et al. 2001. Identification of CpG oligonucleotide sequences with high induction of IFN-alpha/beta in plasmacytoid dendritic cells. *Eur J Immunol* 31: 2154.
39. Izaguirre, A. et al. 2003. Comparative analysis of IRF and IFN-alpha expression in human plasmacytoid and monocyte-derived dendritic cells. *J Leukoc Biol*.
40. Kerkmann, M. et al. 2003. Activation with CpG-A and CpG-B oligonucleotides reveals two distinct regulatory pathways of type I IFN synthesis in human plasmacytoid dendritic cells. *J Immunol* 170: 4465.
41. Honda, K. and Taniguchi, T. 2006. IRFs: master regulators of signalling by Toll-like receptors and cytosolic pattern-recognition receptors. *Nat Rev Immunol* 6: 644.
42. Honda, K. et al. 2004. Role of a transductional-transcriptional processor complex involving MyD88 and IRF-7 in Toll-like receptor signaling. *Proc Natl Acad Sci USA* 101: 15416.
43. Kawai, T. et al. 2004. Interferon-α induction through Toll-like receptors involves a direct interaction of IRF7 with MyD88 and TRAF6. *Nat Immunol* 5: 1061.
44. Hoshino, K. et al. 2006. IκB kinase-α is critical for interferon-α production induced by Toll-like receptors 7 and 9. *Nature* 440: 949.
45. Shinohara, M. L. et al. 2006. Osteopontin expression is essential for interferon-α production by plasmacytoid dendritic cells. *Nat Immunol* 7: 498.
46. Uematsu, S. et al. 2005. Interleukin-1 receptor-associated kinase-1 plays an essential role for Toll-like receptor (TLR)7- and TLR9-mediated interferon-α induction. *J Exp Med* 201: 915.

47. Honda, K. et al. 2005. IRF-7 is the master regulator of type I interferon-dependent immune responses. *Nature* 434: 772.
48. Guiducci, C. et al. 2006. Properties regulating the nature of the plasmacytoid dendritic cell response to Toll-like receptor 9 activation. *J Exp Med* 203: 1999.
49. Honda, K. et al. 2005. Spatiotemporal regulation of MyD88-IRF-7 signalling for robust type-I interferon induction. *Nature* 434: 1035.
50. Kerkmann, M. et al. 2005. Spontaneous formation of nucleic acid-based nanoparticles is responsible for high interferon-α induction by CpG-A in plasmacytoid dendritic cells. *J Biol Chem* 280: 8086.
51. Barrat, F. J. et al. 2005. Nucleic acids of mammalian origin can act as endogenous ligands for Toll-like receptors and may promote systemic lupus erythematosus. *J Exp Med* 202: 1131.
52. Yasuda, K. et al. 2005. Endosomal translocation of vertebrate DNA activates dendritic cells via TLR9-dependent and -independent pathways. *J Immunol* 174: 6129.
53. Means, T. K. et al. 2005. Human lupus autoantibody–DNA complexes activate DCs through cooperation of CD32 and TLR9. *J Clin Invest* 115: 407.
54. Krug, A. et al. 2004. Herpes simplex virus type 1 activates murine natural interferon-producing cells through toll-like receptor 9. *Blood* 103: 1433.
55. Lund, J. et al. 2003. Toll-like receptor 9-mediated recognition of herpes simplex virus-2 by plasmacytoid dendritic cells. *J Exp Med* 198: 513.
56. Krug, A. et al. 2004. TLR9-dependent recognition of MCMV by IPC and DC generates coordinated cytokine responses that activate antiviral NK cell function. *Immunity* 21: 107.
57. Delale, T. et al. 2005. MyD88-dependent and -independent murine cytomegalovirus sensing for IFN-α release and initiation of immune responses *in vivo*. *J Immunol* 175: 6723.
58. Basner-Tschakarjan, E. et al. 2006. Adenovirus efficiently transduces plasmacytoid dendritic cells resulting in TLR9-dependent maturation and IFN-α production. *J Gene Med*.
59. Abe, T. et al. 2005. Involvement of the Toll-like receptor 9 signaling pathway in the induction of innate immunity by baculovirus. *J Virol* 79: 2847.
60. Hochrein, H. et al. 2004. Herpes simplex virus type-1 induces IFN-α production via Toll-like receptor 9-dependent and -independent pathways. *Proc Natl Acad Sci USA* 101: 11416.
61. Malmgaard, L. et al. 2004. Viral activation of macrophages through TLR-dependent and -independent pathways. *J Immunol* 173: 6890.
62. Feyer, G., Drechsel, L., and Freudenberg, M. 2006. Adenovirus infection induces strong early production of type I IFN from splenic myeloid DCs *in vivo* resulting in LPS hypersensitivity. 20th Annual Meeting of EMDS (abstract).
63. Diebold, S. S. et al. 2004. Innate antiviral responses by means of TLR7-mediated recognition of single-stranded RNA. *Science* 303: 1529.
64. Barchet, W. et al. 2005. Dendritic cells respond to influenza virus through TLR7- and PKR-independent pathways. *Eur J Immunol* 35: 236.
65. Kato, H. et al. 2005. Cell type-specific involvement of RIG-I in antiviral response. *Immunity* 23: 19.
66. Lund, J. M. et al. 2004. Recognition of single-stranded RNA viruses by Toll-like receptor 7. *Proc Natl Acad Sci USA* 101: 5598.
67. Cervantes-Barragan, L. et al. 2006. Control of coronavirus infection through plasmacytoid dendritic cell-derived type I interferon. *Blood*.
68. Beignon, A. S. et al. 2005. Endocytosis of HIV-1 activates plasmacytoid dendritic cells via Toll-like receptor-viral RNA interactions. *J Clin Invest* 115: 3265.

69. Fink, K. et al. 2006. Early type I interferon-mediated signals on B cells specifically enhance antiviral humoral responses. *Eur J Immunol* 36: 2094.
70. Siren, J. et al. 2006. Retinoic acid inducible gene-I and Mda-5 are involved in influenza A virus-induced expression of antiviral cytokines. *Microbes Infect* 8: 2013.
71. Rothenfusser, S. et al. 2005. RNA helicase Lgp2 inhibits TLR-independent sensing of viral replication by retinoic acid-inducible gene-I. *J Immunol* 175: 5260.
72. Hornung, V. et al. 2006. 5′ triphosphate RNA is the ligand for RIG-I. *Science.*
73. Pichlmair, A. et al. 2006. RIG-I-mediated antiviral responses to single-stranded RNA bearing 5′ phosphates. *Science.*
74. Gitlin, L. et al. 2006. Essential role of Mda-5 in type I IFN responses to polyriboinosi nic:polyribocytidylic acid and encephalomyocarditis picornavirus. *Proc Natl Acad Sci USA* 103: 8459.
75. Smit, J. J., Rudd, B. D., and Lukacs, N. W. 2006. Plasmacytoid dendritic cells inhibit pulmonary immunopathology and promote clearance of respiratory syncytial virus. *J Exp Med* 203: 1153.
76. Sun, Q. et al. 2006. The specific and essential role of MAVS in antiviral innate immune responses. *Immunity* 24: 633.
77. Loseke, S. et al. 2006. *In vitro*-generated viral double-stranded RNA in contrast to polyinosinic:polycytidylic acid induces interferon-α in human plasmacytoid dendritic cells. *Scand J Immunol* 63: 264.
78. Hornung, V. et al. 2004. Replication-dependent potent IFN-α induction in human plasmacytoid dendritic cells by a single-stranded RNA virus. *J Immunol* 173: 5935.
79. Casanova, J. L. and Abel, L. 2004. The human model: a genetic dissection of immunity to infection in natural conditions. *Nat Rev Immunol* 4: 55.
80. Yang, K. et al. 2005. Human TLR-7-, -8-, and -9-mediated induction of IFN-α/β and -λ is IRAK-4 dependent and redundant for protective immunity to viruses. *Immunity* 23: 465.
81. Schnare, M. et al. 2001. Toll-like receptors control activation of adaptive immune responses. *Nat Immunol* 2: 947.
82. Pasare, C. and Medzhitov, R. 2005. Toll-like receptors: linking innate and adaptive immunity. *Adv Exp Med Biol* 560: 11.
83. Heit, A. et al. 2005. Protective CD8 T cell immunity triggered by CpG-protein conjugates competes with the efficacy of live vaccines. *J Immunol* 174: 4373.
84. Wille-Reece, U. et al. 2006. Toll-like receptor agonists influence the magnitude and quality of memory T cell responses after prime-boost immunization in nonhuman primates. *J Exp Med* 203: 1249.
85. Gaipl, U. S. et al. 2005. Impaired clearance of dying cells in systemic lupus erythematosus. *Autoimmun Rev* 4: 189.
86. Courtney, P. A. et al. 1999. Increased apoptotic peripheral blood neutrophils in systemic lupus erythematosus: relations with disease activity, antibodies to double stranded DNA, and neutropenia. *Ann Rheum Dis* 58: 309.
87. Ren, Y. et al. 2003. Increased apoptotic neutrophils and macrophages and impaired macrophage phagocytic clearance of apoptotic neutrophils in systemic lupus erythematosus. *Arthritis Rheum* 48: 2888.
88. Hanayama, R. et al. 2004. Autoimmune disease and impaired uptake of apoptotic cells in MFG-E8-deficient mice. *Science* 304: 1147.
89. Alarcon-Riquelme, M. E. 2005. The genetics of systemic lupus erythematosus. *J Autoimmun* 25 Suppl: 46.
90. Vallin, H. et al. 1999. Patients with systemic lupus erythematosus (SLE) have a circulating inducer of interferon-α (IFN-α) production acting on leucocytes resembling immature dendritic cells. *Clin Exp Immunol* 115: 196.

91. Bave, U., Alm, G. V., and Ronnblom, L. 2000. The combination of apoptotic U937 cells and lupus IgG is a potent IFN-α inducer. *J Immunol* 165: 3519.
92. Means, T. K. and Luster, A. D. 2005. Toll-like receptor activation in the pathogenesis of systemic lupus erythematosus. *Ann NY Acad Sci* 1062: 242.
93. Boule, M. W. et al. 2004. Toll-like receptor 9-dependent and -independent dendritic cell activation by chromatin-immunoglobulin G complexes. *J Exp Med* 199: 1631.
94. Leadbetter, E. A. et al. 2002. Chromatin–IgG complexes activate B cells by dual engagement of IgM and Toll-like receptors. *Nature* 416: 603.
95. Lau, C. M. et al. 2005. RNA-associated autoantigens activate B cells by combined B cell antigen receptor/Toll-like receptor 7 engagement. *J Exp Med* 202: 1171.
96. Shi, S. N. et al. 1987. Serum interferon in systemic lupus erythematosus. *Br J Dermatol* 117: 155.
97. Bennett, L. et al. 2003. Interferon and granulopoiesis signatures in systemic lupus erythematosus blood. *J Exp Med* 197: 711.
98. Baechler, E. C. et al. 2003. Interferon-inducible gene expression signature in peripheral blood cells of patients with severe lupus. *Proc Natl Acad Sci USA* 100: 2610.
99. Mathian, A. et al. 2005. IFN-α induces early lethal lupus in preautoimmune (New Zealand Black x New Zealand White) F1 but not in BALB/c mice. *J Immunol* 174: 2499.
100. Santiago-Raber, M. L. et al. 2003. Type-I interferon receptor deficiency reduces lupus-like disease in NZB mice. *J Exp Med* 197: 777.
101. Zhuang, H. et al. 2006. Lupus-like disease and high interferon levels corresponding to trisomy of the type I interferon cluster on chromosome 9p. *Arthritis Rheum* 54: 1573.
102. Jego, G. et al. 2003. Plasmacytoid dendritic cells induce plasma cell differentiation through type I interferon and interleukin 6. *Immunity* 19: 225.
103. Le Bon, A. et al. 2001. Type I interferons potently enhance humoral immunity and can promote isotype switching by stimulating dendritic cells *in vivo*. *Immunity* 14: 461.
104. Blanco, P. et al. 2001. Induction of dendritic cell differentiation by IFN-α in systemic lupus erythematosus. *Science* 294: 1540.
105. Banchereau, J. and Pascual, V. 2006. Type I interferon in systemic lupus erythematosus and other autoimmune diseases. *Immunity* 25: 383.
106. Subramanian, S. et al. 2006. A TLR7 translocation accelerates systemic autoimmunity in murine lupus. *Proc Natl Acad Sci USA* 103: 9970.
107. Pisitkun, P. et al. 2006. Autoreactive B cell responses to RNA-related antigens due to TLR7 gene duplication. *Science* 312: 1669.
108. Berland, R. et al. 2006. Toll-like receptor 7-dependent loss of B cell tolerance in pathogenic autoantibody knockin mice. *Immunity* 25: 429.
109. Christensen, S. R. et al. 2006. Toll-like receptor 7 and TLR9 dictate autoantibody specificity and have opposing inflammatory and regulatory roles in murine model of lupus. *Immunity* 25: 417.
110. Christensen, S. R. et al. 2005. Toll-like receptor 9 controls anti-DNA autoantibody production in murine lupus. *J Exp Med* 202: 321.
111. Wu, X. and Peng, S. L. 2006. Toll-like receptor 9 signaling protects against murine lupus. *Arthritis Rheum* 54: 336.
112. Yu, P. et al. 2006. Toll-like receptor 9-independent aggravation of glomerulonephritis in a novel model of SLE. *Int Immunol* 18: 1211.
113. Lartigue, A. et al. 2006. Role of TLR9 in anti-nucleosome and anti-DNA antibody production in LPR mutation-induced murine lupus. *J Immunol* 177: 1349.
114. Moseman, E. A. et al. 2004. Human plasmacytoid dendritic cells activated by CpG oligodeoxynucleotides induce generation of CD4+CD25+ regulatory T cells. *J Immunol* 173: 4433.

115. Werwitzke, S. et al. 2005. Inhibition of lupus disease by anti-double-stranded DNA antibodies of the IgM isotype in the (NZB x NZW)F1 mouse. *Arthritis Rheum* 52: 3629.
116. De Jager, P. L. et al. 2006. Genetic variation in toll-like receptor 9 and susceptibility to systemic lupus erythematosus. *Arthritis Rheum* 54: 1279.
117. Hur, J. W. et al. 2005. Association study of Toll-like receptor 9 gene polymorphism in Korean patients with systemic lupus erythematosus. *Tissue Antigens* 65: 266.
118. Napirei, M. et al. 2000. Features of systemic lupus erythematosus in DNAse I-deficient mice. *Nat Genet* 25: 177.
119. Yasutomo, K. et al. 2001. Mutation of DNAse I in people with systemic lupus erythematosus. *Nat Genet* 28: 313.
120. Shin, H. D. et al. 2004. Common DNAse I polymorphism associated with autoantibody production among systemic lupus erythematosus patients. *Hum Mol Genet* 13: 2343.
121. Yoshida, H. et al. 2005. Lethal anemia caused by interferon-β produced in mouse embryos carrying undigested DNA. *Nat Immunol* 6: 49.
122. Kawane, K. et al. 2003. Impaired thymic development in mouse embryos deficient in apoptotic DNA degradation. *Nat Immunol* 4: 138.
123. Okabe, Y. et al. 2005. Toll-like receptor-independent gene induction program activated by mammalian DNA escaped from apoptotic DNA degradation. *J Exp Med* 202: 1333.
124. Shin, H. D. et al. 2005. DNAse II polymorphisms associated with risk of renal disorder among systemic lupus erythematosus patients. *J Hum Genet* 50: 107.
125. Morita, M. et al. 2004. Gene-targeted mice lacking the Trex1 (DNAse III) 3'-5' DNA exonuclease develop inflammatory myocarditis. *Mol Cell Biol* 24: 6719.
126. Crow, Y. J. et al. 2006. Mutations in the gene encoding the 3'-5' DNA exonuclease TREX1 cause Aicardi–Goutieres syndrome at the AGS1 locus. *Nat Genet* 38: 917.
127. Crow, Y. J. et al. 2006. Mutations in genes encoding ribonuclease H2 subunits cause Aicardi–Goutieres syndrome and mimic congenital viral brain infection. *Nat Genet* 38: 910.
128. Goutieres, F. 2005. Aicardi–Goutieres syndrome. *Brain Dev* 27: 201.
129. Lebon, P. et al. 1988. Intrathecal synthesis of interferon-α in infants with progressive familial encephalopathy. *J Neurol Sci* 84: 201.
130. Akwa, Y. et al. 1998. Transgenic expression of IFN-α in the central nervous system of mice protects against lethal neurotropic viral infection but induces inflammation and neurodegeneration. *J Immunol* 161: 5016.
131. De Laet, C. et al. 2005. Phenotypic overlap between infantile systemic lupus erythematosus and Aicardi–Goutieres syndrome. *Neuropediatrics* 36: 399.
132. Raymond, A. A. et al. 1996. Brain calcification in patients with cerebral lupus. *Lupus* 5: 123.
133. Pashenkov, M. et al. 2002. Recruitment of dendritic cells to the cerebrospinal fluid in bacterial neuroinfections. *J Neuroimmunol* 122: 106.
134. Curtin, J. F. et al. 2006. Fms-like tyrosine kinase 3 ligand recruits plasmacytoid dendritic cells to the brain. *J Immunol* 176: 3566.
135. Kariko, K. et al. 2005. Suppression of RNA recognition by Toll-like receptors: impact of nucleoside modification and the evolutionary origin of RNA. *Immunity* 23: 165.

II

Mechanisms and Therapeutic Applications of Immunomodulatory DNA

6 Natural DNA Recognition by Toll-Like Receptor 9 Does Not Rely upon CpG Motifs

Role of Endosomal Compartmentation

Tobias Haas, Frank Schmitz, Antje Heit, and Hermann Wagner

ABSTRACT

We discuss why evolutionary pressure may have exiled nucleic acid sensing TLRs (TLR3, TLR7/8, and TLR9) to endosomal compartments of immune cells. This expression pattern contrasts with that of TLR1, 2, 4, 15, and 6 and TLR5 known to be anchored on the cell surface. We argue that cell-surface expressed TLRs primarily recognize unique and conserved pathogen-derived structures such as endotoxin that are not present in the host. Thus, a failure of self/non-self discrimination is less probable. Conversely, structural differences among pathogen-derived versus host-derived nucleotides are not as stringent as to exclude the potential for self reactivity. Specifically, we review experimental conditions under which self DNA and single-stranded (ss) DNA devoid of any CpG motifs activate TLR9. The type of endosomal of DNA in plasmacytoid dendritic cells (PDCs) trafficking (early versus late endosomes) rather than the DNA sequence itself determines the type of cytokines produced (type I interferons versus pro-inflammatory cytokines such as tumor necrosis factor (TNF)-α and interleukin (IL)-12). Since enforced endosomal translocation of host (self) DNA activates TLR9, it appears that under homeostatic conditions the poor endosomal translocation of self DNA represents a major control mechanism for self versus non-self (pathogen) DNA discrimination.

INTRODUCTION

Recently, germ line-encoded, thus non-clonal Toll-like receptors (TLRs) have attracted much attention. Upon recognition of distinct microbial components known as pathogen-associated molecular patterns (PAMPs), TLRs activate immune cells

77

including macrophages, dendritic cells, B cells, and certain subsets of T cells.[1-3] TLRs are divided into subfamilies. For example, the subfamily consisting of TLR1, TLR2, and TLR6 recognizes lipopeptides, while TLR3, TLR7/8, and TLR9 recognize nucleic acids.

Certain TLRs (TLR1, 2, 4, 5, and 6) are expressed on the cell surface while the TLRs that recognize nucleic acids (TLR3, 7, 8, 9) are confined to endosomes. It follows that cell surface membrane-expressed TLRs sample the extracellular milieu for their respective ligands, while ligands for endosomal TLRs, i.e., nucleic acids, require internalization (endosomal translocation) before signaling can be initiated. Why nucleic acid-specific TLRs were exiled to endosomes in the course of evolution is unclear.

The current argument holds that TLRs recognize PAMPs, which represent conserved molecular patterns essential for pathogen survival but not present in the host. From this point of view, TLRs appear to be ideal candidates to discriminate self from non-self. However, the ligand specificity of TLRs is still poorly understood. For example, TLR4 senses lipopolysaccharide (LPS) and reportedly also structurally unrelated ligands such as the fusion protein of respiratory syncytial virus (RSV) and heat shock protein (HSP).[1] Information from direct ligand–TLR interaction studies is scarce.

This chapter discusses experimental data on TLR9, which has been shown to get activated by ssDNA rich in unmethylated CpG motifs.[4,5] Since CpG motifs are suppressed in the vertebrate genome and, apart from active promoter regions, cytosines are mostly methylated,[6] self versus non-self (pathogen) DNA discrimination by TLR9 has been thought to rely on such structural differences, although most of the relevant data was generated using phosphorothioate (PS)-modified DNA instead of natural phosphodiester (PD) DNA. Under conditions of "enforced" endosomal translocation, however, TLR9 recognizes both host DNA and ss natural (PD) DNA devoid of CpG motifs. Furthermore, the endosomal compartmentation (early versus late endosomes) rather than the DNA sequence itself controls the type of cytokines (type I interferon versus pro-inflammatory cytokines) produced by PDCs. These data imply that endosomally expressed TLR9 is in principle poor in discriminating self from foreign DNA. The endosomal localization, however, appears to prevent the access of extra cellular self-DNA to TLR9, thus protecting it from constant activation by natural PD self-DNA. On the other hand, this barrier can be breached easily, for example, by infectious DNA viruses. As a consequence, the endosomal localization of TLR9 helps to ensure that its activation is driven primarily by foreign (infectious) DNA.

RESULTS AND DISCUSSION

From a historical perspective, Yamamoto and coworkers[7] and Pisetsky and coworkers[8] made the landmark observations that (1) a palindromic DNA motif centered on a CpG dinucleotide activates mouse spleen cells and human peripheral blood mononuclear cells (PBMCs) to produce cytokines and (2) naked bacterial DNA is mitogenic to B cells. Since DNA and oligodeoxynucleotides (ODNs) containing natural PD linkages become rapidly degraded by ubiquitous DNase activity (its *in vitro* half life is about 20 min), most subsequent studies used PS-modified ODNs to mimic pathogen-derived DNA.[5] By iteratively changing the sequence of immune-stimulating ss (PS)

ODNs, Krieg and associates reported the key observation that a (PS) ODN motif in which an unmethylated CpG dinucleotide is flanked by two 5′ purines and two 3′ pyrimidines displays mitogenicity to B cells.[9] This important observation suggested a link between the immune defense based on the recognition of CpG motifs (enriched in microbial DNA) and poor expression of CpG motifs in vertebrate DNA (CpG-motif suppression). Krieg et al. also noted that at low concentrations of mitogenic CpG ODN, B cell activation was strongly augmented by B cell receptor (BCR) crosslinking, somehow anticipating the situation in "rheumatoid" B cells. In line with Krieg's data on B cells, our own studies revealed that immune-stimulating (PS) CpG ODN sequences specifically activated macrophages and dendritic cells (DCs) to express co-stimulatory molecules (CD40, CD80) and produce an array of pro-inflammatory cytokines.[10]

Our subsequent finding that (PS) CpG ODNs signal via MyD88[11] led to the discovery that (PS) CpG ODNs are sequence specifically recognized by TLR9.[12,13] Later it was shown that TLR9, as with other TLRs that recognize nucleic acids, is expressed within endosomes, but not at the cell surface.[14,15] Why then do we question whether ssDNAs are sequences specifically recognized by endosomally expressed TLR9?

The first caveat arose when we analyzed what we termed *sequence- and species-specific TLR9 activation*. By genetic complementation of TLR9 deficient cells with either human or mouse TLR9, we noted that PS ODN driven activation of human or murine TLR9 required different PS/ODN CpG sequences, respectively.[13] However, when we later repeated these experiments with natural PD CpG ODNs, no such species-specific CpG ODN sequence requirements for TLR9 activation were observed.[16] These results were surprising and raised the question whether the immunobiology of natural PD CpG ODN varies from that of the chemically modified PD CpG ODN. Indeed, many sequence-independent effects have been described for PS ODNs, ranging from avid non-specific protein binding to blockade of immune-stimulatory activities.[5] Although PS CpG ODNs activate TLR9 in a sequence-specific manner,[4,5] we experimentally readdressed the question whether TLR9 mediated recognition of natural PD ssDNA recapitulates these rules.

We used surface plasmon resonance (Biacore™) to determine whether TLR9 physically and directly binds to CpG ODNs in this reductionist system. We were disappointed to note that PS ODN binding was non-specific and thus no informative data could be collected; similar observations were reported by other groups.[15,16] However, PD CpG ODNs bound specifically to the extracellular domain (fused to human Fc) of TLR9 (high affine binding), while binding of methylated CpG ODNs was lower.[17]

Notably, similar results (specific, lower-affine binding to TLR9) were also observed with ODNs displaying no CpG motif.[18] We had thus obtained apparently conflicting experimental data, as the Biacore binding data showed relative CpG-motif-independence of PD-ODN-TLR9 binding, while PS ODN mediated TLR9 activation seemed to be totally CpG-dependent.

Due to the endosomal expression pattern of TLR9, the uptake and endosomal translocation of CpG ODNs may represent a functional bottleneck, in particular for PD ODNs known to be sensitive to DNase-1 mediated degradation. On the other hand, PS CpG ODNs efficiently translocated into endosomes via ill-defined receptor-mediated endocytosis mechanisms.[19] We therefore argued that if we were able to enforce endosomal translocation (and in parallel protect PD ODNs against DNase activity), we might find out about the functional relevance of CpG-motif-independent PD ODN binding by TLR9 observed in the Biacore system. To this end we used the cationic lipid DOTAP™ which forms complexes with ODNs to protect ODNs from DNase degradation and enforces endosomal translocation.

Our experiments provided two distinct pieces of information. First, PS CpG-ODNs complexed to DOTAP displayed enhanced cellular uptake paralleled by enhanced activation of myeloid or plasmacytoid DCs although the CpG motif requirement remained. In other words, PS ODNs lacking CpG motifs failed to activate TLR9 even under conditions of enforced endosomal translocation. Second, the opposite applied to for PD ODNs. Under conditions of DOTAP-mediated enforced endosomal translocation, both PD CpG ODNs as well as PD ODNs lacking CpG motifs activated DCs via TLR9.[20] Overall, these new functional data corresponded well with the Biacore binding studies.

Although the use of DOTAP had allowed us to define ssRNA (vulnerable to RNase) as a ligand for TLR7[21,22] and enabled others to characterize the spatio-temporal regulation of type I interferon production by pDCs as initiated in early endosomes,[23,] we viewed our results on the CpG-independence of TLR9 activation with caution because DOTAP may have confounding effects on DC-activation, as yet poorly defined. We therefore explored whether a relative (or absolute) CpG motif-independent TLR9 activation by natural PD DNA could be reproduced by 3′ extension of the respective ODNs with poly-guanosine (poly-G) tails.

This reasoning was based on reports indicating that 3′ extensions of PD ODNs with poly-G tails protected against DNase activity,[24] and enhanced cellular uptake.[25] While confirming that 3′ poly-G extended PD ODNs translocated to endosomal compartments more efficiently, we could show that 3′ poly-G extended PD ODNs lacking CpG-motifs also activated both enhanced mDCs and pDCs via TLR9.[26] Taken together, these data confirmed that under conditions of enforced endosomal translocation, ss natural DNA activates TLR9 in a CpG motif-independent manner.

CpG ODNs characterized by PS-modified poly-G tails at the 5′ and 3′ ends, and a palindromic phosphodiester CpG motif in the central position (A/D-type ODNs), are potent inducers of type I interferon.[27,28] In contrast, conventional PS-modified CpG ODNs (B/K- type ODNs) fail to stimulate pDCs to produce type I IFNs, although they stimulate production of pro-inflammatory cytokines.[4] Interestingly, DOTAP-mediated retention of B-type ODN in early endosomes allowed pDCs to produce type I IFNs in response to these ODNs.[23] We therefore analyzed whether in pDCs the natural PD version of B-type CpG ODN 1668 extended with a 3′ poly-G tail, which, as we have shown, efficiently induced type I IFN, would also preferentially locate to early, as opposed to late endosomes, where regular PD B-type CpG ODN is found.[23,29]

These experiments revealed that natural phosphodiester ss PD ODNs with or without CpG motifs when 3′ extended with poly-G tails, extended with preferentially locate to early endosomes.[26] Even with ODNs lacking CpG motifs, this differential compartmentation was associated with robust type I interferon but poor pro-inflammatory cytokine production by pDCs. In contrast, PS-modified ODNs lacking poly-G tails translocated preferentially to late endosomes and lysosomes. The latter localization correlated with the production of pro-inflammatory cytokines but not type 1 interferon. It follows that the intracellular compartmentation (early versus late endosomes) of ODNs, but not the sequence control the type of cytokine in pDCs. From these data, we concluded that PS-modified ODNs activate TLR9 primarily in a CpG-dependent manner, whereas natural ss PD ODNs activate TLR9 in a CpG-motif-independent fashion upon enforced endosomal translocation.

Bacterial genomic DNA stimulates immune cells via TLR9.[1] Because TLR9 resides in endosomes, bacterial or viral DNA must be delivered to these intracellular compartments where acidic and reducing conditions lead to denaturation and degradation of dsDNA and subsequent TLR9 ligation. In mammalian genomes (P) CpG motifs are suppressed and cytosines are largely methylated. Since ss PD ODN can nevertheless activate TLR9 independent of its methylation status and CpG motif content, we must correct our view — based on the use of PS-modified ODNs — that methylated mammalian DNA lacking CpG motifs generally fails to activate TLR9. Obviously, the rules governing TLR9 activation by PS ODNs do not apply to PD ODNs.

Compelling evidence for the ability of TLR9 to recognize mammalian DNA has recently been provided by creating a chimeric, cell surface expressed. By exchanging the intracellular/transmembrane portion of TLR9 with the intracellular/transmembrane portion of TLR4, receptor specificity for DNA was retained while the chimeric TLR localized to the cell membrane (as does TLR4).[30] Under these conditions, both mammalian and vertebrate DNA activated immune cells via TLR9, surprisingly without requiring acidic pH. The latter aspect contrasts with our findings that bafilomycin and chloroquine both inhibit CpG DNA-driven signaling in endosomes,[19] and that in vitro (Biacore) TLR9 interacted with CpG DNA more strongly at acidic pH.[17]

Since this chimeric TLR9 relies on the TLR4 signaling pathway,[30] further experiments are required to clarify the role of endosomal acidification in TLR9-dependent DNA recognition. At face value, however, these data support the contention that the endosomal localization of TLR9 regulates the access of TLR9 ligands. Furthermore, to prevent mammalian self DNA recognition under homeostatic conditions, the host ensures degradation of extracellular DNA, for example, by DNase-1. Interestingly, mice deficient in DNase-1 develop features of systemic lupus erythematosus (SLE).[31]

SLE is a prototypical human autoimmune disease in which increased serum levels of type 1 IFN correlate with disease activity and severity. The sera of SLE patients contain immune complexes (ICs) of host DNA and anti-DNA Ig antibodies and/or host RNA and anti-RNA Ig molecules. In vitro, these anti-DNA ICs activate rheumatoid factor positive (RF+) B cells via sequential engagement of B cell receptors (BCRs) and endosomal TLR9. In this system, the RF+ BCRs first recognize the isotype of the anti-DNA autoantibody, and subsequently translocate the ICs into TLR9 expressing endosomes.[32] This mechanism ensures enforced endosomal translocation of its self DNA cargo. Similarly, enforced endosomal translocation of anti-DNA ICs can be mediated by mouse Fcγ RIII or human Fcγ RIIa on DCs and activate the production of cytokines via TLR9.[33,34]

Similar data have been reported for human anti-RNA ICs, which, upon endosomal translocation, activate TLR7.[35-37] These findings support two conclusions: (1) Self-DNA (or -RNA) is recognized by TLR9 (or TLR7) upon enforced endosomal translocation; (2) since recognition of self-DNA or -RNA is potentially harmful to the host, acid recognition, evolutionary pressure may have exiled nucleic acid-specific TLRs to intracellular compartments in order to minimize access of these self-ligands while preserving recognition of pathogen-derived nucleic acids (from DNA or RNA viruses).

We argue that TLR9-mediated "natural" DNA recognition does not rely on CpG motifs. An important question to be addressed now is which structural components of ss PD DNA are recognized by TLR9. It is known that PS-modified CpG ODNs represent powerful immunopharmocological tools.[4] Nevertheless, it is now clear that the rules governing recognition of natural PD DNA by TLR9 differ from those governing PS-modified DNA recognition. We believe that the findings related to PS DNA have clouded our understanding of PD DNA recognition.

REFERENCES

1. Akira, S., Uematsu, S., and Takeuchi, O. 2006, Pathogen recognition and innate immunity. *Cell* 124, 783.
2. Beutler, B. et al. 2006, Genetic analysis of host resistance: Toll-like receptor signaling and immunity at large. *Annu. Rev. Immunol.*, 24, 353.
3. Pasare, C. and Medzhitov, R. 2005, Toll-like receptors: linking innate and adaptive immunity. *Adv. Exp. Med. Biol.*, 560, 11.
4. Krieg, A. M. 2002, CpG motifs in bacterial DNA and their immune effects. *Annu. Rev. Immunol.*, 20, 709.
5. Wagner, H. 1999, Bacterial CpG DNA activates immune cells to signal infectious danger. *Adv. Immunol.*, 73, 329.
6. Bird, A. P. 1980, DNA methylation and the frequency of CpG in animal DNA. *Nucleic Acids Res.*, 8, 1499.
7. Yamamoto, S. et al. 1992, Unique palindromic sequences in synthetic oligonucleotides are required to induce IFN and augment IFN-mediated natural killer activity. *J. Immunol.*, 148, 4072.
8. Pisetsky, D. S. 1996, Immune activation by bacterial DNA: a new genetic code. *Immunity*, 5, 303.
9. Krieg, A. M. et al. 1995, CpG motifs in bacterial DNA trigger direct B-cell activation. *Nature*, 374, 546.
10. Sparwasser, T. et al. 1998, Bacterial DNA and immunostimulatory CpG oligonucleotides trigger maturation and activation of murine dendritic cells. *Eur. J. Immunol.*, 28, 2045.
11. Hacker, H. et al. 2000, Immune cell activation by bacterial CpG-DNA through myeloid differentiation marker 88 and tumor necrosis factor receptor-associated factor (TRAF)6. *J. Exp. Med.* 192, 595.
12. Hemmi, H. et al. 2000, A Toll-like receptor recognizes bacterial DNA. *Nature*, 408, 740.
13. Bauer, S. et al. 2001, Human TLR9 confers responsiveness to bacterial DNA via species-specific CpG motif recognition. *Proc. Natl. Acad. Sci. USA*, 98, 9237.
14. Ahmad-Nejad, P. et al. 2002, Bacterial CpG-DNA and lipopolysaccharides activate Toll-like receptors at distinct cellular compartments. *Eur. J. Immunol.*, 32, 1958.

15. Latz, E. et al. 2004, TLR9 signals after translocating from the ER to CpG DNA in the lysosome. *Nat. Immunol.*, 5, 190.
16. Roberts, T. L. et al. 2005, Cutting edge: species-specific TLR9-mediated recognition of CpG and non-CpG phosphorothioate-modified oligonucleotides. *J. Immunol.*, 174, 605.
17. Rutz, M. et al. 2004, Toll-like receptor 9 binds single-stranded CpG-DNA in a sequence- and pH-dependent manner. *Eur. J. Immunol.*, 34, 2541.
18. Yasuda, K. et al. 2006, CpG motif-independent activation of TLR9 upon endosomal translocation of "natural" phosphodiester DNA. *Eur. J. Immunol.*, 36, 431.
19. Hacker, H. et al. 1998, CpG-DNA-specific activation of antigen-presenting cells requires stress kinase activity and is preceded by non-specific endocytosis and endo-somal maturation. *EMBO J.*, 17, 6230.
20. Yasuda, K. et al. 2005, Endosomal translocation of vertebrate DNA activates dendritic cells via TLR9-dependent and -independent pathways. *J. Immunol.*, 174, 6129
21. Heil, F. et al. 2004, Species-specific recognition of single-stranded RNA via toll-like receptor 7 and 8. *Science*, 303, 1526.
22. Diebold. S. S. et al. 2004, Innate antiviral responses by means of TLR7-mediated rec-ognition of single-stranded RNA. *Science*, 303, 1529.
23. Honda, K. et al. 2005, Spatiotemporal regulation of MyD88-IRF-7 signalling for robust type-I interferon induction. *Nature*, 434, 1035.
24. Bishop, J. S. et al. 1996, Intramolecular G-quartet motifs confer nuclease resistance to a potent anti-HIV oligonucleotide. *J. Biol. Chem.*, 271, 5698.
25. Roberts, T. L. et al. 2005, Differences in macrophage activation by bacterial DNA and CpG-containing oligonucleotides. *J.Immunol.*, 175, 3569.
26. Haas, T. et al. 26. The DNA sugar backbone 2′ deoxyribose determines Toll-like recep-tor 9 activation. *Immunity,* in press.
27. Krug, A. et al. 2001, Identification of CpG oligonucleotide sequences with high induc-tion of IFN-alpha/beta in plasmacytoid dendritic cells. *Eur. J. Immunol.*, 31, 2154.
28. Verthelyi, D. et al. 2001, Human peripheral blood cells differentially recognize and respond to two distinct CPG motifs. *J. Immunol.*, 166, 2372.
29. Guiducci, C. et al. 2006, Properties regulating the nature of the plasmacytoid dendritic cell response to Toll-like receptor 9 activation. *J. Exp. Med.*, 203, 1999.
30. Barton, G. M., Kagan, J. C. and Medzhitov, R. 2006, Intracellular localization of Toll-like receptor 9 prevents recognition of self DNA but facilitates access to viral DNA. *Nat. Immunol.*, 7, 49.
31. Napirei, M. et al. 2000, Features of systemic lupus erythematosus in Dnase1-deficient mice. *Nat. Genet.*, 25, 177.
32. Leadbetter, E. A. et al. 2002, Chromatin-IgG complexes activate B cells by dual engage-ment of IgM and Toll-like receptors. *Nature*, 416, 603.
33. Boulé, M. W. et al. 2004, Toll-like receptor 9-dependent and -independent dendritic cell activation by chromatin-immunoglobulin G complexes. *J. Exp. Med.*, 199, 1631.
34. Means, T. K. et al. 2005, Human lupus autoantibody-DNA complexes activate DCs through cooperation of CD32 and TLR9. *J. Clin. Invest.*, 115, 407.
35. Vollmer, J. et al. 2005, Immune stimulation mediated by autoantigen binding sites within small nuclear RNAs involves Toll-like receptors 7 and 8. *J. Exp. Med.*, 202, 1575.
36. Lau, C. M. et al. 2005, RNA-associated autoantigens activate B cells by combined B cell antigen receptor/Toll-like receptor 7 engagement. *J. Exp. Med.*, 202, 1171.
37. Savarese, E. et al. 2006, U1 small nuclear ribonucleoprotein immune complexes induce type I interferon in plasmacytoid dendritic cells through TLR7. *Blood*, 107, 3229.

7 Discrimination of Self and Non-Self DNAs

Katryn J. Stacey, Francis Clark,
Greg R. Young, and Tara L. Roberts

ABSTRACT

Cellular recognition of pathogen DNA is used as a means to signal infection. Toll-like receptor 9 (TLR9) recognizes DNA containing unmethylated CpG motifs within the endosomes of B cells, dendritic cells, and mouse macrophages. A number of factors appear to contribute to the generation of TLR9 responses to bacterial and viral DNAs, but not self DNA: (1) the CpG dinucleotide occurs at relatively low frequency in vertebrate genomes; (2) cytosine methylation of CpG motifs, prevalent in vertebrates, greatly reduces cellular responses; (3) saturable DNA uptake determines that only DNA with a certain threshold frequency of active sequences will accumulate to sufficient levels to activate cells; and (4) inhibitory DNA motifs may limit activation by self DNA. Apart from TLR9 responses, recent evidence indicates that cells can respond to double-stranded (ds) DNA within the cytoplasm as an indication of viral infection. In this case, the definition of the DNA as foreign or dangerous may rely on its abnormal location. Evidence suggests that discrimination of self from non-self DNA is defective in the autoimmune disease known as systemic lupus erythematosus (SLE), and this may involve both TLR9 and TLR-independent pathways.

INTRODUCTION

From an overall structural viewpoint, the DNA molecules of all species are very similar. It therefore came as a scientific surprise that the immune system utilizes recognition of pathogen DNA as a means to signal infections. Responses to bacterial DNA were first recognized by Tokunaga et al. who showed natural killer (NK) cell activation, interferon induction, and anti-tumor activity.[1,2] Messina et al. subsequently observed B cell proliferation in response to bacterial but not vertebrate DNA.[3] The B cell response was demonstrated to require sequences containing unmethylated CG dinucleotides (CpG motifs),[4] and subsequent work showed macrophages had similar sequence requirements for activation.[5]

Hemmi et al.[6] showed that responses to CpG DNA required TLR9, and Biacore binding studies have shown direct binding of DNA to TLR9.[7] Experiments with tagged TLR9 showed it to reside within the endoplasmic reticulum and move to localize with incoming DNA in the endosomal and lysosomal compartments.[8,9]

Responses to CpG DNA appear to require maturation of the endosomal compartment, since agents such as chloroquine and bafilomycin A that prevent endosomal acidification are inhibitory.[10,11]

Discrimination of self from foreign DNA requires recognition of a distinctive structural or sequence motif common to numerous pathogen DNAs and rarely found in self DNAs. When CpG motifs were elucidated, this discrimination was proposed to be based on (1) the suppression of CpG motifs in the vertebrate genome, and (2) the fact that remaining CpG sequences are predominantly methylated on the cytosine residue, rendering them inactive.[4] In addition, CCGG sequences that are frequently unmethylated in CpG islands in the promoters of active vertebrate genes presented little activity.[12]

The tendency of 5-methyl cytosine to spontaneously deaminate, yielding a thymine residue, is thought to be responsible for the low frequency of CpG dinucleotides in vertebrate genomes. The frequency of CpG dinucleotides in mouse DNA is 12.5% of the frequency found in *Escherichia coli* DNA.[15] Approximately 70 to 80% of CpG dinucleotides in vertebrate DNA contain 5-methyl cytosine,[13,14] a modification shown to prevent activity when introduced into bacterial DNA.[4,5] Thus, unmethylated CpG dinucleotides are present in the mouse genome at 2.5 to 3.8% of the frequency at which they occur in *E. coli* DNA. Many of these remaining unmethylated CpG sequences may be relatively inactive CCGG sequences.

In addition to CpG methylation and suppression controlling activity of self DNA, G-rich inhibitory DNA motifs may play a role.[15,16] Finally, the intracellular location of TLR9 means that responses are controlled by the uptake of DNA or pathogens. Limited uptake of DNA is proposed to help control responses to free circulating self DNA.[15,17,18]

This chapter reviews the factors involved in self versus non-self discrimination of DNA via the TLR9 pathway, and suggest that it is not the only pathway involved in recognition of foreign DNA. dsDNA (e.g., viral DNA) within the cytoplasm is under consideration as a further route for cellular activation. In this case, the means of self and non-self discrimination may merely be the abnormal location of the DNA. A knowledge of the normal routes for distinguishing self and foreign DNA is relevant to understanding SLE, which appears to involve inappropriate responses to self DNA that are both TLR and non-TLR-mediated.[19–21]

ACTIVITIES OF VERTEBRATE AND *ESCHERICHIA COLI* DNAs

Unmethylated CpG dinucleotides are present in vertebrate genomes, as described above, at around 3% of the frequency seen in *E. coli* DNA. For this reason, we assessed whether vertebrate DNA activates macrophages at high concentrations. The mouse macrophage cell line RAW264 was stably transfected with the nuclear factor-kappa-B (NF-κB)-dependent endothelial leukocyte adhesion molecule (ELAM) promoter driving green fluorescent protein (GFP),[15] providing a convenient assay of CpG DNA responses.

Figure 7.1 shows that while the activity of *E. coli* DNA was detectable at 0.03 μg/ml, calf thymus DNA showed no activity up to 100 μg/ml. This does not discount the possibility that vertebrate DNA may stimulate other responses. We sometimes observed

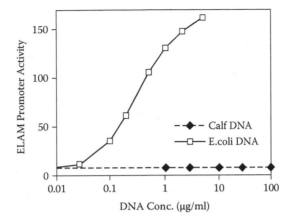

FIGURE 7.1   High fidelity for distinction between vertebrate and bacterial DNA by mouse macrophages. DNA immunostimulatory activity measured using the ELAM mouse cell line, a clone of RAW264 containing an integrated ELAM (E-selectin) promoter driving a green fluorescent protein (GFP) reporter. *E. coli* and calf thymus DNAs were added to cells at indicated concentrations and incubated for 10 hr before measurement of GFP levels by flow cytometry. Results are typical of four experiments. (Source: Stacey, K. J. et al. 2003, *J. Immunol* 170: 2614. With permission.)

low level of nitric oxide production induced by vertebrate DNA in the presence of interferon-γ, although the effect was difficult to reproduce reliably. However, at least for this assay of NF-κB-dependent transcription, the fidelity of the distinction between self and foreign DNA is remarkable.

Reasons for the lack of activity of the residual unmethylated CpG sequences in vertebrate DNA may include: (1) preferential methylation of active sequences, with unmethylated sequences being predominantly the poorly stimulatory CCGG sequences; (2) the presence of immunosuppressive sequences in vertebrate DNA; and (3) limited uptake of DNA imposing a threshold on the density of CpG motifs required for activation. The first possibility is difficult to assess because a genome-wide assessment of the frequency of unmethylated stimulatory sequences would be a difficult undertaking. Evidence in support of roles for the second and third possibilities will be discussed below.

ROLE OF CYTOSINE METHYLATION IN INHIBITING TLR9 RESPONSES

The fact that CpG methylation of bacterial DNA inhibits its immunostimulatory activity[4,5] led to the assumption that this is a major factor involved in the lack of activity of self DNA. However, evidence suggests methylation does not completely prevent the activities of oligodeoxynucleotides (ODNs).[22,23] Figure 7.2 shows that methylation clearly decreased CpG ODN activity at low concentrations, but at higher concentrations, both methylated CpG ODNs and CCGG-containing ODNs considered to be of low activity[24] induced NO production by mouse macrophages to a similar extent as fully active CpG ODNs. As noted earlier, NO production was sometimes observed

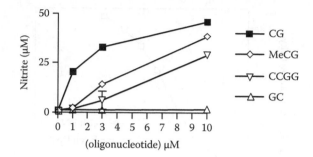

FIGURE 7.2 Methylated CpG and CCGG sequences can display immunostimulatory activities at high concentrations. RAW264 mouse macrophage cells were primed with 20 U/ml interferon-γ for 2 hr followed by ODNs for 24 hr before nitrite was measured in culture supernatant as indication of NO production.[70] Error bars that generally fall within the size of the symbol show the difference in activity between two separate syntheses of each ODN. Natural phosphodiester ODNs based on AO-1[71] were used and sequences varied only in CpG motifs (underlined). MeC = 5-methyl cytosine.

CG = GCTCATGACGTTCCTGATGCTG
MeCG = GCTCATGA^{me}CGTTCCTGATGCTG
CCGG = GCTCATGCCGGTCCTGATGCTG
GC = GCTCATGAGCTTCCTGATGCTG

in response to vertebrate DNA, so it is possible that this response is more readily induced by methylated DNA than some other measures of TLR9 activation.

Methylation of CpG motifs decreased the affinity of an ODN for TLR9 binding but did not prevent binding.[7] TLR9 had measurable binding to all ssDNA sequences tested, although dsDNA binding was barely detectable.[7,25] This suggests that ODN with low affinity for TLR9 may not induce signaling at low concentrations, but at saturating levels of DNA, continuous weak interactions may induce TLR9-mediated signaling. Thus it is possible that methylation does not lead to complete prevention of activation by self DNA as has been commonly assumed.

One issue that has not been examined is the effect of methylation in the context of dsDNA. Evidence suggests that TLR9 recognizes ssDNA but not dsDNA,[7] but how endosomal DNA becomes single stranded is unknown. Helicases or strand-specific nucleases acting at endosomal pH may be involved. It is possible that methylation may be effective in silencing the activity of dsDNA by inhibiting endosomal DNA strand separation. This would explain results from enzymatic methylation of *E. coli* dsDNA that showed a complete loss of activity.[4,5] The possibility that DNA strand separation is another step at which self versus non-self discrimination occurs is worthy of further investigation.

A second finding questioning the role of methylation in silencing activity of self dsDNA was that hypomethylated genomic DNA obtained from ES cells deficient in maintenance methyltransferase Dnmt1 was found to be non-stimulatory.[26] The hypomethylated DNA used in that study was anticipated to retain approximately 18% CpG methylation.[13,15] Since it was possible that the remaining methylation specifically targeted the most active CpG motifs, we investigated the role of methylation by

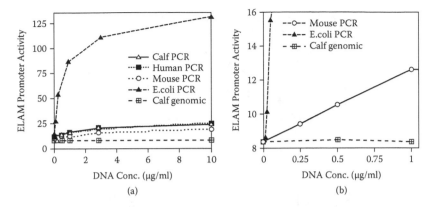

FIGURE 7.3 PCR amplification of vertebrate DNA reveals a role for methylation in preventing DNA activity. A. Calf thymus, human, mouse and *E. coli* DNAs were partially digested with four base cutters and ligated to ds primer sequences. DNAs were then amplified by PCR to remove epigenetic modifications. The abilities of various concentrations of these DNAs and calf thymus genomic DNA to activate the ELAM promoter was assessed after 10-hr incubation with ELAM9 cells (see Figure 7.1). Results are typical of three experiments. B. Data from Figure 7.2A on an expanded scale to allow estimation of demethylated mouse DNA activity relative to *E. coli*. Error bars showing the values obtained with duplicate samples fall within the symbols. (Source: Stacey, K. J. et al. 2003, *J. Immunol* 170: 2614. With permission.)

generating randomly PCR-amplified genomic DNAs devoid of base modifications.[15] Comparison of PCR-amplified DNA with normal genomic DNA in activation of NF-κB-dependent transcription in macrophages showed that unmethylated mouse and human DNA gained limited immunostimulatory activity (Figure 7.3). Thus base modification does play a role in preventing activation by self dsDNA.

LIMITATION OF RESPONSES TO DNA WITH LOW FREQUENCY OF CPG MOTIFS BY SATURABLE DNA UPTAKE

The intracellular location of TLR9 is obviously necessary for recognition of nucleic acids released from phagocytosed and degraded organisms. Exposure of TLR9 to DNA is limited to self or foreign DNA taken up by receptor-mediated endocytosis or DNA released from phagocytosed organisms or apoptotic cells. Endocytic uptake of DNA is clearly saturable in macrophages and dendritic cells.[27,28] Figure 7.3A shows that the responses to both *E. coli* DNA and unmethylated vertebrate DNA reach a plateau despite the fact that the response to unmethylated vertebrate DNA is still far from the maximum possible. To determine whether DNA uptake is limiting for either of these responses, we compared the saturation of uptake of radiolabelled genomic DNA with increasing DNA concentrations to the saturation of immunostimulatory activity. Figure 7.4A shows that although the response to *E. coli* DNA reaches a plateau, it does so at a lower concentration than that where DNA uptake begins to saturate. Thus uptake does not seem to be limiting for DNA with a high frequency of stimulatory motifs.

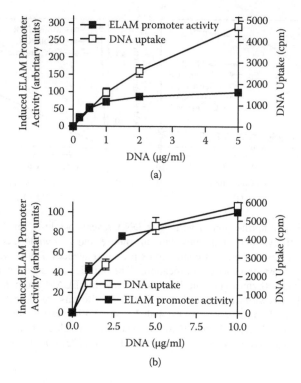

(a)

(b)

FIGURE 7.4   Saturable DNA uptake into macrophages limits response to DNA with a low frequency of stimulatory motifs. A. Responses to *E. coli* DNA are not limited by macrophage DNA uptake. To measure DNA uptake, DNA was labeled by nick translation with ^{33}P-dCTP. RAW264 cells adherent in 96-well plates were incubated with radiolabelled DNA for 30 min, then washed four times with serum-free medium containing 0.2% BSA. Cell-associated radioactivity was assessed by scintillation counting. The experiment was performed twice and gave identical results. Due to differences in DNA concentrations, results of only one experiment are shown. Data points indicate the means and error bars the range obtained with duplicate samples. Immunostimulatory activity of *E. coli* DNA was assessed using the ELAM9 derivative of RAW264 as for Figure 7.1. The low basal expression from the ELAM promoter was subtracted, and results of three separate experiments normalized to the result for the highest concentration. Results shown are the means obtained in three experiments. Error bars show range of values. B. Responses to PCR-amplified CT DNA are limited by macrophage DNA uptake. Experiments showed no difference between uptake of calf thymus and *E. coli* DNA. Results shown here are for *E. coli* DNA as per Figure 7.4A. Activity of PCR-amplified CT DNA was assessed as per Figure 7.4A, and results shown are means of three independent experiments. Error bars show range of values obtained.

However, Figure 7.4B shows that the activity of unmethylated calf thymus DNA follows a similar curve to the amount of DNA taken up, with minimal increase in activity or DNA uptake above 5 μg/ml. Thus DNA uptake does seem to limit the responses to unmethylated calf thymus DNA. DNA with a low density of highly active CpG motifs, such as methylated vertebrate DNA, may never reach a sufficient concentration within the endosomes for a TLR9 response to occur. There then arises the concept of a threshold of CpG motif frequency below which responses will not occur. Since this depends on efficiency of DNA uptake, the threshold may vary from one cell type to another.

Although saturable DNA uptake limits the accumulation of DNA in endosomal compartments, large amounts of self DNA are present in phagosomes following uptake of apoptotic cells. The control of responses to apoptotic DNA is an interesting and unresolved area. Conceivably, TLR9 activation may be inhibited by other signals resulting from recognition of apoptotic cells (e.g., engagement of phosphatidyl serine on apoptotic cell membranes). In addition, the signals leading to trafficking of TLR9 from the ER to incoming endosomes are uncharacterized. It is possible that TLR9 does not traffic to endosomes containing apoptotic cells. However, both these proposals would prevent TLR9 recognition of viral DNA within apoptotic cells, unless viral-induced apoptosis can be distinguished from other forms of apoptosis.

In support of the idea that normal DNA uptake limits responses to self DNA, another group suggested that if vertebrate DNA uptake into endosomes is increased by chemical transfection, TLR9-dependent responses are seen.[17] However, transfected DNA also activates a TLR9-independent pathway for recognition of cytoplasmic dsDNA (discussed below).[29,30]

Another group proposed that the intracellular localization of TLR9 plays a role in restricting responses to self DNA.[18] A fusion protein of TLR9 ligand binding domain–TLR4 transmembrane and cytoplasmic domain localized to the cell surface responded to vertebrate DNA added outside the cell. This result could be interpreted several ways. First, the low pH of the endosome may be required to allow TLR9 to distinguish methylated from unmethylated CpG motifs. Another possibility is that the fidelity of distinction between self and foreign DNA requires appropriate glycosylation. The processing of glycosylated sites on TLR9 that has reached the surface would be different from native TLR9 retained in the ER that does not traffic through the Golgi apparatus.[8,9] Finally, the response to normal methylated vertebrate DNA when TLR9 is expressed at the cell surface may represent a response to a very low density of stimulatory motifs—a response normally prevented by limited DNA uptake. Unfortunately, no comparison of the response of cell surface TLR9 to bacterial DNA and vertebrate DNA was provided,[18] so the relative magnitude of response to self DNA cannot be assessed.

RELATIVE FREQUENCIES OF STIMULATORY MOTIFS IN GENOMIC DNAs

The level of activation achievable with unmethylated vertebrate DNA in no way approached levels seen with bacterial DNA (Figure 7.3A). This leads to consideration of the relative frequencies of activating motifs in bacterial and unmethylated vertebrate DNA. Table 7.1 shows that *E. coli* DNA has 8.4-fold higher frequency of the three best characterized activating CpG hexamers than mouse DNA. If these sequences alone determine the immunostimulatory potential of unmethylated DNA, we might expect that PCR-amplified *E. coli* DNA would be 8.4 times as stimulatory as PCR-amplified mouse DNA.

TABLE 7.1

Frequencies of Stimulatory CpG Motifs in Vertebrate and Bacterial Genomes

	E. coli K12	P. aeruginosa	H. influenzae Rd KW20	Human	Mouse
Genome size	4639670	6264399	1829487	2.858×10^9	2.550×10^9
%GC	50.8	66.6	38.1	40.8	41.7
fAACGTT	3.675	0.506	4.624	0.5675	0.5094
fGACGTT	3.207	1.751	1.311	0.3319	0.3305
fGTCGTT	2.543	2.418	2.001	0.2376	0.2778
Total f	9.425	4.675	7.936	1.137	1.118

Note: Frequency (f) = occurrences in analyzed genome/length of genome analyzed $\times 10^4$. Total f = total frequency of three analyzed CpG motifs

Figure 7.3B shows an expanded view of results of stimulation with PCR-amplified DNAs at low concentration before uptake or CpG responses begin to saturate. The unmethylated mouse DNA at a concentration of 0.5 μg/ml gave a response equivalent to that of amplified *E. coli* DNA at 0.022 μg/ml. Thus in the absence of epigenetic modification, *E. coli* DNA was 23 times more active than mouse DNA. This difference in activity is clearly not completely accounted for by the 8.4-fold difference in frequency of activating sequences, and other factors must limit vertebrate DNA activity.

ROLE FOR INHIBITORY MOTIFS IN SUPPRESSION OF RESPONSES TO SELF DNA?

Evidence indicates that some G-rich DNA sequences can potently suppress responses to CpG motifs.[15,16,21,31,32] While these G-rich sequences administered as ODNs are efficient inhibitors of CpG responses, evidence that such inhibitory sequences act within vertebrate DNA to limit immunostimulation is difficult to obtain. Inhibition by these motifs used as ODNs does not appear to be due to inhibition of DNA uptake,[15,31] and direct competition for binding to TLR9 is a possibility. Inhibitory sequences contain a run of three or four G residues capable of interacting in a planar quartet array to generate four stranded DNA structures.[33] Duramad et al.[34] noted that such structures form with inhibitory ODNs, but all inhibitory activity resides in the single stranded form.

Although most work studied phosphorothioate-modified inhibitory motifs, they do act potently as natural phosphodiesters.[15] Thus it is conceivable that G-rich inhibitory sequences within vertebrate DNA may limit activities of self DNA. Added vertebrate DNA inhibits macrophage responses to CpG ODNs, but the effect is largely a competition for uptake.[15] However, some residual inhibition remains unexplained by effects on uptake. This, along with the fact that unmethylated self DNA has a lower activity than accounted for by CpG suppression, suggests that inhibitory motifs may have some role within vertebrate DNA.

Analysis of the frequencies of inhibitory motifs within bacterial and vertebrate genomes shows that these motifs are less frequent in *E. coli* and *Haemophilus influenzae* genomes than in mouse and human genomes (Table 7.2). On average, the two

TABLE 7.2
Frequencies of Inhibitory Motifs in Vertebrate and Bacterial Genomes

		P. aeruginosa	H. influenzae	E. coli	Human	Mouse
CCNDDNGGGG	f	0.0000829	0.0000339	0.0000608	0.0000790	0.0000843
	f/f exp	0.137	1.08	0.407	1.73	1.64
CCNDDNNGGGG	f	0.000104	0.0000252	0.0000642	0.0000938	0.0000888
	f/f exp	0.172	0.803	0.430	2.05	1.73
GGG	f	0.0162	0.00593	0.01024	0.0132	0.0135
	f/f exp	0.441	0.858	0.625	1.55	1.49
GGGG	f	0.00264	0.00114	0.00188	0.00309	0.00348
	f/f exp	0.215	0.862	0.451	1.79	1.84

Note: Frequency (f) = occurrences in analyzed genome/length of genome analyzed. f exp = expected frequency based on content of each base in genome (see Table 7.1), assuming random distribution of bases. D = A, T, or G in the sequences. Inhibitory sequences based on analysis of base requirements for inhibition by phosphorothioate-modified ODNs.[69] Inhibitory motifs conforming to this consensus are also active as phosphodiester ODNs.[15]

inhibitory sequences analyzed are 1.4-fold higher in mouse and human DNA than *E. coli* DNA. However *Pseudomonas aeruginosa*, with a high GC percent shows no difference from vertebrate genomes in the frequency of GC-rich inhibitory motifs. Of interest, compared to the expected frequency based on the base content of each genome, inhibitory motifs are profoundly suppressed in *P. aeruginosa* and *E. coli* (i.e., f/f exp << 1), and higher than expected in mice and humans. This seems to be attributable to a general suppression of runs of G residues in the bacteria, and an over-representation in mice and humans (Table 7.2).

There is no great suppression of inhibitory motifs in *H. influenzae*, but they are naturally low due to the AT-rich nature of the genome. The reason G runs are unfavorable in bacteria and selected for in vertebrate genomes is not immediately apparent. A broader analysis is needed to determine the range of organisms in which runs of G residues are over- and under-represented. We speculate that if these G runs are unfavorable in bacteria, TLR9 may recognize them as antagonists in order to improve the fidelity of self and non-self recognition.

We previously suggested that the activity of DNA may be determined by the ratio of stimulatory to inhibitory motifs.[15] The three bacterial genomes cited in Table 7.2 were selected because we had assayed the immunostimulatory activities of their DNA and found them of similar magnitude to *E. coli* DNA (results not shown). *P. aeruginosa* has half as many of the CpG motifs analyzed (Table 7.1), and approximately 1.4-fold higher inhibitory motifs (Table 7.2) than *E. coli* DNA. This argues that the analysis of inhibitory and stimulatory motifs is currently too simplistic to be predictive of activity.

It is possible that the wider sequence context in which CpG hexamers occur affects their activity, and that the most potent inhibitory phosphodiester sequences have not yet been defined. In addition, the positions of inhibitory motifs relative to stimulatory motifs may be important; published work suggests that inhibitory motifs placed 5′ but not 3′ to CpG motifs suppressed CpG activity.[32] Most analysis of inhibitory motifs involved single stranded phosphorothioate ODNs, and the effects of inhibitory motifs within dsDNA, acting in cis on CpG motifs, have not been analyzed.

The ends of eukaryotic chromosomes consist of telomeric repeat DNA. In human cells, telomeres generally consist of 500 to 3000 repeats of the TTAGGG hexanucleotide, with a 3′ single stranded extension of 100 to 200 nucleotides.[35,36] Since these structures are unique to eukaryotes and not found in bacterial and viral pathogens, a role for telomeres in suppressing responses to self DNA is an attractive hypothesis. The immunosuppressive action of ODNs consisting of four telomeric repeats has been extensively investigated.[37-39] This work has been done almost exclusively with phosphorothioate-modified ODNs that are proposed to mediate a sequence- or structure-specific inhibition of signaling through STAT proteins.[39] Although early work implicated inhibitory activity of telomeric sequences within genomic DNA,[37] a direct effect of natural telomeric DNA on STAT signaling has not yet been demonstrated.

CPG-INDEPENDENT RECOGNITION OF BACTERIAL DNA?

Further determinants of DNA immunostimulatory activity remain to be character-ized. The modified base N^6-methyladenine (N^6-MeA) is found in bacteria but not vertebrates.[40] ODNs containing N^6-MeA have been reported to induce cytokine production when injected into mice.[41] However, we have shown that plasmid DNA isolated from *E. coli* deficient in the two major methylation enzymes (Dam and Dcm) exerted the same activity in assays of macrophage activation as DNA from wild-type *E. coli*.[27] Dam methylates the N6 position of adenine in GATC sequences, and Dcm methylates the C5 position of the second cytosine in CCA/TGG.[42] Other *E. coli* DNA methylases play lesser roles in generating N^6-MeA because they recognize frequent sites.[27] Thus, while bacterial N^6-MeA probably does not contribute to the macrophage activation phenotypes we analyzed, the *in vivo* cytokine induction observed with N^6-MeA-containing DNA[41] may rely on detection by some other cell type.

Neutrophils have been reported to recognize bacterial but not vertebrate DNA in a TLR9-independent manner.[43,44] Responses to bacterial DNA were seen with ds and ssDNA, and not inhibited by CpG methylation or chloroquine, which inhibit TLR9 responses.[43] Responses were prevented in a knockout of MyD88, the TLR signaling adaptor, but not in a TLR9 knockout.[44] In addition, no effect of Dam or Dcm methy-lation of DNA was observed on neutrophil activation. Experiments used a uniformly high concentration (100 μg/ml) of bacterial DNA, and more subtle effects of CpG or adenine methylation may have been missed. The evidence given that DNA serves as the active component came from DNase digestion followed by heat inactivation of DNase.[44] While this may indicate a heat-sensitive contaminant, earlier work revealed activation by similarly heat denatured DNA.[43] In summary, although TLR9 is clearly the predominant means of response to exogenously added bacterial DNA, some evi-dence suggests collaborating CpG-independent pathways for discrimination between bacterial and self DNA.

ACTIVATION BY INTRACELLULAR DsDNA

TLR9 allows detection of DNA released from bacteria or viruses that are lysed in endosomes. However, a detection system for foreign DNA within a cell would be desirable for a cell infected with a DNA virus. Evidence points to detection of cyto-plasmic dsDNA analogous to the well-characterized responses to viral dsRNA.[29,30,45] More than a decade ago, we observed profound deaths of primary macrophages in response to transfected dsDNA and proposed that this constituted a defense against viral DNA.[45] More recently, induction of type I interferon, chemokines, and cyto-kines, as well as upregulation of major histocompatibility complex (MHC) have been observed in response to transfected DNA.[29,30,46,47]

The activation phenotype and apoptotic responses are typical antiviral reac-tions. Both apoptosis and activation are seen with dsDNA but not ssDNA. Evidence indicates that these responses initiate in the cytoplasm and not the endosomes.[30]

Although there is a higher response to AT-rich DNA,[29] there is no clear specificity for self versus non-self DNA in this pathway. Transfected vertebrate dsDNA activates cells in a similar manner to bacterial and viral DNAs. The assessment that transfected DNA represents a danger to cells is probably based on inappropriate location (e.g., cytoplasm) or lack of association with appropriate cellular proteins.

SYSTEMIC LUPUS ERYTHEMATOSUS: BREAKDOWN OF DISCRIMINATION BETWEEN SELF AND NON-SELF DNA?

The similarity of host and pathogen DNA makes distinctions between them challenging, and some evidence indicates that recognition of self DNA contributes to pathology. Systemic lupus erythematosus (SLE) is an autoimmune disease involving elevated circulating apoptotic debris and characterized by the presence of antibodies against DNA, nuclear proteins, and proteins that complex with RNA. Both DNA and RNA appear to act as adjuvants in the generation of antibodies, activating TLR9 and TLR7, respectively.

Nucleic acid–protein complexes are thought to activate B cells by simultaneous ligation of B cell receptors and TLR7 or TLR9.[19,48] The resulting DNA- and RNA-containing immune complexes are thought to activate plasmacytoid dendritic cells (pDCs) to produce IFNα via simultaneous ligation of Fcγ receptors and TLR7 or TLR9.[20,49–51]

Although controversial,[52–55] recent evidence using TLR knockouts crossed onto a lupus-prone background showed that TLR9 was required for generation of anti-nucleosomal antibodies, and TLR7 was required for generation of antibodies against RNA-binding antigens.[56] The TLR7-knockout mice had less severe disease,[56] while a number of reports showed conversely that TLR9-knockout mice had exacerbated disease.[53,54,56] The reason for this is not apparent, although it is suggested that TLR9 may contribute to the generation of antibodies that enhance clearance of apoptotic debris, or that RNA-containing immune complexes are intrinsically more pathogenic.[57] However, administration of CpG DNA is known to exacerbate SLE.[58] A factor to consider is that knockout mice cannot always produce results equivalent to the acute loss of function of a single gene in a normally developed mouse. TLR9 responses in pDCs are implicated in the development of regulatory T cells.[59] Deficient regulatory compartments in TLR9-knockout mice may explain the observed results.[53]

Whether TLR9 is protective or deleterious in SLE, evidence points to cellular activation by TLR9-mediated recognition of apoptotic self DNA.[60] The basis for this recognition of self DNA is not well understood. The responding B cells are probably only those with receptors that mediate uptake of chromatin, greatly increasing endosomal DNA concentration. This may make recognition of infrequent CpG motifs possible. In addition, the signal through B cell receptors may synergize with TLR9, or decrease selectivity for unmethylated CpG motifs.[22]

Similar arguments can be made about increased uptake and synergistic signaling when immune complexes are taken up via the Fcγ receptors on pDCs. Another consideration is that circulating DNA in SLE may be intrinsically more stimulatory than normal. A number of reports suggest epigenetic changes in SLE, although this has generally been analyzed in terms of changes in T cell gene expression and loss of tolerance.[61] DNA methylation was found to be impaired in T cells from SLE patients[62] perhaps due to lowered levels of a DNA methyl transferase (Dnmt1).[63] The circulating apoptotic DNA of SLE patients may contain sufficient levels of unmethylated CpG motifs to activate via TLR9. Indeed, rheumatoid arthritis and SLE patients had significantly lower percentages of 5-methylcytosine in circulation than healthy controls.[54] In addition, B cells from lupus-prone mice are more sensitive to TLR9 ligands than normal B cells.[65]

The cytoplasmic pathway for recognition of dsDNA may also play a role in SLE. The mouse knockout of DNase II, a lysosomal endonuclease, is embryonic lethal, due to induction of IFNβ in response to accumulation of undegraded DNA in macrophage lysosomes.[66] This response is not mediated by TLR activation.[67] Escape of DNA from endosomes to cytoplasm is well established,[5,68] and undegraded apoptotic DNA may well be detected in the cytoplasm. A similar effect may be seen with endosomal escape of apoptotic DNA in SLE, and evidence has been presented that some DNA-mediated cellular activation is TLR-independent.[20]

CONCLUSION

Suppression of CpG motifs, CpG methylation, and limited DNA uptake all appear to play a role in preventing activation of TLR9 by self DNA. However, unmethylated vertebrate DNA still showed lower immunostimulatory activity than expected on the basis of frequencies of activating motifs. Some of this unexpected low activity may be explained by inhibitory motifs, but the extent to which they act within self DNA is difficult to determine. Conceivably, cells may also discriminate between self and non-self DNA at a number of other stages, including DNA trafficking, recruitment of TLR9 to the endosome, and DNA strand separation to generate single-stranded TLR9 ligands.

The detection of cytoplasmic dsDNA is another means by which potentially dangerous DNA is detected. Inappropriate detection of self DNA via TLR9 or cytoplasmic detection may occur with genetic deficiencies in nucleases that normally clear self DNA. A knowledge of the factors normally restricting activation by self DNA will be useful in the understanding of pathologies such as SLE, which is characterized by inappropriate responses to self nucleic acids.

REFERENCES

1. Tokunaga, T. et al. 1984. Antitumor activity of deoxyribonucleic acid fraction from *Mycobacterium bovis* BCG I. Isolation, physicochemical characterization, and antitumor activity. *J Natl Cancer Inst* 72: 955.
2. Yamamoto, S. et al. 1988. *In vitro* augmentation of natural killer cell activity and production of interferon-α, β, and γ with deoxyribonucleic acid fraction from *Mycobacterium bovis* BCG. *Jpn J Cancer Res* 79: 866.
3. Messina, J. P., Gilkeson, G. S., and Pisetsky, D. S. 1991. Stimulation of *in vitro* murine lymphocyte proliferation by bacterial DNA. *J Immunol* 147: 1759.
4. Krieg, A. M. et al. 1995. CpG motifs in bacterial DNA trigger direct B-cell activation. *Nature* 374: 546.

5. Stacey, K. J., Sweet, M. J., and Hume, D. A. 1996. Macrophages ingest and are activated by bacterial DNA. *J Immunol* 157: 2116.

6. Hemmi, H. et al. 2000. A Toll-like receptor recognizes bacterial DNA. *Nature* 408: 740.

7. Rutz, M. et al. 2004. Toll-like receptor 9 binds single-stranded CpG-DNA in a sequence- and pH-dependent manner. *Eur J Imm*unol 34: 2541.

8. Latz, E. et al. 2004. TLR9 signals after translocating from the ER to CpG DNA in the lysosome. *Nat Immunol* 5: 190.

9. Leifer, C. A. et al. 2004. TLR9 is localized in the endoplasmic reticulum prior to stimulation. *J Immunol* 173: 1179.

10. Hacker, H. et al. 1998. CpG-DNA-specific activation of antigen-presenting cells requires stress kinase activity and is preceded by non-specific endocytosis and endosomal maturation. *EMBO J* 17: 6230.

11. Macfarlane, D. E. and Manzel, L. 1998. Antagonism of immunostimulatory CpG oligodeoxynucleotides by quinacrine, chloroquine, and structurally related compounds. *J Immunol* 160: 1122.

12. Yamamoto, S. et al. 1992. Unique palindromic sequences in synthetic oligonucleotides are required to induce IFN and augment IFN-mediated natural killer activity. *J Immunol* 148: 4072.

13. Ramsahoye, B. H. et al. 2000. Non-CpG methylation is prevalent in embryonic stem cells and may be mediated by DNA methyltransferase 3a. *Proc Natl Acad Sci USA* 97: 5237.

14. Bird, A. 2002. DNA methylation patterns and epigenetic memory. *Genes Dev* 16: 6.

15. Stacey, K. J. et al. 2003. The molecular basis for the lack of immunostimulatory activity of vertebrate DNA. *J Immunol* 170: 3614.

16. Lenert, P. et al. 2001. CpG stimulation of primary mouse B cells is blocked by inhibitory oligodeoxyribonucleotides at a site proximal to NF-κB activation. *Antisense Nucleic Acid Drug Dev* 11: 247.

17. Yasuda, K. et al. 2005. Endosomal translocation of vertebrate DNA activates dendritic cells via TLR9-dependent and -independent pathways. *J Immunol* 174: 6129.

18. Barton, G. M. et al. 2006. Intracellular localization of Toll-like receptor 9 prevents recognition of self DNA but facilitates access to viral DNA. *Nat Immunol* 7: 49.

19. Leadbetter, E. A. et al. 2002. Chromatin-IgG complexes activate B cells by dual engagement of IgM and Toll-like receptors. *Nature* 416: 603.

20. Boule, M. W. et al. 2004. Toll-like receptor 9-dependent and -independent dendritic cell activation by chromatin–immunoglobulin G complexes. *J Exp Med* 199: 1631.

21. Barrat, F. J. et al. 2005. Nucleic acids of mammalian origin can act as endogenous ligands for Toll-like receptors and may promote systemic lupus erythematosus. *J Exp Med* 202: 1131.

22. Goeckeritz, B. E. et al. 1999. Multivalent cross-linking of membrane Ig sensitizes murine B cells to a broader spectrum of CpG-containing oligodeoxynucleotide motifs, including their methylated counterparts, for stimulation of proliferation and Ig secretion. *Int Immunol* 11: 1693.

23. Sparwasser, T. et al. 1997. Macrophages sense pathogens via DNA motifs: induction of tumor necrosis factor-alpha-mediated shock. *Eur J Immunol* 27: 1671.

24. Krieg, A. M. 2002. CpG motifs in bacterial DNA and their immune effects. *Annu Rev Immunol* 20: 709.

25. Yasuda, K. et al. 2006. CpG motif-independent activation of TLR9 upon endosomal translocation of "natural" phosphodiester DNA. *Eur J Immunol* 36: 431.

26. Sun, S. et al. 1997. Mitogenicity of DNA from different organisms for murine B cells. *J Immunol* 159: 3119.

27. Roberts, T. L. et al. 2005. Differences in macrophage activation by bacterial DNA and CpG-containing oligonucleotides. *J Immunol* 175: 3569.

28. Takagi, T. et al. 1998. Involvement of specific mechanism in plasmid DNA uptake by mouse peritoneal macrophages. *Biochem Biophys Res Commun* 245: 729.

29. Ishii, K. J. et al. 2006. A Toll-like receptor-independent antiviral response induced by double-stranded B-form DNA. *Nat Immunol* 7: 40.

30. Stetson, D. B. and Medzhitov, R. 2006. Recognition of cytosolic DNA activates an IRF3-dependent innate immune response. *Immunity* 24: 93.

31. Stunz, L. L. et al. 2002. Inhibitory oligonucleotides specifically block effects of stimulatory CpG oligonucleotides in B cells. *Eur J Immunol* 32: 1212.

32. Yamada, H. et al. 2002. Effect of suppressive DNA on CpG-induced immune activation. *J Immunol* 169: 5590.

33. Keniry, M. A. 2001. Quadruplex structures in nucleic acids. *Biopolymers* 56: 123.

34. Duramad, O. et al. 2005. Inhibitors of TLR-9 act on multiple cell subsets in mouse and man in vitro and prevent death *in vivo* from systemic inflammation. *J Immunol* 174: 5193.

35. Makarov, V. L., Hirose, Y., and Langmore, J. P. 1997. Long G tails at both ends of human chromosomes suggest a C strand degradation mechanism for telomere shortening. *Cell* 88: 657.

36. McElligott, R. and Wellinger, R. J. 1997. The terminal DNA structure of mammalian chromosomes. *EMBO J* 16: 3705.

37. Gursel, I. et al. 2003. Repetitive elements in mammalian telomeres suppress bacterial DNA-induced immune activation. *J Immunol* 171: 1393.

38. Shirota, H., Gursel, M., and Klinman, D. M. 2004. Suppressive oligodeoxynucleotides inhibit Th1 differentiation by blocking IFN-γ and IL-12-mediated signaling. *J Immunol* 173: 5002.

39. Shirota, H. et al. 2005. Suppressive oligodeoxynucleotides protect mice from lethal endotoxic shock. *J Immunol* 174: 4579.

40. Gunthert, U. et al. 1976. DNA methylation in adenovirus, adenovirus-transformed cells, and host cells. *Proc Natl Acad Sci USA* 73: 3923.

41. Tsuchiya, H. et al. 2005. Cytokine induction by a bacterial DNA-specific modified base. *Biochem Biophys Res Commun* 326: 777.

42. Palmer, B. R. and Marinus, M. G. 1994. The dam and dcm strains of *Escherichia coli*: a review. *Gene* 143: 1.

43. Trevani, A. S. et al. 2003. Bacterial DNA activates human neutrophils by a CpG-independent pathway. *Eur J Immunol* 33: 3164.

44. Alvarez, M. E. et al. 2006. Neutrophil signaling pathways activated by bacterial DNA stimulation. *J Immunol* 177: 4037.

45. Stacey, K. J., Ross, I. L., and Hume, D. A. 1993. Electroporation and DNA-dependent cell death in murine macrophages. *Immunol Cell Biol* 71 (Pt 2): 75.

46. Ishii, K. J. et al. 2001. Genomic DNA released by dying cells induces the maturation of APCs. *J Immunol* 167: 2602.

47. Suzuki, K. et al. 1999. Activation of target-tissue immune-recognition molecules by double-stranded polynucleotides. *Proc Natl Acad Sci USA* 96: 2285.

48. Berland, R. et al. 2006. Toll-like receptor 7-dependent loss of B cell tolerance in pathogenic autoantibody knockin mice. *Immunity* 25: 429.

49. Means, T. K. et al. 2005. Human lupus autoantibody-DNA complexes activate DCs through cooperation of CD32 and TLR9. *J Clin Invest* 115: 407.

50. Vollmer, J. et al. 2005. Immune stimulation mediated by autoantigen binding sites within small nuclear RNAs involves Toll-like receptors 7 and 8. *J Exp Med* 202: 1575.

51. Savarese, E. et al. 2006. U1 small nuclear ribonucleoprotein immune complexes induce type I interferon in plasmacytoid dendritic cells through TLR7. *Blood* 107: 3229.

52. Christensen, S. R. et al. 2005. Toll-like receptor 9 controls anti-DNA autoantibody production in murine lupus. *J Exp Med* 202: 321.
53. Wu, X. and Peng, S. L. 2006. Toll-like receptor 9 signaling protects against murine lupus. *Arthritis Rheum* 54: 336.
54. Yu, P. et al. 2006. Toll-like receptor 9-independent aggravation of glomerulonephritis in a novel model of SLE. *Int Immunol.* 18: 1211.
55. Ehlers, M. et al. 2006. TLR9/MyD88 signaling is required for class switching to pathogenic IgG2a and 2b autoantibodies in SLE. *J Exp Med* 203: 553.
56. Christensen, S. R. et al. 2006. Toll-like receptor 7 and TLR9 dictate autoantibody specificity and have opposing inflammatory and regulatory roles in a murine model of lupus. *Immunity* 25: 417.
57. Marshak-Rothstein, A. 2006. Tolling for autoimmunity: prime time for 7. *Immunity* 25: 397.
58. Anders, H. J. et al. 2004. Activation of toll-like receptor-9 induces progression of renal disease in MRL-Fas(lpr) mice. *FASEB J* 18: 534.
59. Moseman, E. A. et al. 2004. Human plasmacytoid dendritic cells activated by CpG oligodeoxynucleotides induce the generation of CD4+CD25+ regulatory T cells. *J Immunol* 173: 4433.
60. Marshak-Rothstein, A. 2006. Toll-like receptors in systemic autoimmune disease. *Nat Rev Immunol* 6: 823.
61. Ballestar, E., Esteller, M., and Richardson, B. C. 2006. The epigenetic face of systemic lupus erythematosus. *J Immunol* 176: 7143.
62. Richardson, B. et al. 1990. Evidence for impaired T cell DNA methylation in systemic lupus erythematosus and rheumatoid arthritis. *Arthritis Rheum* 33: 1665.
63. Deng, C. et al. 2001. Decreased Ras-mitogen-activated protein kinase signaling may cause DNA hypomethylation in T lymphocytes from lupus patients. *Arthritis Rheum* 44: 397.
64. Corvetta, A. et al. 1991. 5-Methylcytosine content of DNA in blood, synovial mononuclear cells and synovial tissue from patients affected by autoimmune rheumatic diseases. *J Chromatogr* 566: 481.
65. Brummel, R. and Lenert, P. 2005. Activation of marginal zone B cells from lupus mice with type A(D) CpG-oligodeoxynucleotides. *J Immunol* 174: 2429.
66. Yoshida, H. et al. 2005. Lethal anemia caused by interferon-beta produced in mouse embryos carrying undigested DNA. *Nat Immunol* 6: 49.
67. Okabe, Y. et al. 2005. Toll-like receptor-independent gene induction program activated by mammalian DNA escaped from apoptotic DNA degradation. *J Exp Med* 202: 1333.
68. Bergsmedh, A. et al. 2006. DNase II and the Chk2 DNA damage pathway form a genetic barrier blocking replication of horizontally transferred DNA. *Mol Cancer Res* 4: 187.
69. Ashman, R. F. et al. 2005. Sequence requirements for oligodeoxyribonucleotide inhibitory activity. *Int Immunol* 17: 411.
70. Sweet, M. J. et al. 1998. IFN-gamma primes macrophage responses to bacterial DNA. *J Interferon Cytokine Res* 18: 263.
71. Sester, D. P. et al. 2000. Phosphorothioate backbone modification modulates macrophage activation by CpG DNA. *J Immunol* 165: 4165.

8 Therapeutic Potential of Immunosuppressive Oligonucleotides Expressing TTAGGG Motifs

Dennis M. Klinman, Debbie Currie,
Chiaki Fujimoto, Igal Gery,
and Hidekazu Shirota

ABSTRACT

Synthetic oligodeoxynucleotides (ODNs) expressing immunosuppressive TTAGGG motifs downregulate the production of pro-inflammatory and Th1 cytokines. We are interested in the ability of these suppressive ODNs to slow or prevent the development of diseases characterized by pathologic levels of immune stimulation including autoimmune diseases and septic shock. In murine models of arthritis, lupus, LPS-induced toxic shock, and inflammatory diseases of several organs, treatment with suppressive ODNs significantly reduced disease severity. These beneficial effects were accompanied by reductions in serum autoantibody and/or inflammatory cytokine levels. While the mechanism underlying these protective effects is incompletely understood, suppressive ODNs bind to and prevent the phosphorylation of STAT1 and STAT4, thereby blocking the signaling cascade central to the initiation and/or perpetuation of some inflammatory disease states. These findings suggest that suppressive ODN may find use in the treatment of acute and chronic diseases characterized by excessive immune stimulation.

INTRODUCTION

DNA exerts multiple and complex effects on the immune system. Bacterial DNA contains immunostimulatory CpG motifs that trigger protective innate immune responses.[1-3] CpG-driven immune activation can exacerbate inflammatory tissue damage, promote the development of autoimmune disease, and increase sensitivity to toxic shock.[4-9] Other immune responses designed to protect the host can have deleterious consequences if inadequately regulated.

Immune-mediated destruction of host tissue can result in the release of organelles and molecules usually restricted to the intracellular compartment. Our studies suggest that the release of host DNA into the systemic circulation serves to down-modulate over-exuberant immune responses. This effect is mediated, in large part, by specialized structures known as telomeres that are composed of large numbers of single-stranded hexanucleotide repeats (in mice and humans, the repeat sequence is TTAGGG), and primarily serve to protect chromosomes from degradation.[10,11] When activated immune cells are exposed to telomeric DNA, the production of pro-inflammatory and Th1 cytokines is downregulated.[12-14]

Synthetic ODNs designed to express multiple TTAGGG motifs mimic the immunoinhibitory activity of telomeric DNA.[12] Although initially identified by their ability to block CpG-induced immune activation, suppressive ODNs were subsequently shown to block multiple forms of immune stimulation[15-18] and to be effective in the prevention and/or treatment of a variety of pathologic autoimmune responses.[19,20]

This chapter reviews the ability of suppressive ODNs to prevent and/or treat disease states characterized by pathologic autoimmune or inflammatory reactions.[5,13,19,21] Findings support the conclusion that TTAGGG-containing ODNs effectively prevent or slow the progression of diseases characterized by hyperactivation of the immune system.

EFFECTS OF SUPPRESSIVE ODNS ON COLLAGEN-INDUCED ARTHRITIS

Collagen-induced arthritis (CIA) is a well established murine model of rheumatoid arthritis that has helped clarify the pathogenesis of RA and examine potential treatments.[12,22] CIA is elicited by injecting DBA/1 mice intradermally with type II bovine collagen (CII) in complete Freund's adjuvant (FA) followed by a boost 3 weeks later of CII in incomplete Freund's adjuvant.[12,22] Arthritis typically develops shortly after the second CII injection, with swelling and inflammation of joints persisting for many weeks.

PROTECTION OF DBA/1 MICE FROM COLLAGEN-INDUCED ARTHRITIS

The ability of suppressive ODNs to block the induction of CIA was examined. A single 300-µg dose of suppression ODN was administered before or after the initial injection of CII. As seen in Table 8.1, this intervention had no effect on the frequency or severity with which DBA/1 mice developed arthritis. When two doses of suppressive ODNs were administered, a modest reduction in the frequency of disease ($p < 0.05$) and a substantial reduction in the level of inflammation ($p < 0.01$, Table 8.1) occurred. By comparison, when suppressive ODNs were delivered twice per week starting 3 days prior to initial CII administration, the incidence of arthritis was reduced by more than half and disease severity by nearly 80% ($p < 0.01$ for both outcomes). This beneficial effect was mediated uniquely by suppressive ODNs, since control ODNs had no effect on the frequency or severity of CIA (Table 8.1).

Joints from mice with CIA manifested histologic evidence of inflammation including severe lymphocytic infiltration, erosion of bone and cartilage, active proliferation

TABLE 8.1
Reduction of Frequency and Severity of Collagen-Induced Arthritis

Treatment	Timing	Disease Incidence (%)	Severity
PBS	Once or twice weekly	100	6.9 ± 0.8
Control ODN (300 µg)	Once or twice weekly	95	6.2 ± 1.1
Suppressive ODN (300 µg):			
Once	Day –3 or +3	90	5.3 ± 0.6
Twice	Day –3/+3 or 0/+18	83[a]	3.9 ± 0.5[a]
Multiple	Twice weekly starting day 3	45[b]	1.5 ± 0.4[b]

Note: DBA/1 mice were injected intradermally with type II collagen emulsified in complete Freund's adjuvant (day 0), then incomplete Freund's adjuvant (day 21). N = 20 to 30 mice per group. Disease severity evaluated on scale of 0 to 12.

[a] $p < 0.05$.

[b] $p < 0.01$ versus combination of PBS and control ODN-treated mice.

of synovial lining cells, and disruption of joints. Consistent with a beneficial effect on disease, joints from mice treated with suppressive ODNs showed only mild cellular infiltration without disruption of the joint architecture.[19]

The effect of suppressive ODNs on the immunologic abnormalities that accompany CIA were examined. These include the production of IgG anti-CII autoantibodies, and the activation of antigen-specific T cells secreting Th1 and pro-inflammatory cytokines.[12,23] Treatment with suppressive ODNs reduced serum autoantibody levels by nearly 3-fold, a reduction that included all isotypes of IgG anti-CII Ab ($p < 0.01$).[19] Similarly, treatment reduced the number of T cells that responded to CII exposure by secreting IFN by 0.3-fold ($p < 0.05$).[19]

EFFECTS OF SUPPRESSIVE ODNS ON INFLAMMATORY ARTHRITIS

Reactive arthritis presents as an asymmetric, oligoarticular inflammatory condition that typically develops several weeks after bacterial infection of the gastrointestinal or genitourinary tract.[24,25] In mice, a similar form of arthritis can be induced by injecting bacterial DNA or immunostimulatory CpG ODN directly into the knee joints of mice.[9,13]

PREVENTION OF CPG-INDUCED INFLAMMATORY ARTHRITIS

Inflammatory arthritis develops when BALB/c mice are injected intraarticularly with 25 µg of CpG ODN. Co-administration of an equal amount of suppressive ODN significantly reduces both inflammation and swelling (Table 8.2). This effect is specifically mediated by the suppressive ODN, since control ODN has no effect on joint swelling or pathology.

Inflammatory arthritis affects multiple joints. Thus, it would be of considerable benefit if suppressive ODN delivered systemically could reduce arthritis in all locations. To examine this possibility, mice were injected intraperitoneally with suppres-

TABLE 8.2

Reduction of Severity of Inflammatory Arthritis

Treatment	Site and Time	Severity (0 to 3 Scale)
PBS	Local/simultaneous	2.3 ± 0.3
Control ODN	Local/simultaneous	2.1 ± 0.2
Suppressive ODN	Local/simultaneous	0.2 ± 0.2[a]
PBS	Systemic/day −3	2.4 ± 0.4
Control ODN	Systemic/day −3	2.7 ± 0.8
Suppressive ODN	Systemic/day −1	1.4 ± 0.4
Suppressive ODN	Systemic/day −3	0.3 ± 0.2[a]

Note: Arthritis was induced in BALB/c mice by intraarticular injection of 25 μg CpG ODNs. In some
        cases, mice were treated simultaneously with intraarticular (local) control or suppressive ODNs
        (25 μg). In other cases, 300 μg of systemic ODNs were delivered intraperitoneally 1 to 3 days
        earlier. Disease severity was evaluated by histologic changes in knees 4 days after challenge. Val-
        ues represent mean + SEM for 6 to 9 mice per group.
[a] $p < 0.05$ versus PBS-treated controls.

sive ODN and challenged intraarticularly with CpG DNA. Administering suppressive
ODN a day prior to CpG challenge had limited impact on disease severity, whereas
treatment 3 days prior to challenge significantly reduced the development of inflammatory arthritis ($p < 0.03$, Table 8.2). In contrast, animals treated systemically with
control ODNs or PBS consistently developed severe arthritis when injected with
CpG ODN. These findings suggest that suppressive ODNs reduce host susceptibility
to inflammatory challenge.

Cell transfer experiments were conducted to clarify the cellular basis of the
effect.

BALB/c mice were treated with suppressive or control ODN. Three days later,
20 million spleen cells were transferred into recipient mice. When the recipients
were challenged with CpG ODN, spleen cells from donors treated with suppressive
(but not control) ODN conferred resistance to arthritis.[5] This resistance was selectively abrogated by removal of CD11c+ (but not B or NK) cells from the transferred
splenocytes.[5] These findings suggest that suppressive ODNs act on CD11c+ cells,
altering subsequent sensitivity to CpG-induced inflammatory signals.

EFFECTS OF SUPPRESSIVE ODNs ON OCULAR INFLAMMATION

Animal models of experimental autoimmune uveitis (EAU) facilitate the study of
a variety of intraocular inflammatory diseases broadly categorized as uveitis. EAU
is elicited by immunizing mice with eye-specific antigens, such as interphotorecep-
tor retinoid-binding protein (IRBP).[26] Alternatively, the efferent limb of the immune
response can be studied by adoptively transferring Th1 cells sensitized against hen
egg lysozyme (HEL) into Tg mice expressing HEL in their retinas.[27]

Both model systems were used to examine the potential therapeutic benefit of
suppressive ODNs in autoimmune uveitis. EAU was induced by immunizing B10.A

TABLE 8.3
Reduction in Severity of Autoimmune Uveitis (Histologic Scores)

	IRBP-Induced	Adoptive Transfer
PBS	1.4 ± 0.1	3.5 ± 0.6
Control ODN	1.4 ± 0.1	3.5 ± 0.6
Suppressive ODN	0.5 ± 0.1[a]	1.3 ± 0.3[a]

Note: Uveitis was induced by immunizing B10.A mice with 40 µg IRBP or adoptively transferring 2×10^5 HEL-specific Th1 cells into HEL-Tg recipients. Animals were treated with 300 µg suppressive or control ODNs. Histologic eye changes were evaluated on day 14 (IRBP) or day 7 (adoptive transfer). Results are averages of three independent experiments.

[a] $p < 0.01$, Mann-Whitney U test.

mice with IRBP. Treating these mice with suppressive (but not control) ODNs on days 0, 3, 7 and 10 significantly reduced the magnitude of the resulting ocular inflammation (Table 8.3, $p < 0.01$). This was associated with a 2- to 3-fold reduction in the number of lymphocytes that responded to IRBP exposure by proliferating or secreting IFN_γ, TNF_α or IL-6 ($p < 0.05$).[28] Ocular inflammation was also examined in the adoptive transfer model. Administering suppressive ODNs 0 and 3 days after the transfer of HEL-specific Th1 cells into HEL Tg mice reduced the magnitude of the resulting uveitis by 3-fold (Table 8.3, $p < 0.01$).[28] Thus, suppressive ODNs significantly reduced the severity of uveitis in two distinct murine models of ocular inflammation.

EFFECTS OF SUPPRESSIVE ODNs ON SYSTEMIC LUPUS ERYTHEMATOSUS

Systemic lupus erythematosus (SLE) is an autoimmune disease characterized by the production of anti-nuclear autoantibodies, immune complex-mediated glomerulonephritis, and multifocal end-organ damage.[29-31] Female NZB/W mice provide a useful model for studying the pathogenesis and treatment of human SLE since they exhibit immunologic and clinical manifestations of disease similar to those present in human lupus patients.[32,33]

SLOWING ONSET OF LUPUS NEPHRITIS IN NZB/W MICE

The ability of suppressive ODNs to impact the development of murine lupus was examined. Female NZB/W mice were treated systemically with 300 µg of suppressive or control ODNs twice monthly starting at 6 weeks of age. By 6 months of age, untreated mice and recipients of control ODNs began to develop proteinuria, an early sign of the kidney damage characteristic of SLE (Table 8.4). The severity and frequency of proteinuria rose with age in both groups. Among mice treated with suppressive ODNs, the onset and magnitude of proteinuria were significantly reduced (Table 8.4, $p < 0.01$). Of particular interest, mice treated with suppressive ODNs survived significantly longer than control animals (Table 8.4, $p < 0.01$).[21] Consistent

TABLE 8.4

Reduction of Onset and Severity of Lupus Nephritis

Timing	Treatment	Proteinuria (0 to 4 scale)	% Survival
Twice monthly starting at 6 weeks of age	PBS	2.4 ± 0.4	20
	Suppressive ODN	0.8 ± 0.3[a]	75[a]
Twice monthly starting at 7 months of age	PBS	3.9 ± 0.1	0
	Suppppressive ODN	2.3 ± 0.3[a]	80[a]

Note: Female NZB/W mice were treated with 300 µg suppressive ODNs. Mean proteinuria ± SEM at 10 months and survival at 12 months are shown for animals treated from 6 weeks of age (N = 20 animals per group), while proteinuria at 9 months and survival at 12 months are shown for animals treated from 7 months of age (N = 5 animals per group).

[a] $p < 0.01$.

with findings in the collagen-induced arthritis model, treatment with suppressive ODNs reduced serum autoantibody titers and the number of cells actively secreting IL-12 and IFNγ in lupus-prone mice by 2- to 8-fold ($p < 0.01$).[21]

To further evaluate the effects of suppressive ODNs in murine lupus, kidneys from mice treated with control or suppressive ODNs were examined for signs of glomerulonephritis. By 8 months of age, kidneys from untreated controls showed evidence of severe inflammation (mean glomerular activity scale of 10.4 ± 0.4 on the NIH 12-point pathology scale). Similarly, kidneys from mice treated with control ODNs developed extensive pathology (9.6 ± 0.8). In contrast, glomerulonephritis was mild among the animals treated with suppressive ODNs (2.8 ± 0.8, $p < 0.01$). Consistent with these findings, the deposition of immune complexes characteristic of autoimmune-mediated glomerulonephritis was significantly reduced in animals treated with suppressive ODNs ($p < 0.01$).[21]

SLOWING PROGRESSION OF ESTABLISHED LUPUS NEPHRITIS

To examine whether suppressive ODNs may prevent the worsening of established autoimmune disease, treatment was initiated in a cohort of older NZB/W mice with established 2+ proteinuria (at approximately 7 months of age). Treating these animals twice weekly for 3 months slowed (but did not halt) the progression of proteinuria and glomerulonephritis ($p < 0.05$) and significantly prolonged their survival (Table 8.4, $p < 0.02$).[21]

EFFECTS OF SUPPRESSIVE ODNs ON INFLAMMATORY LUNG DISEASE

Bacterial DNA can support the induction of inflammation associated with pulmonary infection.[34] For example, immunostimulatory CpG motifs present in bacterial DNA trigger the extravasation of neutrophils into the lungs[27] and the production of cytokines and chemokines (such as TNF and MIP-2), when administered intratracheally to normal

TABLE 8.5
Reduction of CpG-Mediated Lung Inflammation (BAL Content)

Treatment	TNF-α[a]	MIP-2[b]	Neutrophils[c]
PBS	8.2 ± 2.4	4.6 ± 6.9	11 ± 8
CpG ODN	46.9 ± 18.0	53.0 ± 8.2	124 ± 36
CpG + suppressive ODN	11.8 ± 2.7[d]	13.1 ± 6.5[d]	40 ± 31[d]
CpG + control ODN	55.3 ± 9.6	51.9 ± 7.4	143 ± 35

Note: 30 ug of CpG, suppressive and/or control ODN was instilled into the lungs of normal BALB/c mice. BAL was collected 16 h later. Results represent the mean + SD of 8 independently studied mice/group.

[a] pg/ml TNF-α in BAL fluid.
[b] pg/ml MIP-2 in BAL fluid.
[c] Neutrophil count/μl of BAL.
[d] $p < 0.05$ compared to mice treated with CpG ODNs.

BALB/c mice. These effects can be mimicked *in vitro*: alveolar macrophages exposed to CpG DNA become activated to secrete large amounts of TNFα and MIP-2.[27,35,36]

To determine whether the production of these inflammatory mediators is inhibited by suppressive ODNs, RAW 264.7 macrophages were cultured with CpG plus suppressive or control ODNs. Results show that suppressive ODNs uniquely and significantly reduce TNFα and MIP-2 production.[27] Of greater clinical relevance, pulmonary administration of suppressive ODNs significantly reduces neutrophil accumulation and the release of cytokines and chemokines into the bronchoalveolar fluids of mice exposed to CpG DNA (Table 8.5, $p < 0.01$).[27] Thus, suppressive ODNs may be of therapeutic benefit in limiting the inflammatory reactions induced by bacterial infection of the lungs.

EFFECTS OF SUPPRESSIVE ODNs ON LPS-INDUCED TOXIC SHOCK

LPS-induced toxic shock is a major cause of septic shock in humans.[37] LPS binds to Toll-like receptor 4 expressed on macrophages and monocytes, triggering a cascade of cytokine and chemokine production that culminates in the death of the host.[37–39] IFN is one of the cytokines associated with the development of toxic shock, and antibodies that block IFN production can inhibit disease progression.[40–42]

PROTECTION OF MICE FROM LPS-INDUCED TOXIC SHOCK

BALB/c mice challenged with 200 μg of *E. coli* LPS uniformly succumb to endotoxic shock within 3 days. Treating these mice with suppressive ODN immediately prior to challenge led to the survival of all LPS-challenged animals (Table 8.6, $p < 0.001$).[43] Of interest, this was associated with a concomitant reduction in the production of IFNγ by the LPS-challenged mice ($p < 0.001$).[43] Delaying ODN delivery until after challenge was of benefit in significantly prolonging life span, although delaying treatment rarely prevented eventual death (Table 8.6, $p < 0.05$).

TABLE 8.6
Improved Survival Following LPS-Induced Toxic Shock

Treatment	Time of Rx	% Survival	MTD
PBS	Any time	0	26 ± 4
Control ODN	Any time	0	28 ± 5
Suppressive ODN	–3 hr	100[a]	
Suppressive ODN	1 hr	20	42 ± 6[a]

Note: BALB/c mice were treated with 300 µg of suppressive or control ODN 3 hours
before or 1 hour after challenge with 200 µg *E. coli* LPS. The percent of mice
surviving challenge and mean times to death are shown for two to four independent experiments involving 10 to 16 mice per group.

[a] $p < 0.05$.

MECHANISMS OF ACTION OF SUPPRESSIVE ODNs

BLOCKING OF STAT1 AND STAT4 PHOSPHORYLATION

Based on the observation that suppressive ODNs selectively inhibit the production of
IFN following LPS challenge in mice, elements of the regulatory pathway known to
mediate IFN production were examined. Results showed that LPS-mediated IFN release
was regulated (at least in part) by the phosphorylation of STAT1 and STAT4.[40,44–48]
Of considerable interest, this phosphorylation was significantly reduced in a dose- and
sequence-dependent manner by the addition of suppressive ODNs.[43]

To clarify the mechanism by which suppressive ODNs blocked STAT phosphorylation, ligand binding studies were performed. Results showed that suppressive ODNs
selectively bound to STAT1 and STAT4. This interaction was highly specific, since
suppressive ODNs did not bind to other molecules in the NF-κB and MAPK regulatory
cascade, and control ODNs did not bind to STAT1 or STAT4.[43]

DISCUSSION

DNA exerts multiple and complex effects on the immune system. This work reviews
studies supporting the conclusion that repetitive TTAGGG motifs present in mammalian telomeres have immunosuppressive properties of potential therapeutic benefit. Early research showed that telomeric DNA and suppressive ODNs containing
TTAGGG motifs derived from mammalian telomeres down-regulated the production
of Th1 and pro-inflammatory cytokines.[12,14,49] When evaluated in murine models of
arthritis, uveitis, lupus, pulmonary inflammation and LPS-induced toxic shock, suppressive ODNs reduced the frequency and/or severity of disease,[13,19,21,28,43] raising the
possibility that suppressive ODNs may represent novel and broadly effective treatment regimes for autoimmune and inflammatory diseases.

Suppressive ODNs were optimally effective when administered early in the disease process. For example, treating lupus-prone mice starting at 6 weeks of age significantly delayed the onset of renal disease and prolonged their survival.[21] Similarly,

suppressive ODNs reduced mortality when administered prior to LPS challenge in a murine toxic shock model,[43] and prevented the development of arthritis when delivered prior to intraarticular challenge with bacterial DNA.[13,19] By comparison, delayed administration of suppressive ODNs was less effective. Treatment of mice with established lupus nephritis or LPS-induced toxic shock slowed the rate of disease progression but did not improve survival.[21]

Suppressive ODNs were effective when administered locally (e.g., to an inflamed arthritic knee) or systemically. When administered systemically, they broadly reduced host susceptibility to pro-inflammatory stimuli. Thus, local treatment may be of benefit when a site and source of inflammation can be identified. Systemic treatment may be required when the nature or location of the pathologic stimulus is unknown.

The mechanism by which suppressive ODNs influence disease susceptibility is still under investigation. Suppressive ODNs inhibit the production of multiple Th1 and pro-inflammatory cytokines and chemokines including IFNγ, IL-12, IL-6, TNFα, and MIP-2.[14] Suppressive ODN treatment was also found to reduce serum autoantibody levels in several models of autoimmune disease.[19,21] Taken together, these findings suggest that suppressive ODNs inhibit the production of cytokines, chemokines, and antibodies by self-reactive lymphocytes.

The activity of suppressive ODNs is linked with their ability to form poly-G-mediated intra- and inter-chain Hoogsteen hydrogen bonds.[12,50,51] Eliminating the poly-G motifs abrogates the capacity of the ODNs to down-regulate immune activation. Work by Shirota et al. indicates that suppressive ODNs bind to and prevent the phosphorylation of STAT1 and STAT4.[43,49] This is consistent with the findings of Jing et al., who reported that G-rich ODNs block the phosphorylation of STAT1 in cancer cell lines.[52] By preventing STAT phosphorylation, suppressive ODNs presumably inhibit the signal transduction cascade needed to maintain inflammatory and autoimmune conditions.

The studies reviewed in this work suggest that suppressive ODN may be of broad value in the prevention and/or treatment of diseases associated with the development of a pro-inflammatory immune milieu. This activity mimics the ability of telomeric DNA to reduce over-exuberant autoimmune and inflammatory responses. Presumably, the release of TTAGGG-rich self DNA by injured host cells functions to down-regulate pathologic autoimmune responses. The studies reviewed herein analyzed the effects of administering 50 to 300 μg of suppressive ODNs to disease-prone animals for a few days to a few weeks. Additional research is needed to determine whether more aggressive therapy may be even more beneficial and to identify disease states most susceptible to suppressive ODN therapy.

DISCLAIMER

The assertions of the authors are not to be construed as official or as reflecting the views of the National Cancer Institute.

REFERENCES

1. Krieg, A. M. 2002. CpG motifs in bacterial DNA and their immune effects. *Ann Rev Immunol* 20: 709.
2. Krieg, A. M. et al. 1998. CpG DNA induces sustained IL-12 expression *in vivo* and resistance to *Listeria monocytogenes* challenge. *J Immunol* 161: 2428.
3. Zimmermann, S. et al. 1998. CpG oligodeoxynucleotides trigger protective and curative Th1 responses in lethal murine leishmaniasis. *J Immunol* 160: 3627-3630.
4. Krieg, A.M. 1995. CpG DNA: A pathogenic factor in systemic lupus erythematosus? *J Clin Immunol* 15: 284.
5. Zeuner, R. A. et al. 2003. Influence of stimulatory and suppressive DNA motifs on host susceptibility to inflammatory arthritis. *Arthritis Rheum* 48: 1701.
6. Sparwasser, T. et al. 1997. Bacterial DNA causes septic shock. *Nature* 386: 336.
7. Cowdery, J. S. et al. 1996. Bacterial DNA induces NK cells to produce IFNγ *in vivo* and increases the toxicity of lipopolysaccharides. *J Immunol* 156: 4570.
8. Heikenwalder, M. et al. 2004. Lymphoid follicle destruction and immunosuppression after repeated CpG oligodeoxynucleotide administration. *Nat Med* 10: 187-192.
9. Deng, G. M. et al. 1999. Intra-articularly localized bacterial DNA containing CpG motifs induces arthritis. *Nat Med* 5: 702.
10. de Lange, T. and T. Jacks. 1999. For better or worse? Telomerase inhibition and cancer. *Cell* 98: 273.
11. Blasco, M. A., S. M. Gasser, and J. Lingner. 1999. Telomeres and telomerase. *Genes Dev* 13: 2353.
12. Gursel, I. et al. 2003. Repetitive elements in mammalian telomeres suppress bacterial DNA-induced immune activation. *J Immunol* 171: 1393.
13. Zeuner, R. A. et al. 2002. Reduction of GpG-induced arthritis by suppressive oligodeoxynucleotides. *Arthritis Rheum* 46: 2219.
14. Yamada, H. et al. 2002. Effect of suppressive DNA on CpG-induced immune activation. *J Immunol* 169: 5590.
15. Halpern, M. D. and D. S. Pisetsky. 1995. *In vitro* inhibition of murine IFNγ production by phosphorothioate deoxyguanosine oligomers. *Immunopharmacology* 29: 47.
16. Pisetsky, D. S. 2000. Inhibition of murine macrophage IL-12 production by natural and synthetic DNA. *Clin Immunol* 96: 198.
17. Zhu, F. G., C. F. Reich, and D. S. Pisetsky. 2002. Inhibition of murine macrophage nitric oxide production by synthetic oligonucleotides. *J Leukoc Biol* 71: 686.
18. Zhu, F. G., C. F. Reich, and D. S. Pisetsky. 2002. Inhibition of murine dendritic cell activation by synthetic phosphorothioate oligodeoxynucleotides. *J Leukoc Biol* 72: 1154.
19. Dong, I. et al. 2004. Suppressive oligodeoxynucleotides protect against development of collagen-induced arthritis in mice. *Arthritis Rheum* 50: 1686.
20. Ho, P. P. et al. 2003. An immunomodulatory GpG oligonucleotide for the treatment of autoimmunity via the innate and adaptive immune systems. *J Immunol* 171: 4920.
21. Dong, I. et al. 2004. Suppressive oligodeoxynucleotides delay onset of glomerulonephritis and prolong the survival of lupus-prone NZB/W mice. *Arthritis Rheum* 52: 651.
22. Myers, L. K. et al. 1997. Collagen-induced arthritis, an animal model of autoimmunity. *Life Sci* 61: 1861.
23. Terato, K. et al. 1992. Induction of arthritis with monoclonal antibodies to collagen. *J Immunol* 148: 2103.
24. Braun, J. et al. 2000. On the difficulties of establishing a consensus on the definition of and diagnostic investigations for reactive arthritis. *J Rheumatol* 27: 2185.
25. Sieper, J., J. Braun, and G. H. Kingsley. 2000. Report on Fourth International Workshop on Reactive Arthritis. *Arthritis Rheum* 43: 720.

26. Caspi, R. R. et al. 1988. A new model of autoimmune disease. Experimental autoimmune uveoretinitis induced in mice with two different retinal antigens. *J Immunol* 140: 1490.

27. Foxman, E. F. et al. 2002. Inflammatory mediators in uveitis: differential induction of cytokines and chemokines in Th1- versus Th2-mediated ocular inflammation. *J Immunol* 168: 2483.

28. Fujimoto, C. et al. A suppressive oligodeoxynucleotide inhibits ocular inflammation. *Clin Immunol Immunopathol* (submitted).

29. Asselin-Paturel, C. et al. 2001. Mouse type I IFN-producing cells are immature APCs with plasmacytoid morphology. *Nat Immunol* 2: 1144.

30. Klinman, D. M. and A. D. Steinberg. 1995. Inquiry into murine and human lupus. *Immunol Rev* 144: 157.

31. Hahn, B. H. 1998. Antibodies to DNA. *New Engl J Med* 338: 1359.

32. Lambert, P. H. and F. J. Dixon. 1968. Pathogenesis of the glomerulonephritis of NZB/W mice. *J Exp Med* 127: 507.

33. Theofilopoulos, A. N. and F. J. Dixon. 1985. Murine models of SLE. *Adv Immunol* 37: 269.

34. Schwartz, D. A. et al. 1997. CpG motifs in bacterial DNA cause inflammation in the lower respiratory tract. *J Clin Invest* 100: 68.

35. Sparwasser, T. et al. 1997. Macrophages sense pathogens via DNA motifs: induction of tumor necrosis factor α-mediated shock. *Eur J Immunol* 27: 1671.

36. Takeshita, S. et al. 2000. CpG oligodeoxynucleotides induce murine macrophages to up-regulate chemokine mRNA expression. *Cell Immunol* 206: 101.

37. Alexander, C. and E. T. Rietschel. 2001. Bacterial lipopolysaccharides and innate immunity. *J Endotoxin Res.* 7: 167.

38. Dobrovolskaia, M. A. and S. N. Vogel. 2002. Toll receptors, CD14, and macrophage activation and deactivation by LPS. *Microbes Infect* 4: 903.

39. Akira, S. and K. Takeda. 2004. Toll-like receptor signalling. *Nat Rev Immunol* 4: 499.

40. Rendon-Mitchell, B. et al. 2003. IFNγ induces high mobility group box 1 protein release partly through a TNF-dependent mechanism. *J Immunol* 170: 3890.

41. Car, B. D. et al. 1994. Interferon γ receptor-deficient mice are resistant to endotoxic shock. *J Exp Med* 179: 1437.

42. Freudenberg, M. A. et al. 2002. Cutting edge: a murine, IL-12-independent pathway of IFNγ induction by Gram-negative bacteria based on STAT4 activation by Type I IFN and IL-18 signaling. *J Immunol* 169: 1665.

43. Shirota, H. et al. 2005. Suppressive oligodeoxynucleotides protect mice from lethal endotoxic shock. *J Immunol* 174: 4579.

44. Fujihara, M. et al. 1994. Role of endogenous interferon-β in lipopolysaccharide-triggered activation of the inducible nitric-oxide synthase gene in a mouse macrophage cell line J774. *J Biol Chem* 269: 12773.

45. Cassatella, M. A. et al. 1990. Molecular basis of interferon-γ and lipopolysaccharide enhancement of phagocyte respiratory burst capability: studies on the gene expression of several NADPH oxidase components. *J Biol Chem* 265: 20241.

46. Gupta, J.W. et al. 1992. Induction of expression of genes encoding components of the respiratory burst oxidase during differentiation of human myeloid cell lines induced by tumor necrosis factor and γ-interferon. *Cancer Res* 52: 2530.

47. Calandra, T. et al. 1994. The macrophage is an important and previously unrecognized source of macrophage migration inhibitory factor. *J Exp Med* 179: 1895.

48. Thierfelder, W. E. et al. 1996. Requirement for STAT4 in interleukin-12-mediated responses of natural killer and T cells. *Nature* 382: 171.

49. Shirota, H., Gursel, M., and Klinman, D.M. 2004. Suppressive oligodeoxynucleotides inhibit Th1 differentiation by blocking IFNγ- and IL-12-mediated signaling. *J Immunol* 173: 5002.
50. Murchie, A.I. 1994. Tetraplex folding of telomere sequences and the inclusion of adenine bases. *EMBO J* 13: 993.
51. Han, H. 2000. G-quadruplex DNA: a potential target for anti-cancer drug design. *Trends Pharmacol Sci* 21: 136.
52. Jing, N. et al. 2003. Targeting STAT3 with G-quartet oligodeoxynucleotides in human cancer cells. *DNA Cell Biol* 22: 685.

9 Structure/Function of IFNα-Inducing CpG ODNs

Daniela Verthelyi and Montserrat Puig

ABSTRACT

Three different types of CpG oligodeoxynucleotides (CpG ODNs) induced striking different immune responses via the activation of a TLR9-dependent pathway. CpG ODNs that selectively induce production of IFNα (Type D and A ODNs) are characterized by a predominantly phosphorodiester backbone, a CpG motif flanked by self-complementary bases, and a poly-guanosine (poly-G) tail on the 3′ end (or 3′ and 5′ for A ODN) that facilitates G tetrad formation and ODN multimerization. Binding and uptake of poly-G bearing ODNs are facilitated by cell membrane scavenger receptors such as CXCL16 that bind the poly-G strand. Inside cells, these ODNs are retained in early or recycling endosomes where they interact with TLR9 and signal via MyD88 and IRF7 to induce plasmacytoid dendritic cells (pDCs) to secrete high levels of IFNα. Unlike other types of CpG ODNs (K/B or C), ODN types D and A do not stimulate pDC maturation or induce polyclonal B cell activation directly.

In primate studies, D ODN sequences were used effectively with hepatitis B virus and *Leishmania* antigens to promote protective immunoglobulin levels and to protect against infectious challenges, respectively. Moreover, when administered alone around the time of infection, D ODNs improved the clinical outcomes of healthy and immunocompromised macaques challenged with *Leishmania* sp. To date, the clinical development of these ODNs has been hindered by the poly-G associated tetramer formation, which, although critical for the induction of IFNα, leads to increased lot-to-lot variability and causes aggregation and precipitation. Recent development of second-generation type D ODNs that have unique modifications to prevent tetrad formation during manufacture and storage promise to overcome these problems. Clinical studies should follow to determine their immunoprotective and therapeutic efficacies.

INTRODUCTION

Pathogen exposure rapidly triggers an innate immune response characterized by increased phagocytosis, the activation of lytic cells, and the production of immunostimulatory cytokines, chemokines, and polyreactive IgM antibodies that limit pathogen spread and foster development of an appropriate adaptive immune response.[1] In vertebrates, this innate immune response is triggered by a network of germline-encoded, non-clonal pattern recognition receptors (PRRs). These PRRs recognize conserved pathogen-associated molecular patterns (PAMPs) that are highly conserved among classes of pathogens such as components of bacterial cell walls (lipopolysaccharides, lipoteichoic acid, and peptidoglycans), specialized bacterial or parasite proteins (flagellin and profilin), or nucleic acid structures (viral RNA and DNA).[2] The recognition that purified or synthetic PAMP homologues can be used to stimulate and direct the innate and adaptive immune system has opened new areas of vaccine development and immunotherapy.[3,4]

Among the PRRs that recognize nucleic acids, Toll-like receptor 9 (TLR9) is activated by sequences of DNA that contain unmethylated CpG dinucleotides within certain sequence contexts (CpG motifs) present at high frequency in microbial DNA but less frequently in vertebrate DNA.[5,6] Studies in mice show that the synthetic oligonucleotides reproducing these motifs (CpG ODNs) hold promise as vaccine adjuvants and in treating asthma, allergies, infections, and cancers.[3,4]

Early studies showed that although TLR9 recognition of CpG motifs is highly conserved in vertebrates, it diverged through evolution as evidenced by the differential cellular distributions of TLR9 among species.[7,8] For example, TLR9 is expressed by multiple cell types in mice, but is restricted to B cells and pDCs in humans and other primates.[9] Moreover, human TLR9 are poorly activated by CpG ODN sequences that are highly immunostimulatory in mice despite very minor sequence divergence, indicating TLR9 has a high degree of sequence specificity[7,8] (see Table 9.1).

Despite the specificity of microbial DNA recognition, studies using synthetic ODNs identified a large number of sequences that stimulate TLR9 with variations in the numbers and locations of the CpG dimers and in the precise base sequences flanking the CpG dimers.[8,10,11] This led to the classification of synthetic CpG ODNs that exhibited immunostimulatory activity on human peripheral blood mononuclear cells (PBMCs) into three general types that reflect their structural and biological effect on B cells and pDCs (Figure 9.1):[12]

- Type D (and A) ODNs induce pDCs to secrete high levels of interferon-α (IFN-α), but do not stimulate pDC maturation or polyclonal B cell activation.
- Type C CpG ODNs induce polyclonal B cell activation and IFN-α secretion by pDCs. Type K (also known as B) ODNs contain several CpG motifs and phosphorothioate (PS) backbones; they stimulate polyclonal B cell activation and pDC differentiation, but little to no IFN-α secretion.

This chapter focuses on the IFN-α-inducing ODNs.

TABLE 9.1

Selectivity of DNA Recognition

Species specificity of TLR9

Cellular distribution of TLR9

Co-factors and chaperone proteins

Intracellular compartment where TLR9–DNA interaction occurs

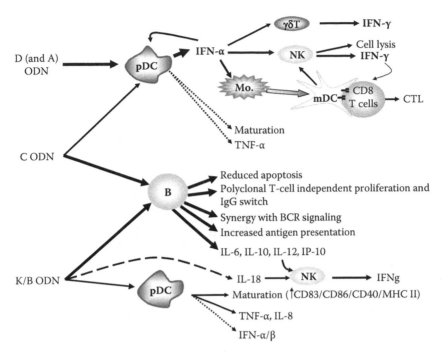

FIGURE 9.1 Three major types of CpG ODNs (D, K, and C) stimulate immune cells. Type D (also known as type A) ODNs do not stimulate B cells but are potent inducers of IFN-α by pDCs. The induced IFN-α leads to maturation of monocytes into mDCs and activation of NK and γ/δ T cells. Type K (also known as type B) ODNs induce polyclonal B cell stimulation and maturation of pDCs. Type C ODNs induce both polyclonal B cell activation and IFN-α secretion by pDCs.

IFN-α-INDUCING CPG DNA SEQUENCES

STRUCTURE

IFN-α plays a key role in responses to viral infections, activates natural killer (NK) cell-mediated cytotoxicity, stimulates IL-15 production and IFN-γ expression, and is key for antigen cross-presentation and memory CD8+ T cell induction.[13–15] Plasmacytoid DCs are thought to be the mains source of IFN-α.[16,17] CpG ODNs were the first synthetic molecules described to induce pDCs to secrete high levels of IFNα.[18] *In vitro*, D and A ODNs foster strong cellular immune responses by inducing high and sustained levels of IFN-α and β that mediate monocyte activation and maturation into functionally active myeloid DCs, along with NK and γδ T cell activation with increased cell lysis and IFN-γ and IP-10 secretion (Figure 9.1).[8,18–22] Unlike CpG ODN type K/B or C, these ODNs do not induce the maturation of pDCs (defined by upregulation of CD80, CD86, and MHCII) and exert no direct effect on B cells.[8,19]

Type D CpG ODNs, originally described by our laboratory,[8] and type A described by Krug et al.,[18] have similar but not identical structures (Table 9.2). The CpG motifs of both types of sequences are flanked on both sides by three to six self-complementary

TABLE 9.2

Structural Differences among IFNα-Inducing CpG ODNs

Characteristic	D ODN	A ODN	C ODN
CpG motif	PuPyCpGPuPy	TCpGT	TCG
Poly-G tail	3' end	3' and 5' ends	No poly-G tail
Number of phosphorothioate-modified bases	4 (2 on each end)	7 to 10	All PS
G-tetrad formation	Parallel	Parallel and anti-parallel	No G tetrad
Tertiary structures formed	Particulate	Particulate and nanowires	No tertiary structures
Intracellular localization	Early endosomes	Early endosomes	Early and late endosomes
Sample sequence	GGTGCATCGATGCAGGGGGG	GGGGGACGATCGTCGGGGG	TCGTCGTCGTTCGAACGACGTTGAT

bases and a poly-G tail. As with other types of CpG ODNs, inversion, replacement, or methylation of the CpG dinucleotide reduces or abrogates ODN activity. Computer simulations suggest that the active CpG motif is exposed at the apex of a stem–loop structure formed by the pairing of self-complementary bases. Changes that alter the self-complementary flanking sequences significantly reduce their immunostimulatory activity.[8] The core bases encoding the CpG hexamer and flanking regions typically have phosphodiester (PD) rather than PS links, as the latter lack the flexibility to form stem–loop structures. Instead, they have two to five bases at both ends, with PS modifications to reduce their susceptibility to nuclease digestion. The poly-G strands at the 3′ end modify cellular uptake and trafficking.[23–25] ODNs that lack poly-G strands and do not form multimers have reduced activity, acting as antagonists in cultures.[23,25]

In addition to these similarities, D and A ODNs have structural differences that do not appear to impact their IFN-α-inducing activity but may influence their overall effects. As outlined in Table 9.2, a D ODN typically has a single central purine–pyrimidine–CpG–purine–pyrimidine motif exposed at the apex of the stem–loop, while the A ODN typically has three CpG dimers with different flanking bases. Further, the D type has a single 3′ poly-G tail that self-associates via Hoogsteen base pairing to form parallel quadruplex structures.[8,26,27] In contrast, A ODNs have poly-G tails at both ends. Sequences with G-strands at both ends tend to form dimeric structures by association of hairpin pairs and assemble into a heterogeneous mixture of parallel and anti-parallel quadruplex structures and G wires in the presence of cations, or they form large globular aggregates in aqueous solution.[24,25,27,28] Moreover, these two poly-G strands allow the formation of quadruplex structures at lower ODN concentrations as compared to D ODNs.[25,27,28] The significance of the secondary and tertiary structures of D and A ODNs will be addressed below.

Finally, C type CpG ODNs, which do not have poly-G strands and therefore do not form large aggregates, do induce pDCs to secrete significant IFN-α (but less than type A or D). These ODNs have self-complementary sequences encoding multiple CpG motifs on a PS backbone.[29,30] These sequences, along with the lack of poly-G strands, favor the formation of dimers but not multimers.[31] In addition, similar to type B/K, type C CpG ODNs stimulate pDC maturation and polyclonal B cell activation with increased Ig secretion and increased pro-inflammatory cytokine secretion.[29,30]

BINDING AND UPTAKE

Early studies in TLR9[−/−] mice showed that cellular responses to bacterial DNA and CpG ODNs are mediated by TLR9.[5] Since functionally significant structural polymorphisms of TLR9 have not been identified, the mechanism by which different DNA sequences acting on the same receptor induce different responses is currently the subject of intense investigation.

Unlike most TLRs that are expressed mainly on cell surfaces, TLR9 has a 14-amino acid sequence on the intracellular domain that dictates its intracellular localization.[32] In resting cells, TLR9 resides in the endoplasmic reticulum and translocates to the endosomal compartment upon activation.[33,34] The mechanism (receptor-mediated endocytosis versus pinocytosis versus caveolin-mediated endocytosis versus DNA binding protein) by which CpG ODNs enter immune cells and reach the endosomes where they signal through TLR9 is still unclear, largely because studies have used widely disparate types of DNAs and target cells, and detailed studies in pDCs and B cells are lacking. Studies show that entry rate is independent of the presence of a CpG motif, but the presence of a poly-G strand facilitates uptake.[12,23]

In PBMCs, K and D CpG ODNs are equally internalized by virtually all monocytes, DCs, B cells, and some NK cells, but once inside the cells they localized to different intracellular compartments.[23] This differential localization was recently supported by studies showing that poly-G bearing ODNs form tertiary structures the size of viral particles that are retained in early transferrin receptor (TfR)-positive and recycling endosomes. In contrast, monomeric K ODNs are rapidly transferred to lysozome-associated membrane protein (LAMP)-1-positive late endosomes and lysozomes.[21,24,35,36] Gursel et al. proposed that CXCL16, a membrane-bound scavenger receptor expressed on pDCs but not B cells, mediates the differential uptake and trafficking of D ODNs. Indeed, HEK293 cells co-transfected with TLR9 and CXCL16 are susceptible to D ODN stimulation, while those transfected with TLR9 alone are not.[37] In that study, however, the activation of HEK293 cells by D ODNs was measured by the activation of nuclear factor-κB (NF-κB). Since the activation of NF-κB is associated with the stimulation of TLR9 by K type ODNs, while the induction of IFN-α by poly-G bearing ODNs is associated with the translocation of IRF7, the question raised is whether the addition of CXCL16 to HEK293 cells changes their intracellular distribution and immunostimulatory effect to approximate that of a K type ODN. Neither of the studies cited above explains the mechanism by which C ODNs, which lack poly-G tails and do not multimerize, induce IFN-α production in pDCs.

SIGNALING

TLR9 activation by K ODNs in late endosomes follows a conserved TLR family signaling pathway that begins with the recruitment of myeloid differentiation primary response gene 88 (MyD88),[38] followed by the activation and phosphorylation of interleukin-1 receptor-activated kinase (IRAK1) and IRAK4,[39] the phosphorylation of TNF receptor-associated factor 6 (TRAF6),[40,41] and the ubiquination of transforming growth factor-β-activating kinase 1 (TAK1), which in turn results in the ubiquitination and subsequent degradation of the inducible IκB kinase (IKK). This releases NF-κB that translocates to the nuclei and regulates their target genes, including those that encode pro-inflammatory cytokines.

In pDCs, B and K ODNs induce cell maturation with upregulation of CD80, CD83, and CD86 on cell surfaces and TNFα production.[14] In contrast, type D and A CpG ODNs are retained (>90 minutes) in early TfR-positive endosomes of pDCs (Figure 9.2). Like K/B type ODNs, D and A CpG ODN initiate a signaling cascade that involves the activation and recruitment of MyD88, IRAK1, IRAK4, and TRAF6. The IRAK–TRAF6 complex physically associates with IRF7 instead of TAK1.[42,43] IRF7, which is constitutively highly expressed in pDC, is phosphorylated by IRAK-1, and then homodimerizes and translocates to nuclei to induce the expression of IFN-α.[21] The role of IRAK1 in signal transduction by TLR9 was recently demonstrated in

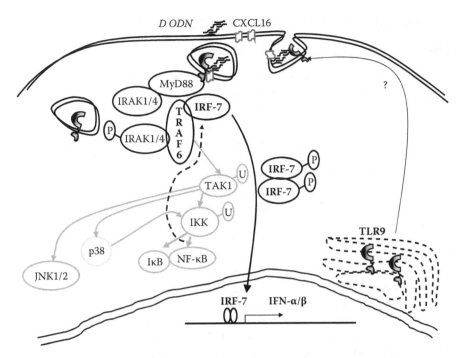

FIGURE 9.2 TLR9 signaling upon stimulation with D or A type CpG ODNs. In resting cells, TLR9s are primarily localized in the endoplasmic reticulum. The signal for TLR9s to relocate to the endosomes remains unclear. Upon internalization, D ODN multimers are retained in the early and recycling endosomes where they interact with TLR9s. Targeting and retention in early endosomes is aided by CXCL16, a cell surface scavenger receptor that binds poly-G strands. TLR9 signaling in early endosomes involves the activation and recruitment of MyD88 that interacts with IRF7 via its death domain and with TRAF6. IRF7 is phosphorylated via the IRAK4–IRAK1–IKKα kinase cascade known to be operational in the NF-κB activation pathway. Upon phosphorylation, IRF7 homodimerizes and translocates to nuclei where it interacts with Smad3 to activate type I IFN gene transcription. Prolonged signaling achieved by retention of ODNs in early endosomes may allow the phosphorylation of *de novo* synthesized IRF7 and the continuous activation of the positive feedback system to induce a robust type I IFN production. Gray area is alternative signaling pathway triggered by monomeric K type ODN signaling via TLR9s in late endosomes.

IRAK$^{-/-}$ pDCs, which showed severe impairment of IFN-α induction in response to TLR9 agonists but normal induction of pro-inflammatory cytokines.[43] From these data emerges a picture in which the response to CpG ODNs is defined by cell type and the cellular compartment where the interaction with TLR9 takes place. Alternatively, the differential activation of TLR9 by K and D ODNs may depend on the aggregation of the receptor as suggested by the need for D ODN multimerization and the lack of cellular activation with monomeric sequences.[25] Future studies are needed to identify the signal that triggers the relocation of TLR9 upon cell activation, and whether the differential signaling initiated by TLR9 in early and late endosomes is due to an intrinsic change in TLR9, differences in receptor–ligand affinity, the kinetics of the interaction, or the additional molecules that act as co-factors or chaperones.

PREVENTION AND TREATMENT OF INFECTIOUS DISEASES

IFN-α-INDUCING ODNs AS VACCINE ADJUVANTS

The development of adjuvants that foster strong Th1-type cellular responses is a high priority in the effort to generate vaccines to prevent diseases such as leishmaniasis, malaria, and HIV infections. Extensive studies in mice established that CpG DNAs act as potent adjuvants when combined with peptide or protein antigens, live or heat-killed viruses, polysaccharide conjugates, DNA vaccines, and autologous cell vaccines.[4,44] Unlike TLR9 activities in mice, TLR9s in non-human primates (including chimpanzees, aotus monkeys, rhesus macaques, and cynomolgus macaques) have the same cellular distribution and recognize the same ODN sequences as human PBMCs, and therefore sever as animal models of choice for preclinical studies.[45]

Studies in non-human primates showed that K ODNs were effective in increasing vaccine immunogenicity against malaria, hepatitis B, and anthrax antigens by accelerating and augmenting antibody titers, even in animals that exhibited poor responses to other vaccines.[46-51] This suggests that using CpG ODNs as adjuvants may allow induction of protective immunity in humans more quickly and with fewer vaccine doses. Studies assessing adjuvant effects of D ODNs showed that when administered with a hepatitis B vaccine, D ODNs increased antibody titers similarly to K CpG ODNs.[49] Notably, D ODNs failed to improve antibody responses of macaques immunized with anthrax recombinant protective antigen.[52]

CpG ODNs that induced IFNα increased the number of CD4+ and CD8+ CD45RA–CD95+ T cells producing IFNγ and IL-2 secreting cells in peripheral blood of primates immunized with HIV Gag protein or with heat killed *Leishmania* parasites.[52,53] The generation of antigen-specific T cells did not necessarily correlate with protection from challenge. For example, when rhesus macaques were immunized with heat-killed *Leishmania major* (HKLV) parasites, the addition of type K or D ODNs resulted in significantly increased numbers of IgG antibodies to soluble *Leishmania* antigen or antigen-specific IFNγ secreting cells. However, upon challenge, only the macaques immunized with HKLV and type D CpG ODNs were protected from live infection 14 weeks after immunization. Those that received HKLV alone, HKLV plus K, and control ODNs were not protected.[45] These studies suggest that different CpG ODN types may be used to obtain optimal adjuvant effects, depending on the pathogen and type of immune response required for protection.

TYPE D ODNs AS IMMUNOPROTECTIVE AGENTS

Evolutionary conservation of CpG recognition suggests that immune responses elicited by CpG DNA provide a survival advantage to a host. Studies in mouse models confirmed that treatment with TLR9 agonists around the time of infection facilitated survival and host clearance for a large group of pathogens including *Leishmania major, L. monocytogenes, Bacillus anthracis, Francisella tularensis,* arenaviruses, and herpes viruses.[54-58] Animals that received CpG ODNs around the time of infection developed protective adaptive immune responses that reduced their susceptibility to re-infection with the same microorganism.[55] CpG ODN-mediated protection involves the activation of Thy1.2+ CD11c+ dendritic cells and is associated with increased levels of antigen-specific antibodies, t-bet mediated induction of pro-inflammatory and Th1-type cytokines (TNF-α, IL-12, IFN-γ), and upregulation of iNOS expression.[55,57,59–62]

Optimal times and routes of treatment vary for individual pathogens. For example, CpG ODN-mediated protection against arenavirus has a very short therapeutic window; animals are protected when treated between days 0 and 3 post-infection. In contrast, optimal protection for Listeria-infected animals is achieved 3 to 6 days before challenge.[63] However, this window of immunoprotection can be extended by repeated exposure to CpG ODNs or by modifications to the ODN backbone or its delivery.[62,64,65] These factors and the differential immunomodulatory effects of D, C, and K CpG ODNs have limited the success of immunoprotection in primate models to date.

Several studies have shown that administration of CpG ODNs at the time of infection elicits local and systemic immune responses that would be expected to protect animals from infection. For example, topical administration of type K or D CpG ODNs increased the levels of IFN-α and IFN-β in the cervicovaginal mucosa of non-human primates, but failed to provide protection from challenges with SIV, suggesting that the development of an antigen-specific cytokine response is not necessarily predictive of protection.[66]

The immunoprotective effects of CpG ODNs were first demonstrated in a series of studies using a rhesus macaque model of cutaneous leishmaniasis. The clinical progression of this model is known to resemble human cutaneous leishmaniasis and to require a Th1-type immune response for protection.[67,68] Local or systemic administration of D ODNs reduced parasite loads and lesion severity in macaques challenged intradermally with up to 10[7] live *Leishmania amazonensis* and *Leishmania major*. Protection was evident when animals were treated between 3 days before and 15 days after the infectious challenge.[69,70] Animals treated with D ODNs during infectious challenges were resistant to rechallenges with the same parasite months after the first exposure.[69] Protection was restricted to animals treated with D ODNs and could not be reproduced with K or C ODNs.

The mechanism of immunoprotection in the *Leishmania* model is still unclear. Preliminary data suggest that clinical outcome did not correlate with the number of *Leishmania*-specific IFN-γ secreting cells in peripheral blood. Further, protection did not appear to be mediated by the induction of IFN-α or the maturation of monocytes into antigen-presenting dendritic cells because no clinical improvement was evident in animals treated with ODN type C. Alternatively, it is possible that the pro-inflammatory effects of ODNs with phosphorothioate backbones may interfere with the development of protective responses (Figure 9.1). These studies suggest that the mechanisms of innate protection by CpG ODNs may vary according to pathogen and site of infection. Developing CpG ODNs as immunoprotective agents will require testing the different ODN types against pathogens in appropriate animal models.

IMMUNOPROTECTION IN HIV PATIENTS AND OTHER IMMUNOCOMPROMISED SUBJECTS

Patients with primary or secondary immunodeficiencies that compromise the adaptive immune response have suboptimal responses to vaccines and are more susceptible to infectious diseases. Since CpG ODNs and other TLR agonists stimulate the innate immune system directly, they present attractive possibilities for reducing susceptibility to infections. Studies investigating lymphocyte proliferation and cytokine secretion suggest that patients with hyper-IgE syndrome[71] and chronic granulomatous disease (manuscript in preparation) do not show defects in their responses to CpG ODNs and other TLR ligands.

In HIV, infection produces a progressive reduction in CD4+ T cells, and a decrease in the numbers and functional activities of NK cells and pDCs as viral load rises.[72–75] *In vitro*, K ODNs induced similar levels of polyclonal B cell activation, IL-6, and IP-10 production in PBMCs from HIV-infected patients and healthy controls.[70] In contrast, stimulation with D ODNs resulted in lower IFNα and IFNγ levels in PBMCs from HIV-infected patients. The reduced response correlated with higher viral loads and lower numbers of CD4+ T cells, respectively.[70,76] A similar viral load increase and CD4+ T cell-dependent reduction in the response to D type, but not K type, ODNs was evident when PBMCs from SIV-infected rhesus macaques were stimulated *in vitro*.[70,77]

Despite the impaired immune responses, both D and K ODNs similarly improved the antibody titers attained upon vaccination with a hepatitis B vaccine, achieving protective antibody levels in 75% of the SIV-infected macaques. Of note, the overall antibody titers observed after three doses of the hepatitis B vaccine spaced one month apart were still significantly lower than those in healthy macaques.[49]

In terms of their use as immunoprotective agents in immunocompromised subjects, studies showed that despite the induction of reduced levels of IFN-α and -γ as compared to healthy animals, type D CpG ODNs significantly improved the clinical outcome in SIV-infected macaques challenged with *L. major*.[70] These findings suggest that by improving the activity of the innate immune system with CpG ODN treatment, it may be possible to reduce susceptibility to infection among patients whose adaptive immune systems are compromised.[70]

Of concern are the potential effects of CpG ODNs on underlying HIV infections. Reports are contradictory. In freshly infected PBMCs, K ODNs and to a lesser extent A ODNs were reported to block HIV replication and reduce CD4+ T cell depletion *in vitro*.[78] In latently infected human T cells, K ODNs were shown to directly trigger reactivation of latent HIV.[79] In primates, the administration of 500 µg of CpG ODNs intradermally did not modify viral loads. However, SIV-infected macaques that received 2 mg of PS CpG ODNs intramuscularly experienced transient (3- to 4-week) CpG-independent five-fold increases in viral load. This effect was not observed in macaques treated with similar doses of D ODNs, probably due to the reduced number of phosphorothioate bases and the antiviral effects of IFN-α (Verthelyi, unpublished, and Gurney et al.[80]).

MANUFACTURE AND DELIVERY OF CPG ODNs

The structures of D and A CpG ODNs are characterized by self-complementary core sequences and poly-G tracks at the 3' or 3' and 5' ends, respectively. Both structures favor the formation of dimers and G-tetrads leading to ODN multimerization. Recent studies show that the formation of multimers is required for these ODNs to localize to early TfR-positive endosomes where they signal through TLR9 to induce IFN-α and suggest that monomeric sequences behave as competitive antagonists.[24,25,36]

Indeed, several studies show that disruption of the poly-G motifs by removing them or replacing one or more guanines with adenines or 7-deazaguanosines to impede tetrad formation abrogates the IFN-α-inducing activity.[8,24,25] However, the poly-G motif-mediated formation of tetrads poses formidable challenges for the synthesis, purification, and characterization of CpG ODN type D and A, often leading to significant variations among product lots and hindering the clinical development of these ODNs.[25,27,30] To date, the only clinical trials assessing the therapeutic effects of IFN-α-inducing CpG ODNs in humans used type C ODN for the treatment of hepatitis C.[4] However, C ODNs have phosophorothioate backbones and stimulate polyclonal B cell activation as well as IFN-α production by pDCs, and this may limit the effectiveness of such a therapy.

Different strategies are being investigated to improve the stability of D and A CpG ODNs, including liposomes,[64] cationic nanoparticles,[81] and polylactide–co-glycolide microparticles[82] or virus-like particles.[83,84] However these compounds (1) do not prevent the formation of large aggregates during manufacture, and (2) can alter the intracellular localization of the ODNs. Recently, our group developed a Prodrug form of CpG-ODN type D by incorporating 2-(N-formyl-N-methyl)aminoethyl groups into the backbone of the 3' poly-G end (Pro-D35 ODN). These protecting groups provided a transitory reduction of G-tetrad formation in potassium-free solutions, while allowing tetrad formation when potassium salt concentrations mimicked those of intracellular environments (~100 mM). This modification to the poly-G tail of D ODNs did not significantly affect cellular uptake or modify biological activity. Indeed, these ODNs induced similar levels of IFN-α, IFN-γ, IP-10, and IL-6 in human and non-human primate PBMCs, and induced >60% of monocytes to mature into CD83[hi]CD86[hi]CD40[hi] CD14[lo]mDCs *in vitro*. *In vivo*, these Pro-ODNs were as effective as the parental D ODN sequences in reducing the severity of cutaneous leishmaniasis.[85] The use of Pro-ODNs may also prolong the therapeutic effects of CpG ODNs.[65]

CLINICAL TRIALS AND SAFETY CONSIDERATIONS

Type C CpG ODNs are currently in phase II clinical studies as immunotherapeutics for patients infected with hepatitis C.[4] Studies of non-human primates showed no adverse events in animals treated with up to 1 mg/kg of type D CpG ODNs. The animals' temperatures, weights, serologies, and hematologies did not differ from those of untreated controls.[86] Unlike K type ODNs, D ODNs did not induce local lymphadenopathies or increases in viral loads of SIV-infected animals.[70,86]

However, since type D or A CpG ODNs induced high levels of IFN-α, the development of flu-like symptoms and other IFN-α-associated adverse events could occur. An additional concern is that CpG ODNs may cause or aggravate autoimmune diseases. Type D CpG ODNs do not stimulate polyclonal B cell activation or trigger the production of pro-inflammatory cytokines, but they do accelerate the maturation of professional antigen presenting cells. Experience in humans will be necessary to establish their safety.

CONCLUDING REMARKS

Since first described in 1999,[87] considerable insight has been gained into the modes of action, specificities, and therapeutic potentials of type D CpG ODNs. These ODNs selectively trigger plasmacytoid DCs to secrete IFN-α, leading to accelerated maturation of monocytes into antigen presenting dendritic cells and improved generation of CD8+ T cell responses. Studies in primates, however, suggest that D type ODNs may have unique therapeutic potentials beyond the induction of IFN-α and DC maturation, as evidenced by the prophylactic and therapeutic effects on cutaneous leishmaniasis that could not be reproduced by C ODNs (which also induced IFN-α secretion and DC maturation).

Structural requirements for D type ODN functions such as the poly-G tails have limited their clinical development to date, but second generation D ODNs (Pro-D ODNs) that allow large scale manufacturing promise to open new avenues for clinical development. Ongoing studies aimed at extending *in vivo* activities and targeting their effects to selected tissues will bring important changes to the field.

DISCLAIMER

The assertions of the authors of this chapter are not to be construed as official or as reflecting the views of the U.S. Food and Drug Administration.

REFERENCES

1. Janeway, C. A., Jr. and R. Medzhitov. 2002. Innate immune recognition. *Annu. Rev. Immunol.* 20: 197-216.
2. Iwasaki, A. and R. Medzhitov. 2004. Toll-like receptor control of the adaptive immune responses. *Nat. Rev. Immunol.* 5: 987-995.
3. Klinman, D. M. 2004. Immunotherapeutic uses of CpG oligodeoxynucleotides. *Nat. Rev. Immunol.* 4: 249-258.
4. Krieg, A. M. 2006. Therapeutic potential of Toll-like receptor 9 activation. *Nat. Rev. Drug Discov.* 5: 471-484.
5. Hemmi, H. et al. 2000. A Toll-like receptor recognizes bacterial DNA. *Nature* 408: 740-745.
6. Takeshita, F. et al. 2001. Cutting edge: role of toll-like receptor 9 in CpG DNA-induced activation of human cells. *J. Immunol.* 167: 3555-3558.
7. Bauer, S. et al. 2001. Human TLR9 confers responsiveness to bacterial DNA via species-specific CpG motif recognition. *Proc. Natl. Acad. Sci. USA* 98: 9237-9242.
8. Verthelyi, D. et al. 2001. Human peripheral blood cells differentially recognize and respond to two distinct CpG motifs. *J. Immunol.* 166: 2372-2377.

9. Kadowaki, N. et al. 2001. Subsets of human dendritic cell precursors express diffeent Toll-like receptors and respond to different microbial antigens. *J. Exp. Med.* 194: 863-869.
10. Bauer, M. et al. 1999. DNA activates human immune cells through a CpG sequence-dependent manner. *Immunology* 97: 699-705.
11. Hartmann, G. and A. M. Krieg. 2000. Mechanism and function of a newly identified CpG DNA motif in human primary B cells. *J. Immunol.* 164: 944-952.
12. Verthelyi, D. and R. A. Zeuner. 2003. Differential signaling by CpG DNA in DCs and B cells: not just TLR9. *Trends Immunol.* 10: 519-522.
13. Biron, C. A. 2001. Interferons α and β as immune regulators: a new look. *Immunity* 14: 661-664.
14. Honda, K. et al. 2005. Regulation of the type I IFN induction: a current view. *Int. Immunol.* 17: 1367-1378.
15. Mescher, M. F. et al. 2006. Signals required for programming effector and memory development by CD8+ T cells. *Immunol. Rev.* 211: 81-92.
16. Siegal, F. P. et al. 1999. The nature of the principal type 1 interferon-producing cells in human blood. *Science* 284: 1835-1837.
17. Pulendran, B. et al. 2000. Flt3 ligand and granulocyte colony-stimulating factor mobilize distinct human dendritic cell subsets *in vivo. J. Immunol.* 165: 566-572.
18. Krug, A. et al. 2001. Identification of CpG oligonucleotide sequences with high induction of IFN-α/β in plasmacytoid dendritic cells. *Eur. J. Immunol.* 31: 2154-2163.
19. Gursel, M., D. Verthelyi, and D. M. Klinman. 2002. CpG oligodeoxynucleotides induce human monocytes to mature into functional dendritic cells. *Eur. J. Immunol.* 32: 2617-2622.
20. Hemmi, H. et al. 2003. The roles of Toll-like receptor 9, MyD88, and DNA-dependent protein kinase catalytic subunit in the effects of two distinct CpG DNAs on dendritic cell subsets. *J. Immunol.* 170: 3059-3064.
21. Kerkmann, M. et al. 2003. Activation with CpG-A and CpG-B oligonucleotides reveals two distinct regulatory pathways of type I IFN synthesis in human plasmacytoid dendritic cells. *J. Immunol.* 170: 4465-4474.
22. Rothenfusser, S. et al. 2001. Distinct CpG oligonucleotide sequences activate human γ/δ T cells via interferon-α/β. *Eur. J. Immunol.* 31: 3525-3534.
23. Gursel, M. et al. 2002. Differential and competitive activation of human immune cells by distinct classes of CpG oligodeoxynucleotide. *J. Leukoc. Biol.* 71: 813-820.
24. Kerkmann, M. et al. 2005. Spontaneous formation of nucleic acid-based nanoparticles is responsible for high interferon-α induction by CpG-A in plasmacytoid dendritic cells. *J. Biol. Chem.* 280: 8086-8093.
25. Wu, C. C. et al. 2004. Necessity of oligonucleotide aggregation for Toll-like receptor 9 activation. J. Biol. Chem. 279: 33071-33078.
26. Panyutin, I. G. et al. 1990. G-DNA: twice-folded DNA structure adopted by single-stranded oligo(dG) and its implications for telomeres. *Proc. Natl. Acad. Sci. USA* 87: 867-870.
27. Costa, L. T. et al. 2004. Structural studies of oligonucleotides containing G-quadruplex motifs using AFM. *Biochem. Biophys. Res. Commun.* 313: 1065-1072.
28. Giraldo, R. et al. 1994. Promotion of parallel DNA quadruplexes by a yeast telomere binding protein: a circular dichroism study. *Proc. Natl. Acad. Sci. USA* 91: 7658-7662.
29. Hartmann, G. et al. 2003. Rational design of new CpG oligonucleotides that combine B cell activation with high IFN-alpha induction in plasmacytoid dendritic cells. *Eur. J. Immunol.* 33: 1633-1641.
30. Marshall, J. D. et al. 2003. Identification of a novel CpG DNA class and motif that optimally stimulate B cell and plasmacytoid dendritic cell functions. *J. Leukoc. Biol.* 73: 781-792.
31. Marshall, J. D. et al. 2005. Superior activity of the type C class of ISs *in vitro* and *in vivo* across multiple species. *DNA Cell Biol.* 24: 63-72.

32. Leifer, C. A. et al. 2006. Cytoplasmic targeting motifs control localization of Toll-like receptor 9. *J. Biol. Chem.* 281: 35585-35592.
33. Latz, E. et al. 2004. TLR9 signals after translocating from the ER to CpG DNA in the lysosome. *Nat. Immunol.* 5: 190-198.
34. Leifer, C. A. et al. 2004. TLR9 is localized in the endoplasmic reticulum prior to stimulation. *J. Immunol.* 173: 1179-1183.
35. Honda, K. et al. 2005. Spatiotemporal regulation of MyD88-IRF-7 signalling for robust type-I interferon induction. *Nature* 434: 1035-1040.
36. Guiducci, C. et al. 2006. Properties regulating the nature of the plasmacytoid dendritic cell response to Toll-like receptor 9 activation. *J. Exp. Med.* 203: 1999-2008.
37. Gursel, M. et al. 2006. CXCL16 influences nature and specificity of CpG-induced immune activation. *J. Immunol.* 177: 1575-1580.
38. Ahmad-Nejad, P. et al. 2002. Bacterial CpG-DNA and lipopolysaccharides activate Toll-like receptors at distinct cellular compartments. *Eur. J. Immunol.* 32: 1958-1968.
39. Wesche, H. et al. 1997. MyD88: an adapter that recruits IRAK to the IL-1 receptor complex. *Immunity* 7: 837-847.
40. Cao, Z., W. J. Henzel, and X. Gao. 1996. IRAK: a kinase associated with the interleukin-1 receptor. *Science* 271: 1128-1131.
41. Li, S. et al. 2002. IRAK-4: A novel member of the IRAK family with the properties of an IRAK-kinase. *Proc. Natl. Acad. Sci. USA* 99: 5567-5572.
42. Honda, K. et al. 2004. Role of a transductional-transcriptional processor complex involving MyD88 and IRF-7 in Toll-like receptor signaling. *Proc. Natl. Acad. Sci. USA* 101: 15416-15421.
43. Uematsu, S. et al. 2005. Interleukin-1 receptor-associated kinase-1 plays an essential role for Toll-like receptor (TLR)7- and TLR9-mediated interferon-α induction. *J. Exp. Med.* 201: 915-923.
44. Klinman, D. M. et al. 2004. Use of CpG oligodeoxynucleotides as immune adjuvants. *Immunol. Rev.* 199: 201-216.
45. Verthelyi, D. et al. 2002. CpG oligodeoxynucleotides as vaccine adjuvants in primates. *J. Immunol.* 168: 1659-1663.
46. Gramzinski, R. A. et al. 1998. Immune response to a hepatitis B DNA vaccine in aotus monkeys: comparison of vaccine formulation, route, and method of administration. *Mol. Med.* 4: 109-118.
47. Jones, T. R. et al. 1999. Synthetic oligodeoxynucleotides containing CpG motifs enhance immunogenicity of a peptide malaria vaccine in aotus monkeys. *Vaccine* 17: 3065-3071.
48. Hartmann, G. et al. 2000. Delineation of a CpG phosphorothioate oligodeoxinucleotide for activating primate immune responses *in vitro* and *in vivo*. *J. Immunol.* 164: 1617-1624.
49. Verthelyi, D. et al. 2004. CpG oligodeoxynucleotides improve the response to hepatitis B immunization in healthy and SIV-infected rhesus macaques. *AIDS* 18: 1003-1008.
50. Davis, H. L. et al. 2000. CpG DNA overcomes hyporesponsiveness to hepatitis B vaccine in orangutans. *Vaccine* 19: 413-422.
51. Payette, P. J. et al. 2006. Testing of CpG-optimized protein and DNA vaccines against the hepatitis B virus in chimpanzees. *Intervirology* 49: 144-151.
52. Verthelyi, D. 2006. Adjuvant properties of CpG ODN in primates, in Saltzman, W. M. et al., Eds. *DNA Vaccines: Methods and Protocols.* Humana Press, Totowa, NJ, pp. 139-158.
53. Wille-Reece, U. et al. 2005. HIV Gag protein conjugated to a Toll-like receptor 7/8 agonist improves the magnitude and quality of Th1 and CD8+ T cell responses in nonhuman primates. *Proc. Natl. Acad. Sci. USA* 102: 15190-15194.
54. Zimmermann, S. et al. 1998. CpG oligodeoxynucleotides trigger protective and curative Th1 responses in lethal murine leishmaniasis. *J. Immunol.* 160: 3627-3630.

55. Elkins, K. L. et al. 1999. Bacterial DNA containing CpG motifs stimulates lymphocyte-dependent protection of mice against lethal infection with intracellular bacteria. *J. Immunol.* 162: 2291-2298.

56. Walker, P. S. et al. 1999. Immunostimulatory oligodeoxynucleotides promote protective immunity and provide systemic therapy for leishmaniasis via IL-12 and IFNγ-dependent mechanisms. *Proc. Natl. Acad. Sci. USA* 96: 6970-6975.

57. Pedras-Vasconcelos, J. A. et al. 2006. CpG oligodeoxynucleotides protect newborn mice from lethal challenge with the neurotropic Tacaribe arenavirus. *J. Immunol.* 176: 4940-4949.

58. Gierynska, M. et al. 2002. Induction of CD8 T-cell-specific systemic and mucosal immunity against herpes simplex virus with CpG–peptide complexes. *J. Virol.* 76: 6568-6576.

59. Ishii, K. J. et al. 2005. CpG-activated Thy1.2+ dendritic cells protect against lethal *Listeria monocytogenes* infection. *Eur. J. Immunol.* 35: 2397-2405.

60. Lugo-Villarino, G. et al. 2005. The adjuvant activity of CpG DNA requires T-bet expression in dendritic cells. *Proc. Natl. Acad. Sci. USA* 102: 13248-13253.

61. Pisetsky, D. S. and C. Reich. 1993. Stimulation of *in vitro* proliferation of murine lymphocytes by synthetic oligonucleotides. *Mol. Biol. Rep.* 18: 217-221.

62. Klinman, D. M., J. Conover, and C. Coban. 1999. Repeated administration of synthetic oligodeoxynucleotides expressing CpG motifs provides long-term protection against bacterial infection. *Infect. Immun.* 67: 5658-5663.

63. Klinman,D.M. 2004. Use of CpG oligodeoxynucleotides as immunoprotective agents. *Expert Opin. Biol. Ther.* 4: 937-946.

64. Gursel, I. et al. 2001. Sterically stabilized cationic liposomes improve the uptake and immunostimulatory activity of CpG oligonucleotides. *J. Immunol.* 167: 3324-3328.

65. Grajkowski, A. et al. 2005. Thermolytic CpG-containing DNA oligonucleotides as potential immunotherapeutic prodrugs. *Nucleic Acids Res.* 33: 3550-3560.

66. Wang, Y. et al. 2005. Toll-like receptor 7 (TLR7) agonist, imiquimod, and the TLR9 agonist, CpG ODN, induce antiviral cytokines and chemokines. *J. Virol.* 79: 14355-14370.

67. Amaral, V. F. et al. 1996. *Leishmania amazonensis*: the Asian rhesus macaque (*Macaca mulata*) as an experimental model for the study of cutaneous leishmaniasis. *Exp. Parasitol.* 82: 34-44.

68. Vanloubbeeck, Y. A. and D. E. Jones. 2004. Immunology of leishmania infection and implications for vaccine development. *Ann. NY Acad. Sci.* 1026: 267-272.

69. Flynn, B. et al. 2005. Prevention and treatment of cutaneous leishmaniasis in primates by using synthetic type D/A oligodeoxynucleotides expressing CpG motifs. *Infect. Immun.* 73: 4948-4954.

70. Verthelyi, D. et al. 2003. CpG Oligodeoxynucleotides protect normal and SIV-infected macaques from Leishmania infection. *J. Immunol.* 170: 4717-4723.

71. Renner, E.D. et al. 2005. No indication for a defect in Toll-like receptor signaling in patients with hyper-IgE syndrome. *Clin. Immunol.* 25: 321-328.

72. Azzoni, L. et al. 2002. Sustained impairment of IFN-γ secretion in suppressed HIV-infected patients despite mature NK cell recovery: evidence for a defective reconstitution of innate immunity. *J. Immunol.* 168: 5764-5770.

73. Chehimi, J. et al. 2002. Persistent decreases in blood plasmacytoid dendritic cell number and function despite effective highly active antiretroviral therapy and increased blood myeloid dendritic cells in HIV-infected individuals. *J. Immunol.* 168: 4796-4801.

74. Howell, D. M. et al. 1994. Decreased frequency of functional natural interferon-producing cells in peripheral blood of patients with the acquired immune deficiency syndrome. *Clin. Immunol. Immunopathol.* 71: 223-230.

75. Pacanowski, J. et al. 2001. Reduced blood CD123+ (lymphoid) and CD11c+ (myeloid) dendritic cell numbers in primary HIV-1 infection. *Blood* 98: 3016-3021.

76. Jiang, W. et al. 2005. Impaired monocyte maturation in response to CpG oligodeoxy-nucleotide is related to viral RNA levels in human immunodeficiency virus disease and is at least partially mediated by deficiencies in α/β interferon responsiveness and production. *J. Virol.* 79: 4109-4119.

77. Teleshova, N. et al. 2006. CpG-C ISS-ODN activation of blood-derived B cells from healthy and chronic immunodeficiency virus-infected macaques. *J. Leukoc. Biol.* 79: 257-267.

78. Schlaepfer, E. et al. 2004. CpG Oligodeoxynucleotides block HIV type 1 replication in human lymphoid tissue infected *ex vivo*. *J. Virol.* 78: 12344-12354.

79. Scheller, C. et al. 2004. CpG oligodeoxynucleotides activate HIV replication in latently infected human T cells. *J. Biol. Chem.* 279: 21897-21902.

80. Gurney, K. B. et al. 2004. Endogenous IFN-α production by plasmacytoid dendritic cells exerts an antiviral effect on thymic HIV-1 infection. *J. Immunol.* 173: 7269-7276.

81. Kwon, Y. J. et al. 2005. Enhanced antigen presentation and immunostimulation of dendritic cells using acid-degradable cationic nanoparticles. *J. Control. Rel.* 105: 199-212.

82. Xie, H. et al. 2005. CpG oligodeoxynucleotides adsorbed onto polylactide–co-glycolide microparticles improve the immunogenicity and protective activity of the licensed anthrax vaccine. *Infect. Immun.* 73: 828-833.

83. Storni, T. et al. 2004. Nonmethylated CG motifs packaged into virus-like particles induce protective cytotoxic T cell responses in the absence of systemic side effects. *J. Immunol.* 172: 1777-1785.

84. Wang, H., E. Rayburn, and R. Zhang. 2006. Synthetic oligodeoxynucleotides containing deoxycytidyl deoxyguanosine dinucleotides (CpG ODNs) and modified analogs as novel anticancer therapeutics. *Curr. Pharmac. Des.* 11: 2889-2907.

85. Puig, M. et al. 2006. Use of thermolytic protective groups to prevent G-tetrad formation in CpG ODN type D: structural studies and immunomodulatory activity in primates. *Nucleic Acids Res.* 34: 6488-6495.

86. Verthelyi, D. and D. M. Klinman. 2002. CpG ODN: Safety considerations, in Raz, E., Ed. *Microbial DNA and Immune Modulation*. Humana Press, Totowa, NJ, pp. 385-396.

87. Verthelyi, D. et al. 1999. Relative potency of immunostimulatory DNA motifs in humans, *Keystone Conference on DNA Vaccines: Immune Responses, Mechanisms, and Manipulating Antigen Processing*. Snowbird, UT. Abstract 232: 53.

10 Clinical Development of Oligodeoxynucleotide TLR9 Agonists

Julie L. Himes and Arthur M. Krieg

ABSTRACT

In recent years recognition of the potential clinical benefits from activating the innate immune system has increased. Toll-like receptor 9 (TLR9) can be readily activated with synthetic CpG oligodeoxynucleotides (ODNs) that have become widely used both as research tools and in the development of novel therapeutics for the treatment of cancers and infectious and allergic diseases. This chapter will summarize the immune effects and early clinical results of these novel classes of drug candidates.

INTRODUCTION

The identification of the immune stimulatory effects of unmethylated CpG motifs within bacterial DNA resulted from investigations into the unexpected immune stimulatory effects of certain antisense ODNs.[1,2] Five years later, TLR9 was identified as the mediator of the immune stimulatory activities of CpG ODNs.[3] Many studies later explored the immune effects of TLR9 activation and their role in immune defense, as discussed in other chapters of this volume. One defining characteristic of the TLR9-induced innate immune response is that it promotes the development of remarkably strong Th1 adaptive immune responses, including high levels of both antigen-specific antibody and CD8+ T cell responses.[4] The targeting of TLR9 with CpG ODNs has emerged as a powerful tool in the generation of Th1 adaptive immunity, and shows promise for enhancing the efficacy of cancer vaccination.

Surprisingly, TLR9 activation can produce very different outcomes, depending on the specific structure of the CpG ODN. Three families of immune stimulatory CpG ODNs with distinct structural and biological characteristics have been described.[5-7] A-class CpG ODNs (also known as type D) are potent activators of natural killer cells and IFN-α secretion from pDCs, but only weakly stimulate B cells. A-class ODNs have poly-G motifs at the 5′ and/or 3′ ends that are capable of forming complex higher ordered structures known as G tetrads. These poly-G motifs are required for the strong IFN-α induction characteristic of this ODN class. For optimal induction of IFN-α secretion, A-class ODNs have central phosphodiester regions containing one or more CpG motifs within a self-complementary palindrome.[4]

B-class ODNs (also known as type K) generally have completely phosphorothio-ate backbones, do not form higher ordered structures, and are strong B cell stimulators, but induce relatively little NK activity or IFN-α secretion.[2] C-class CpG ODNs have intermediate immune properties (between the A and B classes) and induce intermediate levels of both B cell activation and IFN-α secretion.[5-7] C-class ODNs have unique structures, with one or more 5′ CpG motifs, and palindromes 3′ to this, allowing formation of duplexes.

All three CpG ODN classes require the presence of TLR9; they produced essentially no immune stimulatory effects in mice deficient in TLR9.[5] The mechanism through which the different classes induce such divergent immune effects through TLR9 is not fully understood. The fact that maximal induction of pDC IFN-α secretion requires an ODN to adopt a secondary structure, apparently dimeric for the C-class and multimeric for the A-class, suggests that these higher ordered structures may induce TLR9 cross-linking or promote the recruitment of one or more additional cofactors or adaptor proteins into the TLR9 signaling complex. Recent studies indicate that the A-class ODNs achieve much of their IFN-α inducing effect through interactions with CXCL16, which is expressed on pDCs, but not B cells, and appears to enhance uptake of A-class ODNs, but not B-class ODNs.[8] The possibility that additional TLR9 cofactors may take part in the recognition of one or more CpG ODN classes has not been excluded. Recent studies pointed to the possibility that the presence of the different ODN classes in distinct intracellular compartments may contribute to their differential immune effects.[9] Transfection of a B-class ODN can induce IFN-α secretion levels similar to those induced by A-class ODNs, providing new insights into the biology of these pathways while raising new questions about the immune effects of the transfection step.[9] Both B- and C-class CpG ODNs have reached clinical development, as will be discussed below.

CPG ODN MONOTHERAPY FOR INFECTIOUS DISEASE PREVENTION AND TREATMENT

The Th1-like immune effects of TLR9 activation appear to have evolved to stimulate protective immunity against intracellular pathogens. Consistent with this role, many studies have demonstrated that prophylactic treatment of mice with a synthetic TLR9 ligand can provide transient protection against a wide range of viral, bacterial, and even parasitic pathogens including lethal challenges with Category A agents or surrogates such as *Bacillus anthracis*, vaccinia virus, *Francisella tularensis*, and Ebola.[10-27] Protection was achieved through multiple administration routes of B-class CpG ODNs, including injection, inhalation, and even oral routes.[18] The duration of protection depends upon the specific model and ranged from two weeks after a single CpG dose in *Listeria monocytogenes* and *F. tularensis* (LVS strain) challenge models,[10,13,14] to only a day in a vaginal HSV challenge model.[28]

In models of acute infectious challenge such as those listed above, post-exposure therapy with CpG ODN was generally ineffective. However, in certain chronic infection models, CpG ODNs exhibited therapeutic benefit, improving survival when given 4 days post-infection in a Friend leukemia virus model[26] and 1 week after an indolent *L. major* challenge.[12] Protective effects of CpG against Leishmania infectious challenge were observed in rhesus macaques that were protected against *L. amazonensis* infection by treatment with 0.5 mg of A-class (D type) ODNs 3 days before and 3 days after challenge, but were not protected by B-class (K type) ODNs.[29] Similar protection was seen in SIV-infected macaques, suggesting possible clinical benefit in settings such as immune compromised patients. HBV transgenic mice treated with a CpG ODN showed significant decreases in viral expression,[30] suggesting potential application of TLR9 ligands in the treatment of HBV-infected humans.

Approximately 170 million people are infected with hepatitis C virus (HCV). This infection clears spontaneously in about 20% of people who become infected, but becomes chronic in the remainder, for whom existing therapy (48 weeks of combination therapy with IFN-α and ribavirin) is only partially effective. The disease is cleared in fewer than 50% of treated patients. Spontaneous viral clearance has been associated with the induction of Th1-like innate and adaptive immune activation leading to the development of strong and diverse Th1 and CD8 cytolytic T cell responses against the virus.[31] This clinical experience demonstrates the potential for a Th1 immune response to control the virus, at least in the setting of acute infection.

Since TLR9 activation can drive a similar pattern of innate and adaptive immune responses, we investigated whether A-, B-, or C-class CpG ODNs induced IFN-α secretions in the PBMCs of HCV-infected subjects. We found that CPG 10101, a C-class CpG ODN, induced particularly high levels of IFN-α secretion,[32] suggesting this agent may act as a monotherapy against HCV. In a Phase Ia dose escalation study of CPG 10101 in healthy volunteers, it induced serum levels of 2'5' OAS, a downstream effector of IFN-α, equivalent to those reported with a single therapeutic dose (1.0 μg/kg) of pegylated-IFN-α-2b.[33,34]

In a 4-week Phase Ib blinded, randomized, controlled trial involving 60 HCV-infected subjects, CPG 10101 monotherapy caused a dose-dependent decrease in blood viral RNA levels.[32] At the highest dose level of 0.75 mg/kg weekly, a mean reduction of 1.69 $\log_{10} \pm 0.62$ (mean $\pm$ SEM, n = 7) in viral RNA from baseline was noted. This acute anti-viral response at 24 hours after dosing was similar to reductions observed following the single administration of pegylated IFN-α-2b.[35,36] These results suggest that the anti-viral effects of CPG 10101 are equivalent to those induced with therapeutic doses of PEG-IFN. Additionally, the antiviral response induced with CPG 10101 correlated with increases in biomarkers of TLR9 activation, including NK cell activation and serum IFN-α and IFN-inducible chemokines.[37,38] Treatment was generally well tolerated; the most common side effects were mild to moderate flu-like symptoms and injection site reactions. The maximal tolerated dose was not reached.

Subsequently, CPG 10101 was evaluated in combination with pegylated interferon and ribavirin in a population of HCV-infected patients who had relapsed to a previous course of therapy [39]. The addition of CPG 10101 to pegylated interferon and ribavirin produced improvement in all standard early virologic response metrics, including rapid virologic response ($\geq$2 $\log_{10}$ decline at 4 weeks), early virologic response ($\geq$2 $\log_{10}$ decline at 12 weeks), early virologic negativity (undetectable HCV RNA levels at 12 weeks), and on-treatment response (undetectable HCV RNA levels at 24 weeks). The mean virologic responses at week 12 were significantly higher with the addition of CPG 10101 to standard of care compared with standard of care alone (3.26 ± 0.96 versus 2.33 ± 1.46 [mean HCV RNA plasma levels compared with baseline ± SD; P <0.05 (Wilcoxon rank sum)]).

Resistance to certain pathogens such as fungi is associated with a Th2 profile of immune activation. In such a setting, the Th1 activation resulting from treatment with a CpG ODN may be detrimental. Indeed, CpG priming shortened survival slightly in a *Candida albicans* challenge model, in which Th1 cytokines are disadvantageous.[40] Furthermore, the immune expansion induced by TLR9 activation in rodents increased the number of susceptible target cells for Friend leukemia virus, resulting in a more aggressive infection following challenge several days later.[41]

CpG motifs within bacterial DNA can induce HIV transcriptional regulatory elements in the LTRs,[42] and HIV-infected humans treated with a B-class ODN containing a CpG motif exhibited dose-dependent increases in plasma HIV bDNA levels.[43] To avoid inducing HIV expression, CpG ODN therapy of HIV-infected individuals should probably be undertaken only during HAART, unless the therapy is part of a clinical trial strategy to induce anti-HIV immunity. Nevertheless, the high level of IFN-α production induced by A-class ODNs suppressed HIV replication in human fetal thymus cells.[44] B-class ODNs can also suppress HIV replication in cultured human cells, albeit in a sequence-independent fashion.[45]

HIV-infected long-term non-progressors showed much stronger NK cell activation in response to A-class CpG ODNs compared to progressors.[46] Whether this difference in CpG responsiveness is a cause or a consequence of patient clinical status is not clear. As will be discussed further below, B-class CpG ODNs have been used as vaccine adjuvants in HIV-infected humans on HAART with no apparent increases in HIV expression,[47] providing support for the cautious application of TLR9-based immunotherapeutic approaches.

CPG ODNS AS VACCINE ADJUVANTS

TLR9 activation enhances both innate immunity and also antigen-specific humoral and cellular responses. Mechanisms contributing to the strong adjuvant activity of B-class CpG ODNs (with which most vaccine studies have been performed, except as noted otherwise) directly on B cells for humoral responses may include (1) synergy between TLR9 and B cell receptors preferentially stimulating antigen-specific B cells[1]; (2) inhibition of B cell apoptosis, improving B cell survival[48]; and (3) enhanced IgG class switch DNA recombination that may enhance maturation of the immune response.[49-51]

Since TLR9 does not appear to be expressed in resting T cells, the ability of CpG ODNs to enhance the development of antigen-specific CD4+ and CD8+ T cell responses is presumably an indirect consequence of the CpG-induced (1) Th1-like cytokine and chemokine milieu, and (2) DC maturation and differentiation that may explain the resulting strong CTL generation, even in the absence of CD4 T cell help.[52,53] CpG ODNs have become the vaccine adjuvants of choice in many experimental models including peptide or protein antigens, live or killed viruses, dendritic cell vaccines, autologous cellular vaccines, and polysaccharide conjugates. Conjugation of a CpG ODN directly to an antigen can enhance antigen uptake and reduce antigen requirements.[54,55] However, cysteine residues in peptides or proteins may also form spontaneous disulfide bonds with the PS linkages in ODNs, resulting in enhanced CTL responses without the difficulties of a separate conjugation step.[56]

No other adjuvant has been shown to induce stronger Th1 responses than CpG ODN in comparisons using mouse models.[57–60] Fewer adjuvant comparison studies have been performed in primates, but here too a C-class CpG ODN mixed with an HIV Gag antigen was superior to a TLR7/8 ligand mixed with antigen, although inferior to a TLR7/8 ligand conjugated to antigen.[61,62] The Th1 biased vaccine response induced by TLR9 stimulation was maintained even in the presence of vaccine adjuvants such as alum or IFA that normally promote Th2 bias.[49,63,64] Likewise, CpG ODNs can overcome the Th2 bias associated with a respiratory syncytial virus vaccine[65] and with vaccinations in both very young and elderly mice.[66–73] The adjuvant activity of CpG ODNs generally improve markedly when they are formulated or co-administered with other adjuvants or in formulations such as microparticles, nanoparticles, lipid emulsions, or similar formulations that are necessary for inducing strong responses when an antigen is relatively weak.[74] CpG ODNs also are effective mucosal vaccine adjuvants for the respiratory tract,[75–78] vaginal mucosa,[79] oral or intrarectal vaccination,[78,80–82] conjunctival vaccination,[83] and even transcutaneous immunization.[84] Vaccination through mucosal routes succeeded in inducing both local and systemic humoral and cellular immune responses including enhanced protection against infectious challenge.[76,85]

Several human clinical trials were performed using B-class CpG ODNs as adjuvants for hepatitis B surface antigen either in combination with alum[47,86] or alone.[87] In a randomized, double-blind controlled Phase I/II dose escalation study, healthy individuals received three intramuscular (IM) injections (using the FDA-approved vaccination regimen of 0, 4 and 24 weeks) of an alum-absorbed HBV vaccine in saline or mixed with CPG 7909, a B-class ODN, at doses of 0.125, 0.5 or 1.0 mg.[86] HBsAg-specific antibody responses (anti-HBs) appeared earlier and had higher titers at all time points from 2 weeks after the initial prime up to 48 weeks in CPG 7909 recipients compared to individuals who received the vaccine alone. Moreover, most subjects who received CPG 7909 as an adjuvant developed protective levels of anti-HBs IgG within only 2 weeks of the priming vaccine dose, compared to none of the subjects receiving the commercial vaccine alone.[86] The addition of the TLR9 agonist also increased the proportion of antigen-specific high-avidity antibodies, suggesting enhancement of the late affinity maturation process in the activated B cells.[88]

A relatively high rate of unresponsiveness to the HBV vaccine occurs, especially among immune suppressed individuals such as those infected with HIV, among whom approximately 50% typically fail to mount protective levels of antibody. A randomized, double-blind, controlled trial in HIV-infected humans who previously failed to respond to Engerix-B™ alone demonstrated that addition of CPG 7909 to the vaccine significantly enhanced both the mean titers of anti-HBs and the antigen-specific T cell proliferative responses.[47] Perhaps of equal import, the proportion of HIV patients who had seroprotective levels at 12 months following vaccination increased from 63% in the controls to 100% in the group receiving CPG 7909.[47] Moreover, more than 85% of the subjects vaccinated with CPG 7909 maintained protective antibody levels for more than 3½ years, compared to fewer than 20% of the controls according to Cooper et al.[47] and Coley Pharmaceutical Group (data on file).

For vaccines against potential bioterror agents such as anthrax, the potential for faster seroconversion is especially important because it could enable post-exposure vaccination of susceptible populations following an attack. It is notable that CPG 7909 dramatically accelerated seroconversion when used as an adjuvant to the approved anthrax vaccine in a randomized controlled trial in healthy volunteers. Control subjects reached their peak titers of toxin-neutralizing antibodies at day 46, but this titer was achieved in the subjects receiving CPG 7909 as soon as day 22—more than three weeks earlier.[89] Furthermore, the addition of CPG 7909 induced a statistically significant 8.8-fold increase in peak titers of toxin-neutralizing antibodies, and increased the proportion of subjects who achieved strong IgG responses to the anthrax protective antigen from 61 to 100%.[89]

In mice, the use of a CpG ODNs as vaccine adjuvant enables the antigen doses to be reduced by approximately two orders of magnitude, with comparable antibody responses to the full-dose vaccine without CpG.[90] Such an effect in humans could be tremendously beneficial, especially in the setting of influenza vaccination where production of adequate quantities of a new vaccine to combat a pandemic strain may be limiting. In a Phase Ib randomized, double-blind, controlled clinical trial, subjects vaccinated with only a tenth of the normal dose of a commercial influenza vaccine showed reduced levels of antigen-specific IFN-α secretion from re-stimulated PBMCs compared to those measured in PBMCs from subjects administered the full-dose vaccine alone.[91] However, the co-administration of CPG 7909 with the one-tenth vaccine dose restored the antigen-specific IFN-α secretion to the level seen with full-dose vaccine.[91] Should these results be reproduced in further studies, they could advance the development of new generations of influenza vaccines.

Because they induce Th1-biased immune responses, CpG ODNs may have a special role in the development of improved allergy vaccines, to reprogram or overcome the detrimental Th2 allergic response. In allergic mice, CpG ODNs are able to redirect the allergic Th2 response and prevent inflammatory disease manifestations, even in mice with established allergic disease.[92,93] A conjugate of a B-class-like CpG ODN to a portion of the ragweed allergen was evaluated in human clinical trials as an allergy vaccine, with encouraging evidence for a selective and specific redirection of the allergic Th2 response toward a non-allergic and non-inflammatory Th1 response, and showing significant clinical benefit with reduced allergic symptoms,[94,95]

CpG ODNs also show great promise as adjuvants for cancer vaccines.[96] In a small Phase I tumor vaccine trial using a 1-mg dose of CPG 7909 as an adjuvant to recombinant MAGE-3 tumor antigen for triweekly vaccinations of six patients with metastatic melanoma, two stable disease and two partial responses beginning after 7 to 10 vaccinations and lasting at least 1 year by RECIST (response evaluation criteria in solid tumors) were noted.[97] In eight melanoma patients, 0.5 mg CPG 7909 mixed with a Melan-A tumor peptide antigen in Montanide (a water in oil emulsion) stimulated strong and rapid CD8 T cell responses, reaching a mean of >1% of Melan-A-specific CD8+ T cells.[98] The addition of CPG 7909 to the Melan-A peptide vaccine induced a high frequency of CD8+ T cells and also promoted effector cell differentiation to levels comparable to those seen in EBV- and CMV-specific T cells.[99]

ENHANCING ANTI-TUMOR ADAPTIVE IMMUNITY VIA TLR9 ACTIVATION WITHOUT VACCINE

The classical approach for inducing antigen-specific anti-tumor immune responses has been vaccination, as described above. However, the level of success using this approach has been disappointing.[100] Despite a number of possible explanations for the inconsistent clinical responses to tumor vaccines, much evidence suggests that the ability of a tumor to suppress the immune system may be a contributing factor. This led to increasing use of combination therapies, especially approaches that may weaken tumors to make them more susceptible to immune-mediated attack.

In mice with relatively small tumors, CpG monotherapy can be sufficient to induce T cell-mediated tumor regression, but inducing rejection of larger tumors generally requires combining the CpG ODN with other effective anti-tumor strategies such as monoclonal antibodies, radiation, surgery, and chemotherapy. TLR9 activation with CPG 7909 and other CpG ODNs has been shown to induce a Th1-like cytokine response in humans with B cell lymphomas, suggesting a potential role for this approach as part of a combination treatment regimen.[101,102] Although CpG ODNs are strong mitogens for normal B cells, no apparent exacerbation of lymphomas were noted in these patients. Potential therapeutic mechanisms proposed for using CpG ODNs to treat such TLR9-expressing malignancies include the induction of increased immunogenicity of the tumor cells, and the preferential induction of apoptosis in tumor cells stimulated through TLR9.[103,104]

In mouse models, CpG ODNs have shown synergy in combinations with essentially all types of anti-tumor therapies including monoclonal antibodies, surgery, radiotherapy, and chemotherapy.[96] Since chemotherapy is known to be immune-suppressive, it may seem counterintuitive to combine it with TLR9 stimulation and surprising that such combinations produce substantial improvements in survival in mouse tumor models using chemotherapy regimens including the topotecan topoisomerase I inhibitor, the cyclophosphamide alkylating agent, and the anti-metabolite 5-fluorouracil.[105–107] Nevertheless, the increased anti-tumor efficacy of these combination approaches requires T cells but not NK cells, and is associated with the induction of stronger anti-tumor T cell responses in mouse models (Weeratna et al., manuscript in preparation).

Humans receiving certain chemotherapy regimens such as taxanes actually show *increased* T cell and NK cell immune competence[108] that may be related to stimulation of TLR4, induction of pro-inflammatory cytokine production, induction of homeostatic leukocyte proliferation, and reversal of the immune-suppressive effects of regulatory T cells (Tregs) that appear to protect tumors against immune rejection.[109,110]

Based on these results, we investigated the effects of adding the B class CpG ODN known as PF-3512676 (formerly called CPG 7909, and also known as CpG 2006) to standard taxane–platinum chemotherapy for first-line treatment of stage IIIb/IV non-small cell lung cancer (NSCLC) in a Phase II randomized controlled human clinical trial. One hundred eleven chemotherapy-naïve patients were randomized to receive four to six 3-week cycles of standard chemotherapy alone or in combination with 0.2 mg/kg subcutaneous PF-3512676 in weeks 2 and 3 of each cycle. The primary endpoint for the trial, response rate (RECIST, intention-to-treat), was

significantly improved (p <0.05) from 19% in patients randomized to standard chemotherapy to 38% in the patients who also received PF-3512676.[111]

The secondary endpoint of this trial, survival, also showed a trend to improvement
from a median survival of 6.8 months in the chemotherapy arm to 12.3 months in the
combination arm, and an improvement in 1-year survival from 33 to 50%.[111] As in the
other clinical trials with TLR9 agonists, the most common side effects were mild to
moderate injection site reactions and transient flu-like symptoms. Grade 3 or 4 neu-
tropenias were more common in the combination arm, and are thought to reflect neutrophil redistribution, but febrile neutropenia and grade 3/4 infections were actually
slightly less common in the combination arm than in the chemotherapy alone arm.

Thrombocytopenia—a known effect of administration of ODNs containing
phosphorothioate backbones—observed in all systemic clinical trials of antisense
ODNs was seen more commonly in the combination arm, but with no apparent
increase in bleeding events. Unfortunately, two controlled Phase III human clinical trials of PF-3512676 combined with doublet chemotherapy in first-line treatment
of unresectable NSCLC failed to demonstrate any survival benefit of combination
therapy over chemotherapy alone. Several additional Phase I and Phase II trials in
which PF-3512676 is combined with other agents to treat cancer are also underway
(www.clinical trials.gov).

SAFETY OF TLR9 ACTIVATION IN RODENTS AND HUMANS

The safety profile of CpG ODNs results from the combination of their TLR9-
mediated immune stimulatory effects and the sequence-independent effects resulting from their phosphorothioate backbones characterized in antisense studies.[112–114]
However, it can be difficult to distinguish the TLR9-related effects from the backbone effects, since even CpG-free phosphorothioate ODN will stimulate TLR9 at
high concentrations.[115]

It is important to note that although antisense ODNs are commonly dosed at levels of several milligrams per kilogram per day, CpG ODNs for stimulating TLR9 are
generally given at doses in the range of 0.2 mg/kg/wk—about 1% of the typical dose
for an antisense ODN. PS-ODNs are rapidly cleared from the circulation into the
liver, kidneys and to a lesser extent, the spleen and bone marrow.[116,117] Chronic dosing of PS-ODNs in rodents produced dose-dependent mononuclear cell infiltrations
in these organs of ODN deposition, but the cell infiltrates did not occur in monkeys
or humans, possibly because of the broader distribution of TLR9 expression within
the immune cells of rodents.[112,118]

Hepatic effects specific to rodents include the activation of Kupffer cells with
cellular hypertrophy and hyperplasia, basophilic granulation (thought to reflect PS-
ODN deposition), and mononuclear cell infiltrates in hepatic sinusoids and periportal
regions.[112,119] In the kidneys, high local ODN concentrations reached after repeated
high doses can induce degenerative lesions and necrosis in proximal tubules.[112,120]

There have been no reports of adverse effects of PS-ODNs on renal function in
humans, despite the extensive clinical experience thus far. Presumably the species-
specific toxicities are consequences of the cellular pattern of TLR9 expression that
determines the cytokines that will be produced in response to administration of a

CpG ODN, and thus the safety profile. Since TLR9 is expressed in a broader range of immune cells in rodents compared to primates, the rodents tend to over-predict toxicities in primates. For example, rodents respond to CpG ODN administration with high serum concentrations of pro-inflammatory cytokines such as TNF-α that can produce lethal "cytokine storms,"[121] but no changes in serum TNF-α were observed in humans and primates.[122]

In primates, the major dose limiting acute toxicity of PS-ODN is acute cardiovascular collapse and death resulting from systemic activation of the alternative complement pathway.[123,124] This toxicity does not occur below a threshold PS-ODN blood concentration of approximately 40 to 50 µg/mL that is typically reached only after relatively rapid IV administration of the ODN.[112,123,125] High concentrations of PS-ODN bind nonspecifically to the tenase complex of thrombin, causing a prolongation of the activated partial thromboplastin time.[112,126,127] However at the time of this review, no clinical complications due to bleeding have been described.

In human clinical trials, the safety profiles of several TLR9 agonists were studied over a dose range from 0.0025 to 0.81 mg/kg. A maximal tolerated dose in humans has not been reported to date. The primary adverse events are dose-dependent local injection reactions (erythema, pain, swelling, induration, pruritus, and warmth at injection site) or systemic flu-like reactions (headache, rigors, myalgia, pyrexia, nausea, and vomiting), and are consistent with the known mechanisms of action of TLR9 agonists. Depending on dose, systemic symptoms typically appear within 12 to 24 hours of dosing and persist for 1 to 2 days, and in rare cases, longer. Even without the addition of a TLR9 agonist, vaccines commonly induce injection site reactions. Some vaccines containing CpG ODN adjuvants appeared to cause a slight increase in the frequency of injection site reactions compared to the frequency observed with the vaccine alone, but these reactions were generally mild.

CpG ODN treatment can exacerbate autoimmunity in some mouse models of lupus,[128] multiple sclerosis,[129] colitis,[130] and arthritis.[131] Nevertheless, TLR9 also seems to exert counter-regulatory influences that limit or reverse the development of autoimmunity, since lupus-prone mice that are genetically deficient in TLR9 showed increased severity of disease in several models.[132–135] TLR9 activation also can prevent colitis.[136] Clinical experience to date indicates that CpG ODN treatment of normal humans, cancer patients, and individuals infected with HIV or HCV does not readily induce autoimmune disease. However, the duration of therapy has usually been shorter than 6 months; only a few patients have received chronic CpG ODN therapy (longer than 3 years). Since TLR9 activation induces the secretion of IFN-α, and since treatment with recombinant IFN-α induces autoimmune diseases in 4 to 19% of chronically treated patients,[137] continued vigilance would be prudent until larger numbers of patients have been treated with TLR9 agonists for longer periods.

CONCLUSION

Since the CpG motif was described in 1995, a half dozen TLR9 agonists have progressed to clinical trials, including several investigational products in Phase III trials and multiple drugs in Phase II development. The early clinical results are encouraging, with strong indications of substantial clinical benefits that indicate that the targeted activation of TLR9 will enhance the treatment of cancer and infectious diseases, and decrease the harmful inflammatory responses of asthma and other allergic diseases. Although longer term follow-ups of larger numbers of patients are needed, the safety of these TLR9 agonists appears good, and their clinical contributions are potentially enormous.

REFERENCES

1. Krieg, A. M. et al. 1995. CpG motifs in bacterial DNA trigger direct B-cell activation. *Nature* 374: 546-549.
2. Krieg, A. M. 2002. CpG motifs in bacterial DNA and their immune effects. *Annu. Rev. Immunol.* 20: 709-760.
3. Hemmi, H. et al. 2000. A Toll-like receptor recognizes bacterial DNA. *Nature* 408: 740-745.
4. Krieg, A. M. 2006. Therapeutic potential of Toll-like receptor 9 activation. *Nat. Rev. Drug Discov.* 5: 471-484.
5. Vollmer, J. et al. 2004. Characterization of three CpG oligodeoxynucleotide classes with distinct immunostimulatory activities. *Eur. J Immunol.* 34: 251-262.
6. Hartmann, G. et al. 2003. Rational design of new CpG oligonucleotides that combine B cell activation with high IFN-α induction in plasmacytoid dendritic cells. *Eur. J. Immunol.* 33: 1633-1641.
7. Marshall, J. D. et al. 2003. Identification of a novel CpG DNA class and motif that optimally stimulate B cell and plasmacytoid dendritic cell functions. *J. Leukoc. Biol.* 73: 781-792.
8. Gursel, M. et al. 2006. CXCL16 influences the nature and specificity of CpG-induced immune activation. *J. Immunol.* 177: 1575-1580.
9. Honda, K. et al. 2005. Spatiotemporal regulation of MyD88-IRF-7 signalling for robust type-I interferon induction. *Nature* 434: 1035-1040.
10. Elkins, K. L. et al. 1999. Bacterial DNA containing CpG motifs stimulates lymphocyte-dependent protection of mice against lethal infection with intracellular bacteria. *J. Immunol.* 162: 2291-2298.
11. Gramzinski, R. A. et al. 2001. Interleukin-12- and γ interferon-dependent protection against malaria conferred by CpG oligodeoxynucleotide in mice. *Infect. Immun.* 69: 1643-1649.
12. Zimmermann, S. et al. 1998. CpG oligodeoxynucleotides trigger protective and curative Th1 responses in lethal murine leishmaniasis. *J. Immunol.* 160: 3627-3630.
13. Krieg, A. M. et al. 1998. CpG DNA induces sustained IL-12 expression *in vivo* and resistance to *Listeria monocytogenes* challenge. *J. Immunol.* 161: 2428-2434.
14. Klinman, D. M. et al. 1999. Repeated administration of synthetic oligodeoxynucleotides expressing CpG motifs provides long-term protection against bacterial infection. *Infect. Immun.* 67: 5658-5663.
15. Klinman, D. M. et al. 1999. Immune recognition of foreign DNA: a cure for bioterrorism? *Immunity* 11: 123-129.
16. Rees, D. G. et al. 2005. CpG-DNA protects against a lethal orthopoxvirus infection in a murine model. *Antiviral Res.* 65: 87-95.
17. Deng, J. C. et al. 2004. CpG oligodeoxynucleotides stimulate protective innate immunity against pulmonary Klebsiella infection. *J. Immunol.* 173: 5148-5155.
18. Ray, N. B. and A. M. Krieg. 2003. Oral pretreatment of mice with CpG DNA reduces susceptibility to oral or intraperitoneal challenge with virulent *Listeria monocytogenes*. *Infect. Immun.* 71: 4398-4404.

19. Klinman, D. M. 2004. Immunotherapeutic uses of CpG oligodeoxynucleotides. *Nat. Rev. Immunol.* 4: 249-259.
20. Weighardt, H. et al. 2000. Increased resistance against acute polymicrobial sepsis in mice challenged with immunostimulatory CpG oligodeoxynucleotides is related to an enhanced innate effector cell response. *J. Immunol.* 165: 4537-4543.
21. Pyles, R. B. et al. 2002. Use of immunostimulatory sequence-containing oligonucleotides as topical therapy for genital herpes simplex virus type 2 infection. *J Virol.* 76: 11387-11396.
22. Ashkar, A. A. et al. 2003. Local delivery of CpG oligodeoxynucleotides induces rapid changes in the genital mucosa and inhibits replication, but not entry, of herpes simplex virus type 2. *J. Virol.* 77: 8948-8956.
23. Walker, P. S. et al. 1999. Immunostimulatory oligodeoxynucleotides promote protective immunity and provide systemic therapy for leishmaniasis via IL-12- and IFN-γ-dependent mechanisms. *Proc. Natl. Acad. Sci. USA* 96: 6970-6975.
24. Cho, J. Y. et al. 2001. Immunostimulatory DNA sequences inhibit respiratory syncytial viral load, airway inflammation, and mucus secretion. *J. Allergy Clin. Immunol.* 108: 697-702.
25. Juffermans, N. P. et al. 2002. CpG oligodeoxynucleotides enhance host defense during murine tuberculosis. *Infect. Immun.* 70: 147-152.
26. Olbrich, A. R. et al. 2002. Effective postexposure treatment of retrovirus-induced disease with immunostimulatory DNA containing CpG motifs. *J. Virol.* 76: 11397-11404.
27. Freidag, B. L. et al. 2000. CpG oligodeoxynucleotides and interleukin-12 improve the efficacy of *Mycobacterium bovis* BCG vaccination in mice challenged with *M. tuberculosis*. *Infect. Immun.* 68: 2948-2953.
28. Sajic, D. et al. 2003. Parameters of CpG oligodeoxynucleotide-induced protection against intravaginal HSV-2 challenge. *J. Med. Virol.* 71: 561-568.
29. Verthelyi, D. et al. 2003. CpG oligodeoxynucleotides protect normal and SIV-infected macaques from Leishmania infection. *J. Immunol.* 170: 4717-4723.
30. Isogawa, M. et al. 2005. Toll-like receptor signaling inhibits hepatitis B virus replication *in vivo*. *J. Virol.* 79: 7269-7272.
31. Rehermann, B. and M. Nascimbeni. 2005. Immunology of hepatitis B virus and hepatitis C virus infection. *Nat. Rev. Immunol.* 5: 215-229.
32. McHutchison, J. G. et al. 2005. Relationships of HCV RNA responses to CPG 10101, a TLR9 agonist. AASLD.
33. Schmalbach, T. et al. 2004. CPG 10101 (Actilon) oligodeoxynucleotide TLR9 agonist: pharmacokinetics and pharmacodynamics in normal volunteers. Presented at 44th Annual Interscience Conference on Antimicrobial Agents and Chemotherapy.
34. Glue, P. et al. 2006. Pegylated interferon-α2b: pharmacokinetics, pharmacodynamics, safety, and preliminary efficacy data. *Clin. Pharmacol. Ther.* 68: 556-567.
35. McHutchison, J. G. et al. 2006. Final results of a multicenter Phase 1b, randomized, placebo-controlled, dose escalation trial of CPG 10101 in patients with chronic hepatitis C virus. European Association for Study of the Liver, Vienna.
36. Formann, E. et al. 2006. Twice-weekly administration of PEG-interferon-α-2b improves viral kinetics in patients with chronic hepatitis C genotype 1. *J. Viral Hepat.* 10: 271-276.
37. McHutchison, J. G. et al. 2005. Relationships of HCV RNA responses to CPG 10101, a TLR9 agonist: pharmacodynamics and patient characteristics. *Hepatology* 42 (Suppl 1): 249A.
38. McHutchison, J. G. et al. 2005. Immunophenotyping profile of CPG 10101 (Actilon), a new TLR9 agonist antiviral for hepatitis C. *Hepatology* 42 (Suppl. 1): 539A.

39. Jacobson, A. et al. 2006. Early viral response and treatment response to CPG 10101 (Actilon) in combination with pegylated interferon and/or ribavirin, in chronic HCV genotype 1-infected patients with prior relapse response. *Hepatology* 44 (Suppl. 1): 24A.

40. Ito, S. et al. 2005. CpG oligodeoxynucleotides increase the susceptibility of normal mice to infection by *Candida albicans. Infect. Immun.* 73: 6154-6156.

41. Olbrich, A. R., S. Schimmer, and U. Dittmer. 2003. Preinfection treatment of resistant mice with CpG oligodeoxynucleotides renders them susceptible to Friend retrovirus-induced leukemia. *J. Virol.* 77: 10658-10662.

42. Equils, O. et al. 2003. Toll-like receptor 2 (TLR2) and TLR9 signaling results in HIV long terminal repeat transactivation and HIV replication in HIV-1 transgenic mouse spleen cells. *J. Immunol.* 170: 5159-5164.

43. Agrawal, S. et al. 2003. Cutting edge: different Toll-like receptor agonists instruct dendritic cells to induce distinct Th responses via differential modulation of extracellular signal-regulated kinase-mitogen-activated protein kinase and c-Fos. *J. Immunol.* 171: 4984-4989.

44. Gurney, K. B. et al. 2004. Endogenous IFN-α production by plasmacytoid dendritic cells exerts antiviral effect on thymic HIV-1 infection. *J. Immunol.* 173: 7269-7276.

45. Schlaepfer, E. et al. 2004. CpG oligodeoxynucleotides block human immunodeficiency virus type 1 replication in human lymphoid tissue infected *ex vivo. J. Virol.* 78: 12344-12354.

46. Saez, R. et al. 2005. HIV-infected progressors and long-term non-progressors differ in their capacity to respond to an A-class CpG oligodeoxynucleotide. *AIDS* 19: 1924-1925.

47. Cooper, C. L. et al. 2005. CPG 7909 adjuvant improves hepatitis B virus vaccine seroprotection in antiretroviral-treated HIV-infected adults. *AIDS* 19: 1473-1479.

48. Yi, A. K. et al. 1998. CpG oligodeoxyribonucleotides rescue mature spleen B cells from spontaneous apoptosis and promote cell cycle entry [erratum: *J. Immunol.* 163: 1093, 1999] *J. Immunol.* 160: 5898-5906.

49. Davis, H. L. et al. 1998. CpG DNA is a potent enhancer of specific immunity in mice immunized with recombinant hepatitis B surface antigen [erratum *J. Immunol.* 162: 3103, 1999]. *J. Immunol.* 160: 870-876.

50. Liu, N. et al. 2003. CpG directly induces T-bet expression and inhibits IgG1 and IgE switching in B cells. *Nat. Immunol.* 4: 687-693.

51. He, B., X. Qiao, and A. Cerutti. 2004. CpG DNA induces IgG class switch DNA recombination by activating human B cells through an innate pathway that requires TLR9 and cooperates with IL-10. *J. Immunol.* 173: 4479-4491.

52. Lipford, G. B. et al. 2000. CpG-DNA-mediated transient lymphadenopathy is associated with a state of Th1 predisposition to antigen-driven responses. *J. Immunol.* 165: 1228-1235.

53. Sparwasser, T. et al. 2000. Bacterial CpG-DNA activates dendritic cells *in vivo*: T helper cell-independent cytotoxic T cell responses to soluble proteins. *Eur. J. Immunol.* 30: 3591-3597.

54. Tighe, H. et al. 2000. Conjugation of protein to immunostimulatory DNA results in a rapid, long-lasting and potent induction of cell-mediated and humoral immunity. *Eur. J. Immunol.* 30: 1939-1947.

55. Hartmann, E. et al. 2003. Identification and functional analysis of tumor-infiltrating plasmacytoid dendritic cells in head and neck cancer. *Cancer Res.* 63: 6478-6487.

56. Wettstein, P. J. et al. 2005. Cysteine-tailed class I-binding peptides bind to CpG adjuvant and enhance primary CTL responses. *J. Immunol.* 175: 3681-3689.

57. Kim, S. K. et al. 2000. Comparison of the effect of different immunological adjuvants on the antibody and T-cell response to immunization with MUC1-KLH and GD3-KLH conjugate cancer vaccines. *Vaccine* 18: 597-603.

58. Chu, R. S. et al. 1997. CpG oligodeoxynucleotides act as adjuvants that switch on T helper 1 (Th1) immunity. *J. Exp. Med.* 186: 1623-1631.
59. Lipford, G. B. et al. 1997. CpG-containing synthetic oligonucleotides promote B and cytotoxic T cell responses to protein antigen: a new class of vaccine adjuvants. *Eur. J. Immunol.* 27: 2340-2344.
60. Roman, M. et al. 1997. Immunostimulatory DNA sequences function as T helper-1-promoting adjuvants. *Nat. Med.* 3: 849-854.
61 Wille-Reece, U. et al. 2005. HIV Gag protein conjugated to a Toll-like receptor 7/8 agonist improves the magnitude and quality of Th1 and CD8+ T cell responses in non-human primates. *Proc. Natl. Acad. Sci. USA* 102: 15190-15194.
62. Wille-Reece, U. et al. 2006. Toll-like receptor agonists influence the magnitude and quality of memory T cell responses after prime-boost immunization in nonhuman primates. *J. Exp. Med.* 203: 1249-1258.
63. Weeratna, R. D. et al. 2000. CpG DNA induces stronger immune responses with less toxicity than other adjuvants. *Vaccine* 18: 1755-1762.
64. Sugai, T. et al. 2005. A CpG-containing oligodeoxynucleotide as an efficient adjuvant counterbalancing the Th1/Th2 immune response in diphtheria–tetanus–pertussis vaccine. *Vaccine* 23: 5450-5456.
65. Oumouna, M. et al. 2005. Formulation with CpG oligodeoxynucleotides prevents induction of pulmonary immunopathology following priming with formalin-inactivated or commercial killed bovine respiratory syncytial virus vaccine. *J. Virol.* 79: 2024-2032.
66. Brazolot-Millan, C. L. et al. 1998. CpG DNA can induce strong Th1 humoral and cell-mediated immune responses against hepatitis B surface antigen in young mice. *Proc. Natl. Acad. Sci. USA* 95: 15553-15558.
67. Weeratna, R. D. et al. 2001. Priming of immune responses to hepatitis B surface antigen in young mice immunized in the presence of maternally derived antibodies. *FEMS Immunol. Med. Microbiol.* 30: 241-247.
68. Schirmbeck, R. and J. Reimann. 2001. Modulation of gene gun-mediated Th2 immunity to hepatitis B surface antigen by bacterial CpG motifs or IL-12. *Intervirology* 44: 115-123.
69. Zhou, X. et al. 2003. T helper 2 immunity to hepatitis B surface antigen primed by gene gun-mediated DNA vaccination can be shifted toward T helper 1 immunity by codelivery of CpG motif-containing oligodeoxynucleotides. *Scand. J. Immunol.* 58: 350-357.
70. Weeratna, R. D. et al. 2001. CpG ODN can re-direct the Th bias of established Th2 immune responses in adult and young mice. *FEMS Immunol. Med. Microbiol.* 32: 65-71.
71. Manning, B. M. et al. 2001. CpG DNA functions as an effective adjuvant for the induction of immune responses in aged mice. *Exp. Gerontol.* 37: 107-126.
72. Maletto, B. et al. 2002. CpG-DNA stimulates cellular and humoral immunity and promotes Th1 differentiation in aged BALB/c mice. *J. Leukoc. Biol.* 72: 447-454.
73. Alignani, D. et al. 2005. Orally administered OVA/CpG-ODN induces specific mucosal and systemic immune response in young and aged mice. *J. Leukoc. Biol.* 77: 898-905.
74. Krieg, A. M. and H. L. Davis. 2001. Enhancing vaccines with immune stimulatory CpG DNA. *Curr. Opin. Mol. Ther.* 3: 15-24.
75. Moldoveanu, Z. et al. 1998. CpG DNA, a novel immune enhancer for systemic and mucosal immunization with influenza virus. *Vaccine* 16: 1216-1224.
76. Gallichan, W. S. et al. 2001. Intranasal immunization with CpG oligodeoxynucleotides as adjuvant dramatically increases IgA and protection against herpes simplex virus-2 in the genital tract. *J. Immunol.* 166: 3451-3457.
77. McCluskie, M. J. and H. L. Davis. 1998. CpG DNA is a potent enhancer of systemic and mucosal immune responses against hepatitis B surface antigen with intranasal administration to mice. *J. Immunol.* 161: 4463-4466.

78. McCluskie, M. J. and H. L. Davis. 2001. Oral, intrarectal and intranasal immunizations using CpG and non-CpG oligodeoxynucleotides as adjuvants. *Vaccine* 19: 413-422.

79. Kwant, A. and K. L. Rosenthal. 2004. Intravaginal immunization with viral subunit protein plus CpG oligodeoxynucleotides induces protective immunity against HSV-2. *Vaccine* 22: 3098-3104.

80. Eastcott, J. W. et al. 2001. Oligonucleotide containing CpG motifs enhance immune response to mucosally or systemically administered tetanus toxoid. *Vaccine* 19: 1636-1642.

81. McCluskie, M. J. et al. 2001. CpG DNA is an effective oral adjuvant to protein antigens in mice. *Vaccine* 19: 950-957.

82. Dong, J. L. et al. 2005. Oral immunization with pBsVP6-transgenic alfalfa protects mice against rotavirus infection. *Virology* 339: 153-163.

83. Nesburn, A. B. et al. 2005. Local and systemic B cell and Th1 responses induced following ocular mucosal delivery of multiple epitopes of herpes simplex virus type 1 glycoprotein D together with cytosine-phosphate-guanine adjuvant. *Vaccine* 23: 873-883.

84. Berry, L. J. et al. Transcutaneous immunization with combined cholera toxin and CpG adjuvant protects against *Chlamydia muridarum* genital tract infection. *Infect. Immun.* 72: 1019-1028.

85. Dumais, N. et al. 2002. Mucosal immunization with inactivated human immunodeficiency virus plus CpG oligodeoxynucleotides induces genital immune responses and protection against intravaginal challenge. *J. Infect. Dis.* 186: 1098-1105.

86. Cooper, C. L. et al. 2004. CpG 7909, an immunostimulatory TLR9 agonist oligodeoxynucleotide, as adjuvant to Engerix-B HBV vaccine in healthy adults: double-blind Phase I/II study. *J. Clin. Immunol.* 24: 693-702.

87. Halperin, S. A. et al. 2003. A phase I study of the safety and immunogenicity of recombinant hepatitis B surface antigen co-administered with an immunostimulatory phosphorothioate oligonucleotide adjuvant. *Vaccine* 21: 2461-2467.

88. Siegrist, C. A. et al. 2004. Co-administration of CpG oligonucleotides enhances the late affinity maturation process of human anti-hepatitis B vaccine response. *Vaccine* 23: 615-622.

89. Rynkiewicz, D. et al. 2005. Marked enhancement of antibody response to anthrax vaccine aAdsorbed with CPG 7909 in healthy volunteers. ICAAC.

90. Weeratna, R., L. Comanita, and H. L. Davis. 2003. CPG ODN allows lower dose of antigen against hepatitis B surface antigen in BALB/c mice. *Immunol. Cell Biol.* 81: 59-62.

91. Cooper, C. L. et al. 2004. Safety and immunogenicity of CpG 7909 injection as an adjuvant to Fluarix influenza vaccine. *Vaccine* 22: 3136-3143.

92. Kline, J. N. et al. 1998. Modulation of airway inflammation by CpG oligodeoxynucleotides in a murine model of asthma. *J. Immunol.* 160: 2555-2559.

93. Jain, V. V. et al. 2002. CpG-oligodeoxynucleotides inhibit airway remodeling in a murine model of chronic asthma. *J. Allergy Clin. Immunol.* 110: 867-872.

94. Creticos, P. S. et al. 2002. Immunotherapy with immunostimulatory oligonucleotides linked to purified ragweed Amb-a-1 allergen: effects on antibody production, nasal allergen provocation, and ragweed seasonal rhinitis. *J. Allergy Clin. Immunol.* 109: 742-743.

95. Simons, F. E. et al. 2004. Selective immune redirection in humans with ragweed allergy by injecting Amb-a-1 linked to immunostimulatory DNA. *J. Allergy Clin. Immunol.* 113: 1144-1151.

96. Krieg, A. M. 2004. Antitumor Applications of Stimulating Toll-like Receptor 9 with CpG Oligodeoxynucleotides. *Curr. Oncol. Rep.* 6: 88-95.

97. van Ojik, H. et al. 2002. Phase I/II study with CpG 7909 as adjuvant to vaccination with MAGA-3 protein in patients with MAGE-3 positive tumors. *Ann. Oncol.* 13: 157.

98. Speiser, D. E. et al. 2005. Rapid and strong human CD8(+) T cell responses to vaccination with peptide, IFA, and CpG oligodeoxynucleotide 7909. *J. Clin. Invest.* 115: 739-746.

99. Appay, V. et al. 2006. New generation vaccine induces effective melanoma-specific CD8+ T cells in the circulation but not in the tumor site. *J. Immunol.* 177: 1670-1678.

100. Rosenberg, S. A., J. C. Yang, and N. P. Restifo. 2004. Cancer immunotherapy: moving beyond current vaccines. *Nat. Med.* 10: 909-915.

101. Link, B. et al. 2006. Oligodeoxynucleotide CPG 7909 delivered as intravenous infusion demonstrates immunologic modulation in patients with previously treated non-Hodgkin's lymphoma. *J. Immunother.* 29: 558-568.

102. Friedberg, J. W. et al. 2005. Combination immunotherapy with a CpG oligonucleotide (1018 ISS) and rituximab in patients with non-Hodgkin's lymphoma: increased interferon-α/βa-inducible gene expression. *Blood* 105: 489-495.

103. Jahrsdorfer, B. et al. 2005. Immunostimulatory oligodeoxynucleotides induce apoptosis of B cell chronic lymphocytic leukemia cells. *J. Leukoc. Biol.* 77: 378-387.

104. Jahrsdorfer, B. et al. 2005. B-cell lymphomas differ in their responsiveness to CpG oligodeoxynucleotides. *Clin. Cancer Res.* 11: 1490-1499.

105. Weigel, B. J. et al. 2003. CpG oligodeoxynucleotides potentiate the antitumor effects of chemotherapy or tumor resection in an orthotopic murine model of rhabdomyosarcoma. *Clin. Cancer Res.* 9: 3105-3114.

106. Balsari, A. et al. 2004. Combination of a CpG-oligodeoxynucleotide and a topoisomerase I inhibitor in the therapy of human tumour xenografts. *Eur. J. Cancer* 40: 1275-1281.

107. Wang, X. S. et al. 2005. CpG oligodeoxynucleotides inhibit tumor growth and reverse the immunosuppression caused by the therapy with 5-fluorouracil in murine hepatoma. *World J. Gastroenterol.* 11: 1220-1224.

108. Carson, W. E., et al. 2004. Cellular immunity in breast cancer patients completing taxane treatment. *Clin. Cancer Res.* 10: 3401-3409.

109. Emens, L. A., R. T. Reilly, and E. M. Jaffee. 2004. Augmenting the potency of breast cancer vaccines: combined modality immunotherapy. *Breast Dis.* 20: 13-24.

110. Lake, R. A. and B. W. Robinson. 2005. Immunotherapy and chemotherapy: a practical partnership. *Nat. Rev. Cancer* 5: 397-405.

111. Manegold, C. et al. 2005. Addition of PF-3512676 (CpG 7909) to a taxane–platinum regimen for first-line treatment of unresectable non-small cell lung cancer. *Eur. J. Cancer* 3: 326.

112. Levin, A. A. et al. 2001. Toxicity of antisense oligonucleotides, in Crooke, S. T., Ed., *Antisense Drug Technology.* Marcel Dekker, New York, pp. 201-267.

113. Levin, A. A. 1999. A review of the issues in the pharmacokinetics and toxicology of phosphorothioate antisense oligonucleotides. *Biochim. Biophys. Acta* 1489: 69-84.

114. Monteith, D. K. and A. A. Levin. 1999. Synthetic oligonucleotides: the development of antisense therapeutics. *Toxicol. Pathol.* 27: 8-13.

115. Vollmer, J. et al. 2004. Oligodeoxynucleotides lacking CpG dinucleotides mediate Toll-like receptor 9 dependent T helper type 2-biased immune stimulation. *Immunology* 113: 212-223.

116. Geary, R. S. et al. 1997. Antisense oligonucleotide inhibitors for the treatment of cancer: 1. Pharmacokinetic properties of phosphorothioate oligodeoxynucleotides. *Anticancer Drug Des.* 12: 383-393.

117. Cossum, P. A. et al. 1993. Disposition of the ^{14}C-labeled phosphorothioate oligonucleotide ISIS 2105 after intravenous administration to rats. *J. Pharmacol. Exp. Ther.* 267: 1181-1190.

118. Henry, S. P. et al. 1997. Evaluation of the toxicity of ISIS 2302, a phosphorothioate oligonucleotide, in a 4-week study in CD-1 mice. *Antisense Nucleic Acid Drug Dev.* 7: 473-481.
119. Heikenwalder, M. et al. 2004. Lymphoid follicle destruction and immunosuppression after repeated CpG oligodeoxynucleotide administration. *Nat. Med.* 10: 187-192.
120. Jason, T. L., J. Koropatnick, and R. W. Berg. 2004. Toxicology of antisense therapeutics. *Toxicol. Appl. Pharmacol.* 201: 66-83.
121. Sparwasser, T. et al. 1997. Bacterial DNA causes septic shock. *Nature* 386: 336-337.
122. Krieg, A. M. et al. 2004. Induction of systemic TH1-like innate immunity in normal volunteers following subcutaneous but not intravenous administration of CPG 7909, a synthetic B-class CpG oligodeoxynucleotide TLR9 agonist. *J. Immunother.* 27: 460-471.
123. Henry, S. P. et al. 2002. Complement activation is responsible for acute toxicities in rhesus monkeys treated with a phosphorothioate oligodeoxynucleotide. *Int. Immunopharmacol.* 2: 1657-1666.
124. Galbraith, W. M. et al. 1994. Complement activation and hemodynamic changes following intravenous administration of phosphorothioate oligonucleotides in the monkey. *Antisense Res. Dev.* 4: 201-206.
125. Monteith, D. K. et al. 1998. Preclinical evaluation of the effects of a novel antisense compound targeting C-raf kinase in mice and monkeys. *Toxicol. Sci.* 46: 365-375.
126. Henry, S. P. et al. 1997. Inhibition of coagulation by a phosphorothioate oligonucleotide. *Antisense Nucleic Acid Drug Dev.* 7: 503-510.
127. Sheehan, J. P. and H. C. Lan. 1998. Phosphorothioate oligonucleotides inhibit the intrinsic tenase complex. *Blood* 92: 1617-1625.
128. Hasegawa, K. and T. Hayashi. 2003. Synthetic CpG oligodeoxynucleotides accelerate the development of lupus nephritis during preactive phase in NZB × NZWF1 mice. *Lupus* 12: 838-845.
129. Ichikawa, H. T., L. P. Williams, and B. M. Segal. 2002. Activation of APCs through CD40 or Toll-like receptor 9 overcomes tolerance and precipitates autoimmune disease. *J. Immunol.* 169: 2781-2787.
130. Obermeier, F. et al. 2002. CpG motifs of bacterial DNA exacerbate colitis of dextran sulfate sodium-treated mice. *Eur. J. Immunol.* 32: 2084-2092.
131. Ronaghy, A. et al. 2002. Immunostimulatory DNA sequences influence the course of adjuvant arthritis. *J. Immunol.* 168: 51-56.
132. Christensen, S. R. et al. 2006. Toll-like receptor 7 and TLR9 dictate autoantibody specificity and have opposing inflammatory and regulatory roles in a murine model of lupus. *Immunity* 25: 417-428.
133. Ma, Z. et al. 2006. Modulation of autoimmunity by TLR9 in the chronic graft-versus-host model of systemic lupus erythematosus. *J. Immunol.* 177: 7444-7450.
134. Yu, P. et al. 2006. Toll-like receptor 9-independent aggravation of glomerulonephritis in a novel model of SLE. *Int. Immunol.* 18: 1211-1219.
135. Wu, X. and S. L. Peng. 2006. Toll-like receptor 9 signaling protects against murine lupus. *Arthritis Rheum.* 54: 336-342.
136. Katakura, K. et al. 2005. Toll-like receptor 9-induced type I IFN protects mice from experimental colitis. *J. Clin. Invest.* 115: 695-702.
137. Ioannou, Y. and D. A. Isenberg. 2000. Current evidence for induction of autoimmune rheumatic manifestations by cytokine therapy. *Arthritis Rheum.* 43: 1431-1442.

11 Prospects for TLR9-Based Immunotherapy for Asthma and Allergy

David Broide

ABSTRACT

Allergic inflammation and asthma are both characterized by increased expression of Th2 cytokines such as interleukin (IL)-4, IL-5, IL-9, and IL-13. Immunostimulatory sequences of DNA (ISS DNA) containing CpG motifs (CpG DNA) bind to TLR9 receptors on cells involved in the innate immune response such as plasmacytoid dendritic cells (pDCs), and as a consequence inhibit adaptive Th2 immune responses. ISS DNA therefore represents a potential novel therapeutic strategy to inhibit allergic inflammation and asthma. Studies in mouse and primate models of allergen-induced asthma have demonstrated that ISS DNA inhibits eosinophilic airway inflammation, Th2 cytokine expression, airway hyperreactivity, and features of airway remodeling. Human study data on the use of ISS DNA in allergy and asthma is limited. In a pilot study of subjects with asymptomatic mild asthma, nebulized ISS DNA did not inhibit immediate or late phase airway responses to inhaled allergen challenge. In contrast, in a small Phase II study in human subjects with symptomatic allergic rhinitis, subcutaneous injection of ISS DNA conjugated to a ragweed protein allergen significantly inhibited allergic rhinitis symptoms during the ragweed season. Further studies of ISS DNA in human subjects with allergy and asthma will assist in determining its safety and effectiveness as a therapeutic strategy for such patients.

INTRODUCTION

Asthma is an inflammatory disease of the airway that affects approximately 7% of the U.S. population.[1] Although asthma is well controlled in most subjects through the use of inhaled beta-2 adrenergic receptor agonists and inhaled corticosteroids, 10 million acute exacerbations, 2 million urgent care visits, 400,000 hospitalizations, and 5,000 deaths result from asthma annually in the U.S. This underscores the fact that asthma is not well controlled in large numbers of patients.[2] In particular, the 10% who have severe asthma constitute approximately 50% of the health care costs associated with asthma.

The estimated annual direct cost for asthma care in the U.S. in 2002 was approximately $6 billion. In addition to experiencing acute episodes of asthma, a subset of asthmatics develops chronic structural changes of their airways (airway remodeling) over time. These structural features include increased mucus expression, increased thickness of the smooth muscle layer, increased subepithelial fibrosis, and angiogenesis.

Exposure of a sensitized allergic asthmatic to airborne allergens such as cat dander or dust mites can trigger wheezing, cough, and shortness of breath due to immediate activation of resident airway mast cells with release of histamine and recruitment of pro-inflammatory leukocytes from the bone marrow and circulation into the airway. The inflammatory changes in the airway include increased expression of Th2 cytokines and accumulation of pro-inflammatory eosinophils, Th2 lymphocytes, and IgE-sensitized mast cells. The importance of Th2 cytokines to allergic inflammation is suggested from their effector functions. IL-4 acts as a switch factor for IgE synthesis; IL-5, an eosinophil growth factor; IL-9, an inducer of mucus expression; and IL-13, a cytokine that can induce airway hyperreactivity.[1]

A variety of novel therapeutic strategies are currently under investigation in asthma patients to determine whether interruption of this cascade of pro-inflammatory cytokines can reduce levels of airway inflammation and associated airway hyperreactivity. In general, therapeutic strategies have focused on antagonizing single downstream cytokines and mediators such as IL-5, LTC_4, TNF, and IL-13, or alternatively targeting upstream cells (Th2 lymphocytes) that regulate the expression of multiple downstream cytokines and mediators.[3] The advantage of targeting a downstream mediator is the likely greater safety margin of such a therapy compared to targeting an upstream cell that regulates the expression of multiple mediators and cytokines, some of which may be important to host defense. However, therapies that target a single cytokine or mediator may be less likely to beneficially impact a disease such as asthma in which multiple cells, cytokines, and mediators play important roles in disease expression.

One novel approach for down-regulating the complex inflammatory responses in asthma is utilizing TLR9-based strategies to inhibit Th2 cells that express several cytokines important to pathogenesis. This chapter reviews results of studies investigating the therapeutic potentials of several TLR9-based strategies in asthma, including the use immunostimulatory sequences of DNA containing a CpG motif (ISS DNA) and ISS DNA conjugated to an allergen protein in animal models of asthma, and results of preliminary studies in humans.

MECHANISMS OF ACTIONS OF TLR9-BASED STRATEGIES

TLR9 is a receptor expressed on cells of the innate immune response. Activation of TLR9 (in particular on pDCs) plays a key role in modulating the adaptive Th2 immune response to an allergen. The profiles of cells expressing TLR9 are more restricted in humans (pDCs and B cells) compared to mice, where it is more broadly expressed on monocytes, myeloid dendritic cells, pDCs, and B cells.[4,5]

The immunostimulatory activity of ISS DNA is contained in a six-base pair CpG DNA sequence that follows the 5′–purine–purine–CpG–pyrimidine–pyrimidine–3′ formula.[4,5] This DNA sequence does not code for a protein, but rather exerts its biologic function by binding to TLR9 expressed on cells of the innate immune system.

The ability of ISS DNA to inhibit Th2 cells—key participants in allergic inflammation—is the rationale for investigating its effectiveness as a novel asthma therapy.

ISS DNA does not inhibit Th2 responses by binding to T cells (that do not express TLR9), but rather acts indirectly on T lymphocytes by inducing cells of the innate immune response (especially pDCs that express high levels of TLR9) to express cytokines that inhibit the Th2 response. The immune response to ISS DNA is characterized by the expression of Th1 cytokines including IFN-γ, IL-6, IL-10, IL-12, IL-18, and IFNs α and β.[4,5] In addition to inducing Th1 responses, CpG DNA inhibits expression of Th2 cytokines including IL-4, IL-5, IL-9, and IL-13 in the lungs, while also reducing levels of expression of Th2 cytokine receptors such as the IL-4 receptor.[4] Later studies demonstrated that ISS DNA also induces indoleamine-2,3-dioxygenase (IDO), an enzyme classically known for its role in the tryptophan degradation pathway.[6] Studies also demonstrated induction of IDO expression by ISS DNA in a mouse model of asthma.[6] Administration of ISS resulted in induction of IDO and inhibition of allergic inflammation, suggesting that IDO expression may play a role in limiting allergic inflammatory responses.[6] Administration of a pharmacologic inhibitor of IDO to mice treated with ISS inhibited IDO activity as assessed by inhibition of kynurenine levels in the lungs.[6] In parallel, the pharmacologic inhibitor of IDO inhibited the anti-allergic effect of ISS. These studies suggest that ISS may mediate its anti-allergic effect via induction of IDO and its metabolites.

ISS DNA AND MOUSE MODELS OF ALLERGEN-INDUCED ASTHMA

Studies in mouse models of allergen-induced asthma indicate that ISS DNA can prevent airway eosinophilic inflammation, Th2 cytokine responses, and increased airway hyperreactivity to methacholine.[7,8] It can also reverse episodes of established allergen-induced airway inflammation.[9] ISS DNA is effective in inhibiting eosinophilic airway inflammation when administered systemically or mucosally to the airway. Administration of a single dose of ISS systemically inhibited allergen-induced eosinophilic inflammation and Th2 cytokine responses for at least 4 weeks.

Interestingly, in preventive models of asthma in mice, a single dose of ISS DNA was shown to reduce airway eosinophilia and hyperreactivity as effectively as systemic administration of corticosteroids administered for 7 days.[7] The combination of ISS DNA and corticosteroid therapy was more effective in reducing airway hyperactivity than monotherapy with either agent alone.[9] Thus if ISS DNA were effective as a therapy in humans with asthma, it could potentially be used as a corticosteroid sparing agent.

ISS DNA AND MOUSE MODELS OF VIRAL-INDUCED ASTHMA

Episodes of asthma can be triggered by both allergens and viruses.[1] Since ISS DNA can activate the innate immune system to generate antiviral cytokines such as IFN-γ, studies have used mouse models of respiratory syncytial virus (RSV) infection, a virus associated with asthma in humans,[10] to determine whether ISS modulates episodes of viral-induced asthma. Administration of ISS to RSV-infected mice induced the expression of the IFN-γ antiviral cytokine in the lungs. This was associated with significantly reduced RSV viral titers (Figure 11.1), peribronchial inflammation (Figure 11.2), and airway mucus (Figure 11.3).[10]

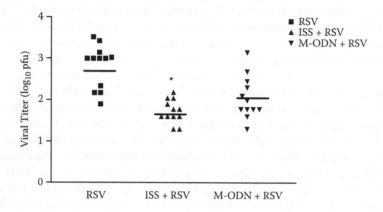

FIGURE 11.1 Effect of ISS DNA on RSV viral load in a mouse model of viral-induced asthma. ISS inhibits the number of RSV plaque-forming units ($\log_{10}$ scale) in the lungs of mice infected with RSV and treated with ISS compared with mice infected with RSV and treated with a control mutated oligodeoxynucleotide (M-ODN) lacking a CpG motif. (*Source:* Cho, J. Y. et al. 2001, *J. Allergy Clin. Immunol.*, 108: 697. With permission.)

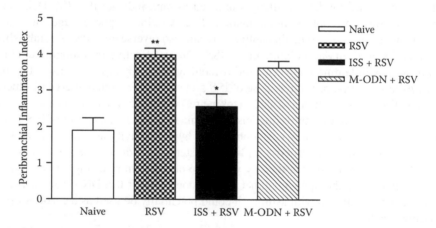

FIGURE 11.2 Effect of ISS DNA on RSV-induced peribronchial inflammation in a mouse model of viral-induced asthma. RSV infection increases the numbers of peribronchial inflammatory cells compared with numbers in uninfected naïve mice. ISS reduces the number of peribronchial inflammatory cells in the lungs of mice infected with RSV compared with mice infected with RSV and treated with a control mutated oligodeoxynucleotide (M-ODN) lacking a CpG motif. (*Source:* Cho, J. Y. et al. 2001, *J. Allergy Clin. Immunol.*, 108: 697. With permission.)

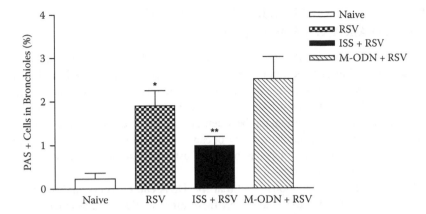

FIGURE 11.3   Effect of ISS DNA on RSV-induced mucus expression in a mouse model of viral-induced asthma. RSV infection increases levels of airway mucus (PAS+ cells) compared with levels in uninfected mice. ISS reduces the number of PAS+ cells in the airways of mice infected with RSV compared with mice infected with RSV and treated with a control mutated oligodeoxynucleotide (M-ODN) lacking a CpG motif. (*Source:* Cho, J. Y. et al. 2001, *J. Allergy Clin. Immunol.*, 108: 697. With permission.)

ISS DNA AND MOUSE MODELS OF ALLERGIC CONJUNCTIVITIS AND ALLERGIC RHINITIS

Allergic conjunctivitis involving itching, redness, and tearing of the eyes is a common manifestation of reactions to airborne allergens. Studies investigated whether administration of ISS DNA inhibits ragweed-induced allergic conjunctivitis in a mouse model.[11,12] Systemic or mucosal administration to the conjunctiva of ISS DNA inhibited both the immediate hypersensitivity response and late-phase cellular infiltration in the conjunctiva, and also induced a ragweed-specific Th1 immune response.[11] ISS administration was more effective than dexamethasone in inhibiting ocular allergic responses. The ISS-induced immunomodulatory effects were abolished when mice were treated with anti-IL-12 neutralizing antibodies, suggesting a role for IL-12 in the inhibition of both the immediate hypersensitivity and the late-phase cellular infiltration of the conjunctiva.

A second study evaluated the effect of ISS on clinical ocular allergy symptoms in a mouse model.[12] Topical administration of ragweed allergen to the conjunctiva in ragweed-sensitized wild-type mice induced symptoms of allergic conjunctivitis including conjunctival redness, eye lid edema and redness, tearing, and discharge. Administration of ISS significantly reduced the total conjunctival clinical response to ragweed and ocular symptoms. Histologic examination of the conjunctiva demonstrated that administration of ISS significantly inhibited conjunctival accumulation of eosinophils and neutrophils.

ISS DNA AND MOUSE MODELS OF ALLERGEN-INDUCED AIRWAY REMODELING

Asthma may be associated with structural changes in the airways known as airway remodeling.[13] Although the importance of airway remodeling to asthma is not yet completely understood, studies show that the rate of loss of lung function as assessed by FEV_1 decline is greater in non-smoking asthmatics compared to non-smoking healthy controls.[14] The relationship between airway remodeling and the loss of lung function in asthma is difficult to study because it would require repeated invasive bronchoscopies over time in large numbers of asthmatic subjects to assess levels of remodeling in airway biopsies.

Recent studies have demonstrated that ISS DNA can prevent[15] and reverse[16] airway remodeling in mice exposed to repetitive airway allergen challenges. Considerable evidence from mouse models of asthma and human studies indicates that airway remodeling is associated with expression of the profibrotic cytokine known as transforming growth factor (TGF)-β1.[13] In mouse models of allergen-induced airway remodeling, remodeling was significantly reduced by depleting eosinophils that express TGF-β1,[13] or by administering an anti-TGF-β antibody.[17]

Interestingly, studies in humans with asthma suggest an important role for eosinophil expression of TGF-β1 in airway remodeling.[18] Asthmatics treated with anti–IL-5 showed significant reductions in airway remodeling paralleled by significant reductions of bronchoalveolar lavage (BAL) eosinophils, BAL TGF-β1, and eosinophil expression of TGF-β1, suggesting that eosinophil expression of TGF-β1 played a significant role in remodeling.[18]

Since ISS inhibits eosinophilic airway inflammation, mouse models were studied to determine whether ISS DNA inhibits allergen-induced airway remodeling. These studies demonstrated that ISS DNA reduces accumulation of peribronchial myofibroblasts, peribronchial fibrosis (peribronchial collagen III and V deposition), thickness of the peribronchial smooth muscle layer, and airway hyperresponsiveness.[15,16] (See Figure 11.4 through Figure 11.6.) This effect on airway remodeling is likely due to the ability of ISS DNA to inhibit expression of profibrotic TGF-β1 by eosinophils and macrophages. Studies in IL-5–deficient mice demonstrated an important role for eosinophil expression of TGF-β1 in airway remodeling.[13] Thus, the ability of ISS DNA to reduce levels of eosinophil TGF-β1 expression may be one key mechanism by which it reduces airway remodeling. Such studies in mouse models of asthma suggest that ISS DNA prevents Th2-mediated airway inflammation in response to acute allergen challenge, and also inhibits TGF-β1 expression and airway remodeling associated with chronic allergen challenge.

ISS also modulates levels of peribronchial angiogenesis[19] and enzymes that degrade extracellular matrices such as matrix metalloproteinase (MMP)-9 and its tissue inhibitor of metalloproteinase (TIMP)-1.[20] Mice challenged with allergen by inhalation develop significantly increased levels of peribronchial angiogenesis (increases in CD31+ peribronchial small blood vessels) and increases in peribronchial vascular area.[19] The allergen-induced peribronchial angiogenesis is associated with

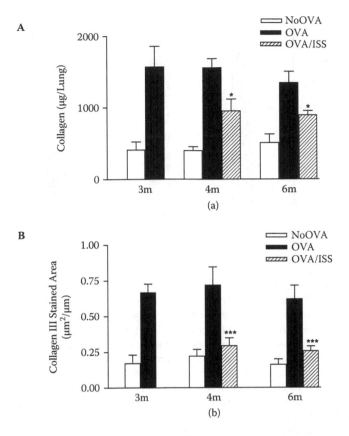

FIGURE 11.4 Effect of ISS DNA on chronic allergen exposure-induced total lung collagen (a) and peribronchial collagen III (b) in a mouse model of airway remodeling. Repetitive OVA challenge for 3 months in untreated mice induced significant increases in lung collagen (a) as well as peribronchial collagen III immunostaining (b) compared with non-OVA-challenged mice. Mice challenged with OVA for 3 months and then followed for a further 1 or 3 months without further OVA challenges continued to show significant increases in levels of lung collagen, as well as peribronchial collagen III compared with non-OVA-challenged mice. Institution of ISS treatment after 3 month of OVA challenges for 1 month (4 months total) or 3 months (6 months total) significantly reduced levels of lung collagen and peribronchial collagen III. (*Source:* Cho, J. Y. et al. 2004, *J. Immunol.,* 173: 7556. With permission.)

increased BAL levels of vascular endothelial growth factor (VEGF) and increased numbers of peribronchial cells expressing VEGF.

Treatment of mice with ISS before repetitive allergen challenge significantly reduced the levels of peribronchial angiogenesis and of BAL VEGF and the number of peribronchial cells expressing VEGF.[19] ISS is unlikely to act directly on endothelial cells to inhibit angiogenesis because lung endothelial cells do not express TLR9, the receptor for ISS as assessed by RT-PCR.[19] The ability of ISS to inhibit angiogenesis *in vivo* is likely to be mediated by several mechanisms, including its reduction

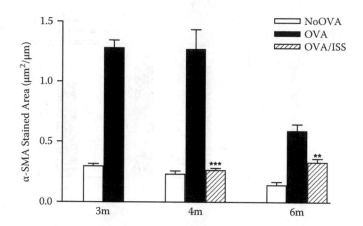

FIGURE 11.5 Effect of ISS DNA on chronic allergen exposure-induced peribronchial smooth muscle levels in a mouse model of airway remodeling. Repetitive OVA challenge for 3 months in untreated mice induced significant increases in areas of peribronchial α-smooth muscle actin immunostaining compared with non-OVA-challenged mice ($p = 0.0001$). Mice challenged with OVA for 3 months and followed without further OVA challenges continued to show significant increases in areas of peribronchial α-smooth muscle actin immunostaining compared with non-OVA-challenged mice at 1 month of follow-up (total 4 months; $p = 0.001$). At 3 months of follow-up, the area of peribronchial α-smooth muscle actin immunostaining was still significantly increased compared with non-OVA-challenged mice (total 6 months; $p = 0.001$), but was significantly less than that noted following 1 month after discontinuing OVA challenges. Institution of ISS treatment after 3 months of OVA challenges significantly reduced the area of peribronchial α-smooth muscle actin immunostaining in mice following 1 month of ISS treatment (total 4 months; $p = 0.0001$), as well as following 3 months of ISS treatment (total 6 months; $p = 0.01$). ^, $p = 0.05$. *, $p = 0.01$. **, $p = 0.001$. ***, $p = 0.0001$. (*Source:* Cho, J. Y. et al. 2004, *J. Immunol.,* 173: 7556. With permission.)

of the number of peribronchial inflammatory cells that express VEGF, inhibition of expression of Th2 cytokines such as IL-13 that promote VEGF expression, and direct effects on macrophages to inhibit VEGF expression.

MMP-9 and its TIMP-1 inhibitor are hypothesized to play a role in the pathogenesis of airway remodeling in asthma. Repetitive allergen challenges in mice induced significant increases in levels of MMP-9, TIMP-1, and peribronchial collagen deposition.[20] The patterns of expression of MMP-9 and TIMP-1 in the remodeled airways of mice are significantly different. MMP-9 but not TIMP-1 is expressed in airway epithelium, whereas both MMP-9 and TIMP-1 are expressed in peribronchial inflammatory cells.[20] ISS significantly reduced levels of expression of MMP-9 in airway epithelium and MMP-9 levels in peribronchial inflammatory cells. The ability of ISS to inhibit the expression of MMP-9 in airway epithelium (where its TIMP-1 inhibitor is not induced by allergen challenge) may be important in determining whether ISS contributes to reductions in airway remodeling by reducing levels of MMP-9.

Other mechanisms of action of ISS (i.e., on TGF-β1 expression) may be more important than its effect on MMP-9 in inhibiting airway remodeling. Studies of MMP-9-deficient mice demonstrated that MMP-9 may play a more modest role in mediating selected aspects of allergen-induced airway remodeling (i.e., modest reductions

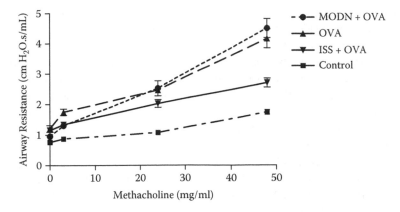

FIGURE 11.6 Effect of ISS DNA on chronic allergen exposure-induced airway responsiveness. Airway resistance was assessed in intubated and ventilated mice. Mice challenged repetitively with OVA for 3 months developed significant increases in airway resistance in response to MCh (48 mg/ml) challenge as compared with non-OVA-challenged control mice ($p = 0.01$). ISS significantly reduced airway resistance to MCh (48 mg/ml) in OVA-challenged mice ($p = 0.05$), whereas M-ODN had no effect on inhibiting OVA-induced increases in airway resistance to MCh. Circles, M-ODN + OVA. Triangles, OVA. Inverted triangles, ISS + OVA. Squares, no OVA. (*Source:* Cho, J. et al. 2004, *Am. J. Resp. Cell Mol. Biol.,* 30: 651. With permission.)

in levels of peribronchial fibrosis),[21] whereas MMP-9 does not play a significant role in mucus expression, smooth muscle thickness, or airway responsiveness.

ISS DNA AND PRIMATE STUDIES OF ALLERGEN-INDUCED ASTHMA AND AIRWAY REMODELING

To determine whether ISS DNA reduces airway hyperresponsiveness and airway remodeling in primates, house dust mite-allergic rhesus monkeys with experimentally induced allergic airway disease were treated with inhaled ISS DNA periodically for 8 months.[22] Airway hyperresponsiveness was reduced significantly in treated versus control allergic monkeys. Airways from ISS DNA-treated monkeys exhibited fewer mucous cells, eosinophils, and mast cells than control monkeys. In addition, the thicknesses of reticular basement membranes were reduced in ISS DNA-treated monkeys, suggesting an effect on airway remodeling. These studies suggest that ISS DNA inhibits Th2 responses and airway remodeling in mice, and inhibits airway inflammation and airway hyperreactivity in primates.

ISS DNA AND HUMAN STUDIES OF ALLERGEN-INDUCED ASTHMA

In humans and mice, TLR9 is the receptor for ISS DNA, and thus cell types that express TLR9 (in particular pDCs) respond to ISS DNA stimulation.[4,5] ISS DNA directly activates human pDCs to express increased levels of co-stimulatory molecules and cytokines. Studies of human PBMCs also demonstrated that ISS DNA inhibits Th2 responses.[4,5]

To determine whether ISS DNA produces immunomodulatory activities in subjects with asymptomatic mild allergic asthma, 40 individuals (ISS, 21; placebo, 19) were enrolled in a randomized, double-blind, placebo-controlled, parallel group study to examine the safety, pharmacologic activity, and efficacy of ISS related to allergen-induced airway responses.[23] Subjects received 36 mg of ISS DNA or placebo weekly via nebulization for 4 weeks. At 2 weeks and at the end of the study (4 weeks), subjects underwent inhalation allergen challenge to determine whether ISS inhibited immediate airway responses and/or late-phase airway responses to allergen inhalation challenge.

Treatment with ISS DNA significantly increased expression of interferon (IFN)-γ and IFN-inducible genes such as IFN-γ-inducible 10-kD protein (IP10), monokine induced by IFN-γ (MIG), IFN-stimulated gene (ISG)-54, monocyte chemotactic protein (MCP)-1, and MCP-2 in induced sputum cells as assessed by TaqMan PCR.[23] No attenuations of early- or late-phase decreases in FEV_1, reductions in allergen-induced sputum eosinophils, or Th2-related cytokine gene expression profiles measured in sputum cells were noted in the ISS versus placebo treated groups. ISS was safe and well tolerated by the inhalation route in these mild asthmatics.

Clinical studies evaluating new asthma therapies are also conducted in patients with persistent asthma symptoms to determine whether the therapy reduces symptoms, medication use, and several other outcome measures, depending upon the clinical trial design and duration. The fact that ISS DNA did not inhibit allergen challenge induced FEV_1 responses in mild asymptomatic asthmatics suggests that ISS DNA may not be effective for persistent asthma. However, clinical responses to ISS DNA in subjects with moderate persistent asthma may be different from responses of asymptomatic asthmatics challenged with allergens. Persistent symptoms are driven by low levels of allergens, viruses, pollutants, and other environmental stimuli that may not be accurately modeled with higher doses of inhaled allergen in an allergen challenge model. Further studies are needed to determine whether ISS is effective in patients with asthma symptoms instead of investigations using models of allergen challenge. More information is required to determine optimal dosing regimen and route of administration. Another determination to be made is whether conjugation of the allergen to ISS (shown to be effective in humans with allergic rhinitis immunized with a ragweed protein ISS DNA conjugate) is required.[24]

EFFECT OF CONJUGATION OF ISS DNA TO PROTEIN ALLERGEN IN MOUSE MODELS OF ALLERGEN-INDUCED ASTHMA

Several studies have investigated whether conjugation of ISS DNA to a protein allergen would enhance the immune response to the protein allergen. The rationale for conjugation of ISS DNA to a protein allergen is based on the hypothesis that the subcutaneously injected ISS DNA and the conjugated protein allergen would be engulfed by the same antigen-presenting cell (APC). The stimulation of TLR9 in APCs by ISS induces a strong Th1-biased immune response to the engulfed allergen. In contrast, when the ISS DNA and protein allergen are injected separately and not conjugated, the ISS DNA and the protein allergen will not always locate to the same APCs.

Studies of a fluorescent-labeled ISS DNA protein conjugate in mice demonstrated its enhanced uptake by APCs. Dendritic cells bound 100-fold more ISS DNA conjugated to protein antigen compared with ISS DNA alone.[25] ISS DNA conjugated to antigen, therefore, potentiates the ability of dendritic cells to promote Th1 cellular immune responses.[25,26] The ability of ISS DNA conjugated to a protein antigen to induce a Th1 immune response is 100-fold higher than that induced by a mixture of equivalent amounts of ISS DNA and protein antigen.[26] In mouse models of asthma, ISS DNA conjugated to an allergen protein significantly reduced airway hyperreactivity to methacholine for at least 2 months.[27] The ISS DNA allergen protein conjugate was significantly more efficient than the unconjugated mixture in reducing airway eosinophilia and bronchial hyperreactivity.

ISS DNA CONJUGATED TO RAGWEED PROTEIN AND ALLERGIC RHINITIS

Based on preclinical studies demonstrating that ISS DNA conjugated to a protein allergen has enhanced immunogenicity, ISS DNA was conjugated to the Amb-a-1 major ragweed protein in a therapeutic product called AIC that is undergoing clinical trials in subjects with ragweed allergies.[24] AIC inhibited Th2 and induced Th1 immune responses in PBMCs derived from human patients with ragweed allergies.[28] In *in vitro* studies using PBMCs, ISS DNA conjugated to a ragweed protein was more effective than ISS DNA alone for enhancing Th1 (e.g., IFN-γ production) and inhibiting Th2 (e.g., IL-4 production) responses.[28]

In *in vivo* studies, ragweed-allergic human subjects were immunized with a series of six weekly out-of-season injections of AIC to determine whether redirection from a Th2 to a Th1 immune response would occur.[29] Peripheral blood was used to assess immune responses to ragweed, streptokinase, and PHA before and after a course of AIC or placebo injections. AIC injections inhibited ragweed-specific Th2 responses and induced Th1 immune responses by PBMCs. AIC reduced levels of IL-5, Th2-dependent TARC (thymus and activation regulated chemokine, also known as CCL17), and MDC (macrophage-derived chemokine, also known as CCL22).[29] AIC induced ragweed-specific Th1 immune responses with significant increases in IFN-γ, the Th1-dependent chemokines Mig (monokine induced by IFN-γ, also known as CXCL9) and IP-10 (a 10-kDA IFN-inducible protein also known as CXCL10). Thus, AIC injected in concentrations approximately 40-fold lower than those used in most murine studies published to date led to a shift from Th2 immune responses to ragweed toward Th1 responses. Cytokine and chemokine responses to the unrelated streptokinase bacterial antigen streptokinase and the general ability to mount immune responses on polyclonal activation with PHA did not change. No clinically significant systemic or local allergic reactions were associated with AIC injections in these short-term studies with a 4-month follow-up.

Preliminary *in vivo* studies suggest that the ISS DNA conjugated to Amb-a-1 protein is less allergenic than the allergen protein based on immediate hypersensitivity skin test reactivity.[30] This reduced allergenicity is probably due to the physical properties of the four ISS DNA sequences in the AIC conjugate impeding binding of the Amb-a-1 allergen in the AIC conjugate to high affinity IgE receptors on mast cells, rather than a direct inhibitory effect of ISS DNA on mast cell degranulation. Although incubation of ISS DNA conjugated to Amb-a-1 protein also induced significantly lower basophil histamine release compared with ragweed alone,[30] this result was likely due to the physical properties of the conjugate. Incubation of ISS DNA *in vitro* with mouse bone marrow-derived mast cells (that express TLR9) did not inhibit allergen-induced mast cell degranulation,[31] suggesting that ISS DNA does not directly inhibit mast cell degranulation.

Preliminary studies with a six-injection regimen of the ISS DNA Amb-a-1 conjugate demonstrated that it reduces symptoms of allergic rhinitis in ragweed-sensitized human subjects followed through the ragweed season in some but not all studies.[24,32] In a study in Montreal, Canada, patients with ragweed allergies and allergic rhinitis were treated with six escalating doses of AIC (0.06 to 12 µg, n = 28) or placebo (n = 29) at weekly intervals immediately before the 2001 ragweed season.[32] Symptom scores and medication use were recorded for the 2001 and 2002 ragweed seasons. Nasal biopsy specimens were taken before AIC immunization (baseline) and twice after immunization (pre- and post-first ragweed season). The post-AIC immunization (pre- and post-season) specimens were taken 24 hours after ragweed nasal allergen challenge. The results of cytokine and inflammatory cell responses noted in the biopsies were compared with the initial unchallenged nasal biopsy specimens via immunocytochemistry and *in situ* hybridization.

The biopsies demonstrated significant decreases in nasal mucosal eosinophils and Th2 cytokines (i.e., IL-4) associated with significant increases in Th1 cytokines (IFN-γ) in AIC-treated compared with placebo-treated subjects.[32]

However, during the first ragweed season, no difference between treatment groups was observed in rhinitis symptom scores or medication use. During the second ragweed season, decreases in chest symptoms and a trend toward reduced nasal symptoms were noted in the AIC-treated group. AIC was safe and well tolerated by all patients. This study suggests that short-course immunotherapy with AIC can modify the response of the nasal mucosa to allergen challenge by decreasing Th2 cytokine production and eosinophilia. This modification of the immune response in the nasal mucosa was not immediate, but was observed 4 to 5 months after completion of AIC immunotherapy, and this may account for the delay in evidence of clinical efficacy to the second ragweed season.

In another study of AIC performed in subjects with allergic rhinitis in Baltimore (n = 25), those receiving AIC demonstrated significant clinical improvement that persisted for two ragweed seasons.[24] Patients received six weekly injections of AIC or a placebo vaccine before the first ragweed season and were monitored during the next two ragweed seasons. No patterns of vaccine-associated systemic reactions or clinically significant laboratory abnormalities were found. Although AIC did not alter the primary endpoint (nasal vascular permeability response to ragweed nasal provocation measured by albumin level in nasal lavage) during the first ragweed season, the AIC group had significantly better peak-season rhinitis scores on the visual analogue scale, peak-season daily nasal symptom diary scores, and mid-season overall quality-of-life scores than the placebo group.[24] Subjects on AIC used fewer rescue allergy medications during the ragweed season compared to subjects on placebo.

AIC induced a transient increase in Amb-a-1-specific IgG antibody, but suppressed the seasonal increase in Amb-a-1-specific IgE antibody. A reduction in the number of IL-4-positive basophils in AIC-treated patients correlated with lower rhinitis visual analogue scores. Clinical benefits of AIC were again observed in the subsequent ragweed season without additional AIC injections, with improvements over placebo in peak-season rhinitis visual analogue scores and peak-season daily nasal symptom diary scores.[24] Seasonal specific IgE antibody responses were again suppressed. In this pilot study, a 6-week regimen of AIC vaccine appeared to offer long-term clinical efficacy in the treatment of ragweed-allergic rhinitis. Further large scale studies are needed to determine the efficacy and safety of this approach.

SUMMARY

A variety of DNA-based vaccine strategies (ISS DNA alone, ISS DNA conjugated to an allergen protein) have demonstrated the potential of ISS DNA to prevent or reverse allergen-induced upper and lower airway inflammation, Th2 cytokine responses, mucus production, airway remodeling, and airway hyperreactivity in animal models. Preliminary studies in humans demonstrated the potential for immunotherapy with a conjugate of ISS DNA and ragweed protein to inhibit Th2 responses in nasal mucosa, as well as clinical responses in subjects with allergic rhinitis. Further ongoing studies in human subjects will determine whether any of these ISS DNA-based strategies are safe and effective for treating patients with allergic inflammation, including allergic rhinitis and asthma.

REFERENCES

1. Busse, W.W. et al. Asthma. *New Eng. J. Med.*, 2001, 344:350.
2. McFadden, E.R. Jr. Acute Severe Asthma. *Am J Respir Crit Care Med.*, 2003, 168:740.
3. Holgate, S.T. et al. New targets for allergic rhinitis—a disease of civilization. *Nat. Rev. Drug Discov.*, 2003, 2:902.
4. Horner, A.A. et al. Toll-like receptor ligands: hygiene, atopy and therapeutic implications. *Curr. Opin. Allergy Clin. Immunol.*, 2004, 4:555.
5. Krieg, A.M. Therapeutic potential of Toll-like receptor 9 activation. *Nat. Rev. Drug Discov.*, 2006, 5:471.
6. Hayashi, T. et al. Inhibition of experimental asthma by indoleamine 2,3-dioxygenase. *J. Clin. Invest.*, 2004, 114:270.
7. Broide, D et al. Immunostimulatory DNA sequences inhibit IL-5, eosinophilic inflammation, and airway hyperresponsiveness in mice. *J. Immunol.*, 1998, 161:7054.
8. Kline, J.N. et al. Modulation of airway inflammation by CpG oligodeoxynucleotides in a murine model of asthma. *J. Immunol.*, 1998, 160:2555.
9. Ikeda, R.K. et al. Resolution of airway inflammation following ovalbumin inhalation: comparison of ISS DNA and corticosteroids. *Am. J. Resp. Cell Mol. Biol.*, 2003, 28:655.
10. Cho, J.Y. et al. Immunostimulatory DNA sequences inhibit respiratory syncytial viral load, airway inflammation, and mucus secretion. *J. Allergy Clin. Immunol.*, 2001, 108:697.
11. Magone, M.T. et al. Systemic or mucosal administration of immunostimulatory DNA inhibits early and late phases of murine allergic conjunctivitis. *Eur J Immunol.*, 2000, 30:1841.
12. Miyazaki, D. et al. Prevention of acute allergic conjunctivitis and late-phase inflammation with immunostimulatory DNA sequences. *Invest. Ophthalmol. Vis. Sci.*, 2000, 41:3850.

13. Cho, J.Y. et al. Inhibition of airway remodeling in IL-5-deficient mice. *J. Clin. Invest.*, 2004, 113:551.

14. Lange, P. et al. A 15-year follow-up study of ventilatory function in adults with asthma. *N. Engl. J. Med.*, 1998, 339:1194.

15. Cho, JY et al. Immunostimulatory DNA inhibits transforming growth factor-beta expression and airway remodeling. *Am. J. Resp. Cell Mol. Biol.*, 2004, 30:651.

16. Cho, J.Y. et al. Immunostimulatory DNA reverses established allergen-induced airway remodeling. *J. Immunol.*, 2004, 173:7556.

17. McMillan, S.J. et al. Manipulation of allergen-induced airway remodeling by treatment with anti-TGF-beta antibody: effect on the Smad signaling pathway. *J. Immunol.*, 2005, 174:5774.

18. Flood-Page, P. et al. Anti-IL-5 treatment reduces deposition of ECM proteins in the bronchial subepithelial basement membrane of mild atopic asthmatics. *J. Clin. Invest.*, 2003, 112:1029.

19. Lee, S.Y. et al. Immunostimulatory DNA inhibits allergen-induced peribronchial angiogenesis in mice. *J. Allergy Clin. Immunol.*, 2006, 117:597.

20. Cho, J.Y. et al. Remodeling associated expression of matrix metalloproteinase 9 but not tissue inhibitor of metalloproteinase 1 in airway epithelium: modulation by immunostimulatory DNA. *J. Allergy Clin. Immunol.*, 2006. 117:618.

21. Lim, D.H. et al. Reduced peribronchial fibrosis in allergen-challenged MMP-9-deficient mice. *Am. J. Physiol. Lung Cell Mol. Physiol.*, 2006, 291:L265.

22. Fanucchi, M.V. et al. Immunostimulatory oligonucleotides attenuate airways remodeling in allergic monkeys. *Am. J. Respir. Crit. Care Med.*, 2004, 170:1153.

23. Gauvreau, G.M. et al. Immunostimulatory sequences regulate interferon genes but not allergic airway responses. *Am. J. Respir. Crit. Care Med.*, 2006, 174:15.

24. Creticos, P.S. et al. Immunotherapy with a Ragweed-Toll-Like Receptor 9 Agonist Vaccine for Allergic Rhinitis. *N. Engl. J. Med.*, 2006, 355:1445.

25. Shirota, H. et al. Novel roles of CpG oligodeoxynucleotides as a leader for the sampling and presentation of CpG-tagged antigen by dendritic cells. *J. Immunol.*, 2001, 167:66.

26. Tighe, H. et al. Conjugation of Immunostimulatory DNA to the short ragweed allergen amb a 1 enhances its immunogenicity and reduces its allergenicity. *J. Allergy Clin. Immunol.*, 2000, 106:124.

27. Santeliz, J.V. et al. Amb a 1-linked CpG oligodeoxynucleotides reverse established airway hyperreponsiveness in a murine model of asthma. *J. Allergy Clin. Immunol.*, 2002, 109:455.

28. Marshall, J.D. et al. Immunostimulatory sequence DNA linked to the Amb a 1 allergen promotes T(H)1 cytokine expression while downregulating T(H)2-cytokine expression in PBMCs from human patients with ragweed allergy. *J. Allergy Clin. Immunol.*, 2001, 108:191.

29. Simons, F.E. et al. Selective immune redirection in humans with ragweed allergy by injecting Amb a 1 linked to immunostimulatory DNA. *J. Allergy Clin. Immunol.*, 2004, 113:1144.

30. Creticos, P.S. et al. Immunostimulatory oligonucleotides conjugated to Amb a 1: Safety, skin test reactivity, and basophil histamine release. *J. Allergy Clin. Immunol.*, 2000, 105:S70.

31. Ikeda, R.K. et al. Accumulation of peribronchial mast cells in a mouse model of ovalbumin allergen induced chronic airway inflammation: modulation by immunostimulatory DNA sequences. *J. Immunol.*, 2003, 171:4860.

32. Tulic, M.K. et al. Amb a 1-immunostimulatory oligodeoxynucleotide conjugate immunotherapy decreases the nasal inflammatory response. *J. Allergy Clin. Immunol.*, 2004, 113:235.

12 Toll-Like Receptors in Development of Systemic Autoimmune Disease

In Vitro *Artifact or* In Vivo *Paradigm?*

Ann Marshak-Rothstein and Mark J. Shlomchik

ABSTRACT

Considerable *in vitro* data over the past decade have pointed to a critical role for TLR9 and TLR7 in the activation of autoantibody-producing B cells and IFN-α-producing plasmacytoid dendritic cells. However, it is only within the past year that the availability of the appropriate TLR-targeted autoimmune-prone strains has made it possible to formally test the Toll hypothesis *in vivo*. As anticipated, these studies have indicated that both TLR9 and TLR7 play an important role in the activation of autoreactive B cells. However, unexpectedly, TLR9 deficiency appeared to exacerbate rather than improve the overall level of clinical disease. Potential mechanisms and implications for further studies will be discussed in this review.

INTRODUCTION

The prevalence of autoantibodies reactive with chromatin and other nuclear and cytoplasmic ribonucleoproteins (RNPs) is a common feature of systemic lupus erythematosus (SLE) and other systemic autoimmune diseases such as scleroderma, Sjögren's syndrome, and even myositis.[1] Mutations that impair the ability to appropriately clear apoptotic debris often result in the development of autoimmunity.[2–7] However, as the common lupus target autoantigens are only some of the molecules that move to the surfaces of apoptotic cells, it has been more difficult to explain why only certain autoantigens are more dominant autoantibody targets in these diseases, or even why some but not all nuclear and cytoplasmic components can provoke such remarkably strong and consistent autoantibody responses.

This latter question has been brought into focus by the remarkable advances in the field of innate immunity that help explain the adaptive immune responses to microbial antigens. It is now well accepted that a wide range of hematopoietic cells express pattern recognition receptors (PRRs) that are capable of rapidly detecting pathogen-associated molecular patterns expressed by infectious agents.[8] These receptors serve as immune system sentinels by both activating the innate immune system to contain infections, and alerting the more flexible adaptive immune system to mount more comprehensive protective responses.[9] Indeed, it is generally thought that productive responses by the adaptive immune system require signals generated by the innate immune system. This paradigm provides a useful perspective for reexamining the adaptive immune responses to self that are the hallmarks of systemic autoimmunity.

Key to this issue are members of the Toll-like receptor (TLR) gene family that are among the best described PRRs. They include, TLR4, the receptor for endotoxin, and approximately 10 other family members specific for microbial products, or at least displaying relative specificity for microbial forms of such products.[10] Of particular relevance to this review is the recent discovery that some TLRs bind to products that may also be expressed in metazoan cells, even if they selectively bind microbial versions of these molecules. This is especially true for nucleic acid-specific TLRs. TLR9 was originally identified as the receptor specific for the hypomethylated CpG motifs commonly found in bacterial DNA, and is also activated by certain viral DNAs.[11] By comparison, TLR3 and TLR7 are thought to detect relatively unmodified viral ssRNA, and dsRNA, respectively.[12–15] These nucleic acid-specific TLRs are localized in cytoplasmic compartments, not on cell surfaces, and their activation is inhibited by agents such as chloroquine and bafilomycin A that interfere with late endosome/early lysosome acidification.[16,17]

While it was initially presumed that TLRs would maintain a rigid discrimination between self and non-self determinants, it has become increasingly apparent that at least some TLR family members respond to endogenous ligands associated with tissue injuries or damaged cells.[18–20] Moreover, as outlined below, *in vitro* studies have clearly shown that TLR9 and TLR7 can respond to mammalian nucleic acids, and that this reactivity can at least partly account for the activation of autoreactive B cells specific for nucleic acid-associated autoantigens.[21,22] These insights provide a novel perspective for the study of autoimmune disease. While autoimmunity was assumed for many years to represent a breakdown in the discriminatory properties of the adaptive immune system, it is also possible that autoimmune diseases result from overexuberant responses of the innate immune system. The analysis of TLR-deficient autoimmune-prone mice is the best way to test such a premise.

AM14 BCR TRANSGENIC MODEL AS PROTOTYPE FOR AUTOREACTIVE B CELLS

A major proportion of the B cells activated spontaneously by the autoimmune disease process in Fas-deficient mice express a B cell receptor specific for autologous IgG.[23] Representative examples of these rheumatoid factor (RF) expressing cells were captured as hybridoma cell lines,[24] and the heavy and light chains of the IgG2a$^{a/j}$-reactive AM14 line were used to create B cell receptor transgenic mouse lines.[25] AM14 B cells recognize monomeric IgG2a with relatively low affinity and, on a non-autoimmune-prone genetic background, apparently avoid tolerance induction and differentiate into mature B cells that produce little if any autoantibody.[25,26] However, on an autoimmune-prone background, AM14 B cells convert to plasmablast phenotype and actively secrete autoantibody.[27,28]

To determine the factors contributing to the activation of low affinity autoreactive B cells such as AM14, a variety of ligands were used to stimulate AM14 B cells *in vitro*. AM14 B cells responded normally to anti-IgM F(ab')$_2$ and other common TLR ligands such as lipopolysaccharide (LPS), CpG oligodeoxynucleotide 1826, and the synthetic compound R848, and ligands for TLR4, 9 and 7, respectively, and therefore they were not anergic by any detectable criteria. AM14 B cells were next assayed for their responses to cognate antigen (IgG2a) by using a panel of monoclonal antibodies (mAbs). Although AM14 B cells failed to proliferate in response to monomeric hapten-specific IgG2a mAbs, they proliferated robustly in response to IgG2a mAbs known to react with DNA, histones, and nucleosomes such as monoclonal antibody PL2-3. The unexpected response to the monoclonal autoantibodies most likely depended on the ability of these antibodies to bind DNA-associated autoantigens released into the culture medium by the purified B cells, as the capacity of these autoantibody/autoantigen immune complexes to stimulate AM14 B cells was blocked by the addition of DNAse to the culture medium.[21] It was initially assumed that the higher avidity of the complexed autoantibodies compared to the limited valency of the anti-hapten antibodies promoted BCR cross-linking that then triggered the proliferative response. However, if this were the case, it should have been possible to convert the anti-hapten antibodies into stimulatory ligands by simply increasing their valency. This was not the case. The anti-hapten antibodies failed to stimulate AM14 B cells, even when they were heat-aggregated or complexed to highly derivatized haptenated proteins.

The potential contribution of a second receptor was then considered. Based on the reported TLR dependency of a macrophage response to dead cell debris, the AM14 heavy and light chain genes were crossed onto a MyD88-deficient background, and the resulting AM14 MyD88-deficient B cells were compared to cells from MyD88-sufficient littermates. Remarkably, AM14 MyD88-deficient B cells completely failed to respond to PL2-3, thereby strongly linking TLRs to the recognition of endogenous ligands.

Subsequent studies involving endosome acidification inhibitors and TLR-deficient mouse strains further implicated TLR9, and to a lesser extent TLR7, in the response to PL2-3. The response of AM14 TLR9-deficient B cells was routinely decreased about 60 to 70% compared to AM14 TLR-sufficient cells; while the response of AM14 3D (Unc93-deficient and incapable of responding to TLR3, TLR7, or TLR9 ligands) mice was reduced to baseline.[29] These data suggest that in the absence of TLR9, other TLRs, and most likely TLR7, can contribute signals. Possible explanations for TLR7 involvement in the PL2-3 response include the potential association of this anti-histone antibody with histone-bound RNA or with apoptotic bodies that incorporate both histone-bound DNA and other RNA-associated autoantigens.

These studies suggested a novel paradigm. BCR recognition of an autoantigen results in the BCR-directed delivery of the autoantigen to a cytoplasmic compartment containing TLR9 (and/or TLR7), and a combination of signals resulting from the dual engagement of the BCR and TLR9 (and/or TLR7) is required to cross the signaling threshold and fully activate the B cell.

To further validate this hypothesis, it was necessary to extend the initial observations to additional B cell reactivities and better defined autoantigens. Fragments of conventional phosphodiester dsDNA approximately 600 bp in length were isolated from plasmids kindly provided by Dr. A. Krieg. The sequence of a fragment designated CG50 incorporated 50 CpG motifs; a second fragment designated CGneg had comparable C+G content but its sequence lacked CG dinucleotides and canonical CpG motifs. When B cells expressing the 3H9 transgene-encoded heavy chain that encodes DNA-reactive BCRs in combination with a variety of L chains[30,31] were stimulated with the CG50 and CGneg fragments, only the CG50 fragment induced a proliferative response. Similarly, hapten-conjugated NP-CG50, but not NP-CGneg, stimulated B cells expressing an NP-specific receptor. Importantly, the response to CG50 was entirely dependent on receptor-mediated uptake and TLR9, and therefore supported the BCR/TLR paradigm.[32]

To determine whether the BCR/TLR paradigm also applied to RNA and RNA-associated autoantigens, AM14 B cells were stimulated with mAbs reactive with U1RNA-binding protein SmD. These antibodies modestly activated AM14 B cells, but their relatively weak activity could be significantly improved by exogenously provided SmRNP (protein-bound RNA) as well as type 1 interferon (IFN). AM14 B cells also proliferated in response to a mAb directly reactive with RNA, and again this response was markedly increased by type 1 interferon. As predicted by the paradigm, the responses to the RNA-containing immune complexes were sensitive to RNAse and blocked by inhibitors of endosome acidification. These responses were also TLR-dependent, but here the relevant TLR was TLR7.[22] It seems likely that the major effect of type 1 interferon on the responses to RNA ICs resulted from the ability of type 1 interferons to dramatically upregulate TLR7 expression in B cells.[33]

DENDRITIC CELL ACTIVATION BY DNA AND RNA IMMUNE COMPLEXES

In most patients afflicted with SLE, the production of IFN-α by plasmacytoid dendritic cells (pDCs) appears to be a major pathogenic mechanism. Apart from its effects on TLR7 expression, IFN-α directly or indirectly promotes monocyte and dendritic cell maturation and antigen presentation, T cell cytotoxic activity, and B cell plasmablast differentiation.[34–36] DNA and RNA containing immune complexes were originally identified as critical inducers of pDC IFN production by Ronnblöm and colleagues.[37–39] Better defined RNA-containing ICs formed by combining purified small nuclear RNA protein (snRNP) macromolecular particles such as U1 RNA and its associated proteins with Sm-reactive very effectively activated pDCs.[40,41] As in the case of B cells, it is likely that cytokine production depends on receptor-mediated uptake of the immune complexes and delivery to an internal compartment containing a nucleic acid-specific receptor. For human pDCs, the relevant receptor is FcγRIIa (CD32).[42,43] FcγRIII has been shown to serve this function in murine conventional DCs.[44] *In vitro* studies utilizing TLR-deficient mouse strains and inhibitors specific for TLR9 or TLR7 linked TLR9 as well as TLR7 and TLR8 in the detection of these DNA- and RNA-containing ICs.[40,45]

YAA: A MUTATION THAT INCREASES COPY NUMBER OF TLR7

The Y-linked Yaa mutation was first described in 1979 as an accelerator of autoimmune disease that arose spontaneously during the development of the BXSB mouse line.[46] Despite many years of study, the genetic basis for this mutation was only recently determined. Yaa is the result of the duplication and Y chromosome translocation of a 4-Mb segment of the X chromosome pseudoautosomal region.[47,48] Importantly, TLR7 is one of the approximately 15 expressed genes included in this genetic interval. The extra copy of TLR7 results in a two-fold increase in the level of TLR7 protein expression and increased sensitivity to RNA-associated autoantigens.

As an autoimmune accelerator, Yaa dramatically exacerbated the clinical disease of SLE-prone B6 FcγRIIb-deficient mice, as evidenced by dramatically shortened lifespans and the rapid onsets of proteinuria and splenomegaly.[49] The impact of the Yaa mutation was also evident in B6 mice inheriting the sle1 SLE susceptibility locus. B6.sle1 mice spontaneously produced anti-nuclear antibodies, but never developed overt signs of clinical disease. By contrast, B6.Sle1.Yaa mice developed severe glomerulonephritis with a greater than 50% mortality rate by 9 months of age.[48] As might be anticipated by the association with TLR7, the autoantibody repertoires of both the FcγRIIb−/− Yaa and sle1 Yaa mice were more skewed toward the recognition of RNA-associated autoantigens than the corresponding non-Yaa strains.

The effect of the Yaa mutation has also been examined in BCR transgenic lines in which the BCR-encoded transgene can recognize specific autoantigens. The results have been variable. Yaa did not appear to change the phenotypes of mice expressing the Sp6 transgene, a receptor that binds DNA.[50] This observation was consistent with the premise that the amplification of TLR7 would be most likely to influence responses to RNA or RNA-associated proteins. However, the Yaa mutation produced a very dramatic effect when crossed to mice that expressed 4C8, a transgene-encoded receptor specific for red blood cells.[50] The 4C8.Yaa mice developed rapid onset lethal anemias and did not survive past 8 weeks of age. Non-Yaa 4C8 littermates had much more prolonged lifespans. The reason for elevated TLR7 expression triggering such a dramatic effect on anti-RBC antibody production is puzzling. Possible explanations include: (1) a yet-to-be-identified cross-reactivity between the 4C8 receptor and an RNA-associated molecule, and (2) a role in the development of autoimmune disease for a gene other than TLR7 in the Yaa interval.

EFFECT OF TLR DEFICIENCY ON *IN VIVO* PRODUCTION OF AUTOANTIBODIES

If the activation of B cells and pDCs by DNA- and RNA-containing immune complexes is dependent on TLR9 and TLR7, then TLR deficiency should exert a major effect on autoantibody production and disease progression in animal models of systemic autoimmune disease. Mice inheriting the lpr mutation are among the most commonly-studied models of SLE because they spontaneously produce autoantibodies with specificities similar to those found in human patients, particularly when crossed onto the MRL genetic background.[51,52]

Early studies designed to examine the role of TLRs in SLE evaluated autoantibody production in lpr/lpr mice lacking the MyD88 adaptor protein by immunofluorescent staining of HEp2 cells. Consistent with a major role for functional TLR7 and TLR9 signaling cascades in the production of autoantibodies reactive with DNA or RNA associated molecules, sera of the lpr/lpr MyD88-deficient mice did not stain HEp2 cells with a homogeneous or speckled nuclear pattern.[22] Such patterns are considered hallmarks of autoantibodies reactive with chromatin or snRNP, respectively. B6 FcγRIIb-deficient mice also developed an SLE-like disease associated with the production of antinuclear antibodies (ANAs), but when crossed to MyD88-deficient mice, the double knockouts showed much lower ANA titers.[53]

TLR9-deficient lpr/lpr mice have been used to more directly evaluate the contribution of TLR9 to the production of autoantibodies. They exhibited much more restricted defects in autoantibody production than MyD88-deficient lpr/lpr mice.[54,55] Sera from these mice failed to develop the homogeneous nuclear pattern typically found in TLR-sufficient lpr/lpr mice, and did not stain chromosomes in mitotic plates. In addition, these sera weakly stained or did not stain kinetoplasts of *Crithidia luciliae*. A kinetoplast is a giant mitochondrion located at the base of the trypanosome flagella, and kinetoplast staining is generally considered a definitive assay for antibodies reactive with dsDNA. However, many of the TLR9-deficient sera showed the speckled nuclear staining pattern of HEp-2 cells, and remarkably, many exhibited cytoplasmic staining patterns not generally seen in TLR9-sufficient mice. Overall, these staining data indicated that the TLR9-deficient mice produced autoantibodies specific for cytoplasmic and nuclear ribonucleoproteins, but failed to make dsDNA- or chromatin-specific antibodies.

Several groups used ELISA-based assays to quantify anti-DNA titers in TLR9-deficient autoimmune-prone mice and reached a different conclusion. TLR9-sufficient and -deficient MRL/lpr/lpr, B6/lpr/lpr, MRL/+/+ and Ali5 mice were found to produce comparable titers of antibodies reactive with plate-bound DNA.[55-58] The lack of homogeneous nuclear staining in these same samples, however, suggests that DNA ELISA detects a broader range of autoantibodies than dsDNA.

The role of TLR9 was further explored with mice that expressed the transgene-encoded 56R heavy chain associated with high affinity DNA-binding antibodies. 56R FcγRIIb-deficient TLR9-deficient mice produced much lower titers of IgG2a DNA-reactive autoantibodies than their TLR9-sufficient counterparts, consistent with a role for TLR9 in the production of T-bet, a transcription factor implicated in switching to the IgG2a isotype.[53] In contrast to the TLR9-deficient mice, TLR7-deficient lpr/lpr mice still produced autoantibodies that stained HEp-2 cells with the homogeneous nuclear pattern indicative of dsDNA or chromatin reactivity. These sera also strongly stained *C. luciliae* kinetoplasts and mitotic chromosomes. However, they failed to react with Sm/RNP and therefore appeared to have lost the capacity to bind RNA-associated autoantigens.[55] The impact of TLR7 deficiency was also shown by the analysis of a mouse strain that has targeted insertions of heavy and light chain genes that generate a BCR reactive with both ssDNA and RNA. Antibodies spontaneously produced by these B6 564Igi mice stained the cytoplasm and nucleoli of HEp-2 cells. These autoantibodies were not found in the sera of TLR7-deficient 564Igi mice.[59]

TLR9 LOSS OF FUNCTION: *IN VIVO* CONUNDRUM

Autoantibodies reactive with dsDNA and immune complexes containing these autoantibodies are strongly linked to the pathogenesis of SLE. Therefore the loss of antibodies reactive with dsDNA in TLR9-deficient autoimmune-prone mice was expected to reduce the extent of clinical disease in these mice. However, studies involving MRL/+, MRL/lpr/lpr, B6/lpr/lpr, and Ali5 mice have shown that this is not the case. Rather, the TLR9-deficient lines had higher numbers of activated T cells and antigen-presenting cells, in some cases developed more extensive renal disease, and had, at least for MRL/lpr/lpr, significantly reduced lifespans.[55-58] By contrast, TLR7-deficient MRL/lpr/lpr mice did not develop more severe disease than TLR-sufficient mice, and in fact showed evidence of decreased lymphocyte activation, reduced IgG titers, and ameliorated clinical disease.[55]

Overall, this *in vivo* analysis indicates a role for TLRs in the development of systemic autoimmune disease. Importantly, TLR9-deficiency appears to promote the development of autoimmune disease while TLR7-deficiency appears to limit the development of autoimmune disease. These results are particularly intriguing based on a recent clinical report correlating Type I IFN-responsive genes, SLE disease activity, and the presence of antibodies specific for any of the common RNP autoantigens.[60] Similarly, in TLR9-deficient MRL/lpr/lpr mice, disease severity and activation of pDCs did correlate with the presence of antibodies to RNA-containing autoantigens.[55] Moreover, TLR7 ligands (and not TLR9 ligands) were recently shown to stimulate higher levels of IFN-α production in females than males.[61] While further studies with additional autoimmune strains are needed to establish the generality of this TLR9/TLR7 dichotomy, the accumulated data point to TLR7 as a specific therapeutic target in systemic diseases such as SLE.

It is particularly puzzling that TLR7 and TLR9 exert such opposite effects on disease progression, considering that they generally are expressed on the same cell types, signal by the same pathway, and are highly homologous members of the TLR family.[62-65] This dichotomy may reflect the different specificities of TLR7 and TLR9 and the *in vivo* metabolisms of the corresponding autoantigens. Perhaps TLR9 plays a key role in the removal of cell debris because it contributes to the production of autoantibodies that promote such clearance.[6,66-68] If so, these TLR9-dependent auto-antibodies may be protective to a degree by removing the major autoantigenic trigger for the autoimmune response. Although in principle RNA-specific autoantibodies may also promote cell clearance, they may not be as effective because of the densities of autoantigens or their stability to ubiquitous RNAses.

Related to this is the potential for TLR7 and TLR9 cross-talk. Since both receptors signal via MyD88, signals through one TLR may be able to induce counter-regulators that affect both TLRs. A number of counter-regulators of TLR signaling are known, but less is known about TLR7 and TLR9 regulation in particular. Among potential pathways are IRF4, IRAK-M, SIKE, Fli-1, SOCS family members, and even possibly TGF-β in an IFN-α-dependent regulatory loop.[69-75] Perhaps TLR9 is subject to constitutive low level signals because of the constant presence of extracellular chromatin[76-78] and TLR7 is not.

The tuning of these receptors may well have evolved to take this into account. In the absence of TLR9, induction of counter-regulation may be diminished, leading to more potent signals through TLR7, thus exacerbating disease. Alternatively, TLR9 and TLR7 may directly compete for downstream signaling components, again with the result that TLR7 signals would be stronger in the absence of TLR9. In fact, one very recent report describes cross-inhibition of TLR7 by TLR9, although the mechanism has not been determined.[79] With regard to autoimmunity, the phenotypes of TLR7/9 double knockouts will be of interest to determine whether a genetic interaction between these two receptors controls disease severity.

At least in some cell types, the quality of the signal resulting from TLR9 ligands depends on the ligand as a result of spatiotemporal differences in ligand metabolism inside cells.[80] A similar situation may apply to TLR7, with its signaling perhaps of a different spatiotemporal nature, leading to a different biological outcome, despite using generally the same signaling mediators.

The expressions of TLR7 and TLR9, as mentioned above, are also differentially controlled upon cellular activation. While this has not been thoroughly studied, we know that TLR7 is more inducible by Type 1 IFN than TLR9.[22,33] Thus, signals through TLR7 may more efficiently engage a positive feedback regulatory loop involving Type 1 IFN secretion (by pDCs) as a result of TLR7 ligation.

Hematopoietic cells other than pDCs may contribute to disease pathogenesis and may preferentially express TLR7 or TLR9. Activated neutrophils and eosinophils have been reported to respond more strongly to TLR7 ligands than TLR9 ligands,[81,82] and therefore may theoretically respond more vigorously to RNA-containing immune complexes. Finally, regulatory function may be attributed to TLR9 via preferential induction of a regulatory cell or inhibitory cytokines such as IL-10 by TLR9 ligands.[83,84] Based on the importance of TLRs in regulating systemic autoimmunity and the mysterious paradoxical phenotype of TLR9-deficient autoimmune-prone mice, this is an area that requires and deserves much more effort to unravel.

REFERENCES

1. Tan, E. 1989. Antinuclear antibodies: diagnostic markers for autoimmune diseases and probes for cell biology. *Adv. Immunol.* 44: 93.
2. Lu, Q. and Lemke, G. 2001. Homeostatic regulation of the immune system by receptor tyrosine kinases of the Tyro 3 family. *Science* 293: 306.
3. Bickerstaff, M. C. et al. 1999. Serum amyloid P component controls chromatin degradation and prevents antinuclear autoimmunity. *Nat. Med.* 5: 694.
4. Hanayama, R. et al. 2004. Autoimmune disease and impaired uptake of apoptotic cells in MFG-E8-deficient mice. *Science* 304: 1147.
5. Maekawa, Y. and Yasutomo, K. 2001. Defective clearance of nucleosomes and systemic lupus erythematosus. *Trends Immunol.* 22: 662.
6. Ogden, C. A. et al. 2005. IgM is required for efficient complement mediated phagocytosis of apoptotic cells *in vivo. Autoimmunity* 38: 259.
7. Taylor, P. R. et al. 2000. A hierarchical role for classical pathway complement proteins in the clearance of apoptotic cells *in vivo. J. Exp. Med.* 192: 359.

8. Medzhitov, R. and Janeway, C. A. J. 2000. Innate immune recognition: mechanisms and pathways. *Immunol. Rev.* 173: 89.
9. Akira, S., Takeda, K., and Kaisho, T. 2001. Toll-like receptors: critical proteins linking innate and acquired immunity. *Nat. Immunol.* 2: 675.
10. Akira, S., Uematsu, S., and Takeuchi, O. 2006. Pathogen recognition and innate immunity. *Cell* 124: 783.
11. Hemmi, H. et al. 2000. A Toll-like receptor recognizes bacterial DNA. *Nature* 408: 740.
12. Alexopoulou, L. et al. 2001. Recognition of double-stranded RNA and activation of NF-kappaB by Toll-like receptor 3. *Nature* 413: 732.
13. Diebold, S. S. et al. Innate antiviral responses by means of TLR7-mediated recognition of single-stranded RNA. *Science* 303: 1529.
14. Lund, J. M. et al. 2004. TLR7: A new sensor of viral infection. *Proc. Natl. Acad. Sci. USA* 101: 6835.
15. Heil, F. et al. 2004. Species-specific recognition of single-stranded RNA via Toll-like receptor 7 and 8. *Science* 303: 1526.
16. Hacker, H. et al. 1998. CpG-DNA-specific activation of antigen-presenting cells requires stress kinase activity and is preceded by non-specific endocytosis and endosomal maturation. *EMBO J.* 17: 6230.
17. Ahmad-Nejad, P. et al. 2002. Bacterial CpG-DNA and lipopolysaccharides activate Toll-like receptors at distinct compartments. *Eur. J. Immunol.* 32: 1958.
18. Li, M. et al. 2001. An essential role of the NF-kappa-B/ Toll-like receptor pathway in induction of inflammatory and tissue-repair gene expression by necrotic cells. *J. Immunol.* 166: 7128.
19. Jiang, D. et al. 2005. Regulation of lung injury and repair by Toll-like receptors and hyaluronan. *Nat. Med.* 11: 1173.
20. Johnson, G. B., Brunn, G. J., and Platt, J. L. 2004. An endogenous pathway to systemic inflammatory response syndrome (SIRS)-like reactions through Toll-like receptor 4. *J. Immunol.* 172: 20.
21. Leadbetter, E. A. et al. 2002. Chromatin-IgG complexes activate autoreactive B cells by dual engagement of sIgM and Toll-like receptors. *Nature* 416: 603.
22. Lau, C. M. et al. 2005. RNA-associated autoantigens activate B cells by combined BCR/Toll-like receptor 7 engagement. *J. Exp. Med.* 202: 1171.
23. Shan, H. et al. 1994. The mechanism of autoantibody production in an autoimmune MRL/lpr mouse. *J. Immunol.* 153: 5104.
24. Wolfowicz, C. B. et al. 1988. Oligoclonality of rheumatoid factors arising spontaneously in lpr/lpr mice. *Clin. Immunol. Immunopathol.* 46: 382.
25. Shlomchik, M. J. et al. 1993. A rheumatoid factor transgenic mouse model of autoantibody regulation. *Int. Immunol.* 5: 1329.
26. Hannum, L. G. et al. 1996. A disease-related rheumatoid factor autoantibody is not tolerized in a normal mouse: implications for the origins of autoantibodies in autoimmune disease. *J. Exp. Med.* 184: 1269.
27. William, J. et al. 2002. Evolution of autoantibody responses via somatic hypermutation outside of germinal centers. *Science* 297: 2066.
28. William, J. et al. 2005. Visualizing the onset and evolution of an autoantibody response in systemic autoimmunity. *J. Immunol.* 174: 6872.
29. Busconi, L. et al. 2006. DNA and RNA autoantigens as autoadjuvants. *J. Endotoxin Res.* 12: 379.
30. Radic, M. Z. et al. 1991. Ig H and L chain contributions to autoimmune specificities. *J. Immunol.* 146: 176.
31. Radic, M. Z. and Weigert, M. 1994. Genetic and structural evidence for antigen selection of anti-DNA antibodies. *Annu. Rev. Immunol.* 12: 487.

32. Viglianti, G. A. et al. 2003. Activation of autoreactive B Cells by CpG dsDNA. *Immunity* 19: 837.
33. Bekeredjian-Ding, I. B. et al. 2005. Plasmacytoid dendritic cells control TLR7 sensitivity of naive B cells via type I IFN. *J. Immunol.* 174: 4043.
34. Ronnblom, L. and Alm, G. V. 2001. An etiopathogenic role for the type I IFN system in SLE. *Trends Immunol.* 22: 427.
35. Jego, G. et al. 2003. Plasmacytoid dendritic cells induce plasma cell differentiation through type I interferon and interleukin 6. *Immunity* 19: 225.
36. Banchereau, J., Pascual, V., and Palucka, A. K. 2004. Autoimmunity through cytokine-induced dendritic cell activation. *Immunity* 20: 539.
37. Vallin, H. et al. 1999. Anti-double-stranded DNA antibodies and immunostimulatory plasmid DNA in combination mimic the endogenous IFN-alpha inducer in systemic lupus erythematosus. *J. Immunol.* 163: 6306.
38. Ronnblom, L. and Alm, G. V. 2001. A pivotal role for the natural interferon a-producing cells (plasmacytoid dendritic cells) in the pathogenesis of lupus. *J. Exp. Med.* 194: F59.
39. Lovgren, T. et al. 2004. Induction of interferon-alpha production in plasmacytoid dendritic cells by immune complexes containing nucleic acid released by necrotic or late apoptotic cells and lupus IgG. *Arthritis Rheum.* 50: 1861.
40. Vollmer, J. et al. 2005. Autoantigen binding sites within small nuclear RNAs induce innate immunity through Toll-like receptors 7 and 8. *J. Exp. Med.* 202: 1575.
41. Lovgren, T. et al. 2006. Induction of interferon-alpha by immune complexes or liposomes containing systemic lupus erythematosus and Sjogren's syndrome autoantigen-associated RNA. *Arthritis Rheum.* 54: 1917.
42. Bave, U. et al. 2003. Fc gamma RIIa is expressed on natural IFN-alpha-producing cells (plasmacytoid dendritic cells) and is required for the IFN-alpha production induced by apoptotic cells combined with lupus IgG. *J. Immunol.* 171: 3296.
43. Means, T. K. et al. 2005. Human lupus autoantibody-DNA complexes activate DCs through cooperation of CD32 and TLR9. *J. Clin. Invest.* 115: 407.
44. Boule, M. W. et al. 2004. Toll-like receptor 9-dependent and -independent dendritic cell activation by chromatin–immunoglobulin G complexes. *J. Exp. Med.* 199: 1631.
45. Barrat, F. J. et al. 2005. Nucleic acids of mammalian origin can act as endogenous ligands for Toll-like receptors and may promote systemic lupus erythematosus. *J. Exp. Med.* 202: 1131.
46. Murphy, E. D. and Roths, J. B. 1979. A Y chromosome associated factor in strain BXSB producing accelerated autoimmunity and lymphoproliferation. *Arthritis Rheum.* 22: 1188.
47. Pisitkun, P. et al. 2006. Autoreactive B cell responses to RNA-related antigens due to TLR7 gene duplication. *Science* 312: 1669.
48. Subramanian, S. et al. 2006. A Tlr7 translocation accelerates systemic autoimmunity in murine lupus. *Proc. Natl. Acad. Sci. USA* 103: 9970.
49. Bolland, S. et al. 2002. Genetic modifiers of systemic lupus erythematosus in Fc-gamma-RIIB(–/–) mice. *J. Exp. Med.* 195: 1167.
50. Moll, T. et al. 2005. Differential activation of anti-erythrocyte and anti-DNA autoreactive B lymphocytes by the Yaa mutation. *J. Immunol.* 174: 702.
51. Pisetsky, D. S. et al. 1982. lpr gene control of the anti-DNA antibody response. *J. Immunol.* 128: 2322.
52. Eisenberg, R. A. et al. 1989. The genetics of autoantibody production in MRL/lpr lupus mice. *Clin. Exp. Immunol.* 7/S-3: 35.
53. Ehlers, M. et al. 2006. TLR9/MyD88 signaling is required for class switching to pathogenic IgG2a and 2b autoantibodies in SLE. *J. Exp. Med.* 203: 553.
54. Christensen, S. R. et al. 2005. Toll-like receptor 9 controls anti-DNA autoantibody production in murine lupus. *J. Exp. Med.* 202: 312.

55. Christensen, S. R. et al. 2006. TLR7 and TLR9 dictate autoantibody specificity and have opposing inflammatory and regulatory roles in lupus. *Immunity* 25: 417.
56. Wu, X. and Peng, S. L. 2006. Toll-like receptor 9 signaling protects against murine lupus. *Arthritis Rheum.* 54: 336.
57. Yu, P. et al. 2006. Toll-like receptor 9-independent aggravation of glomerulonephritis in a novel model of SLE. *Int. Immunol.* 18: 1211.
58. Lartigue, A. et al. 2006. Role of TLR9 in anti-nucleosome and anti-DNA antibody production in lpr mutation-induced murine lupus. *J. Immunol.* 177: 1349.
59. Berland, R. et al. 2006. TLR7-dependent loss of B cell tolerance in pathogenic autoantibody knock-in mice. *Immunity* 25: 429.
60. Hua, J. et al. 2006. Functional assay of type I interferon in systemic lupus erythematosus plasma and association with anti-RNA binding protein autoantibodies. *Arthritis Rheum.* 54: 1906.
61. Berghofer, B. et al. 2006. TLR7 ligands induce higher IFN-alpha production in females. *J. Immunol.* 177: 2088.
62. Dunne, A. and O'Neill, L. A. 2005. Adaptor usage and Toll-like receptor signaling specificity. *FEBS Lett.* 579: 3330.
63. Hemmi, H. et al. 2003. Role of Toll-like receptor 9, MyD88, and DNA-dependent protein kinase catalytic subunit in the effects of two distinct CpG DNAs on dendritic cell subsets. *J. Immunol.* 170: 3059.
64. Kawai, T. and Akira, S. 2005. Toll-like receptor downstream signaling. *Arthritis Res. Ther.* 7: 12.
65. Kawai, T. and Akira, S. 2006. TLR signaling. *Cell Death Differ.* 13: 816.
66. Mevorach, D. et al. 1998. Systemic exposure to irradiated apoptotic cells induces autoantibody production. *J. Exp. Med.* 188: 387.
67. Werwitzke, S. et al. 2005. Inhibition of lupus disease by anti-double-stranded DNA antibodies of the IgM isotype in the NZB/NZW F1 mouse. *Arthritis Rheum.* 52: 3629.
68. Witte, T. et al. 1998. IgM anti-dsDNA antibodies in systemic lupus erythematosus: negative association with nephritis. *Rheumatol. Int.* 18: 85.
69. Chow, E. K. et al. 2005. TLR agonists regulate PDGF-B production and cell proliferation through TGF-beta/type I IFN crosstalk. *EMBO J.* 24: 4071.
70. Honma, K. et al. 2005. Interferon regulatory factor 4 negatively regulates the production of proinflammatory cytokines by macrophages in response to LPS. *Proc. Natl. Acad. Sci. USA* 102: 16001.
71. Huang, J. et al. 2005. SIKE is an IKK epsilon/TBK1-associated suppressor of TLR3- and virus-triggered IRF-3 activation pathways. *EMBO J.* 24: 4018.
72. Kobayashi, K. et al. 2002. IRAK-M is a negative regulator of Toll-like receptor signaling. *Cell* 110: 191.
73. Negishi, H. et al. 2005. Negative regulation of Toll-like-receptor signaling by IRF-4. *Proc. Natl. Acad. Sci. USA* 102: 15989.
74. Wang, T. et al. 2006. Flightless I homolog negatively modulates the TLR pathway. *J. Immunol.* 176: 1355.
75. Yoshimura, A. et al. 2004. Regulation of TLR signaling and inflammation by SOCS family proteins. *J. Leukoc. Biol.* 75: 422.
76. Tan, E. M. et al. 1966. Deoxybonucleic acid (DNA) and antibodies to DNA in the serum of patients with systemic lupus erythematosus. *J. Clin. Invest.* 45: 1732.
77. Bjorkman, L., Reich, C. F., and Pisetsky, D. S. 2003. Use of fluorometric assays to assess the immune response to DNA in murine systemic lupus erythematosus. *Scand. J. Immunol.* 57: 525.
78. Jiang, N. et al. 2003. The expression of plasma nucleosomes in mice undergoing in vivo apoptosis. *Clin. Immunol.* 106: 139.

79. Wang, J. et al. 2006. Functional effects of physical interactions among Toll-like receptors 7, 8, and 9. *J. Biol. Chem.* 281: 37427.
80. Honda, K. et al. 2005. Spatiotemporal regulation of MyD88-IRF-7 signalling for robust type-I interferon induction. *Nature* 434: 1035.
81. Nagase, H. et al. 2003. Expression and function of Toll-like receptors in eosinophils: activation by Toll-like receptor 7 ligand. *J. Immunol.* 1771: 3977.
82. Hayashi, F., Means, T. K., and Luster, A. D. 2003. Toll-like receptors stimulate neutrophil function. *Blood* 102: 2660.
83. Moseman, E. A. et al. 2004. Human plasmacytoid dendritic cells activated by CpG oligodeoxynucleotides induce the generation of CD4+CD25+ regulatory T cells. *J. Immunol.* 173: 4433.
84. Lenert, P. et al. 2005. TLR-9 activation of marginal zone B cells in lupus mice regulates immunity through increased IL-10 production. *J. Clin. Invest.* 25: 29.

13 Impacts of Nucleoside Modification on RNA-Mediated Activation of Toll-Like Receptors

Katalin Karikó and Drew Weissman

ABSTRACT

DNA and RNA stimulate the mammalian innate immune system by triggering a variety of nucleic acid-responsive sensors including Toll-like receptors (TLRs). TLR9 present in the endosomal compartments of immune cells responds to DNA. TLR3 is activated by dsRNA, while TLR7 and human TLR8 signal when exposed to single-stranded RNA. Most DNA and RNA from natural sources contain modified nucleosides. In eukaryotes, modification of DNA is limited to 5-methylcytidine of CpG motifs; whereas in RNA more than 100 different types of modified nucleosides have been identified.

The extent and quality of these modifications depend on RNA subtype and correlate directly with the evolutionary level of the organism from which the RNA is isolated. Accordingly, eukaryotic RNAs contain much higher levels of modified nucleosides than prokaryotic RNAs. In general, the biological significance of nucleoside modifications present in DNA and RNA is not well understood. However, one role for 5-methylcytidines in CpG motifs of DNA is well established. DNA that contains such modifications does not activate TLR9. The question of whether an analogous effect can be attributed to nucleoside modifications in RNA has been addressed only recently. This chapter characterizes a few naturally occurring nucleoside modifications of RNA and their effects on the capacity of RNA to activate immune cells and TLR-transformed cell lines.

INTRODUCTION

It has been known for decades that "foreign" nucleic acids such as bacterial DNA and viral double-stranded (ds) RNA, are potent adjuvants of the mammalian immune system.[1] However, the molecular mechanisms underlying recognition of nucleic acids by the immune system were not understood until recently. We now know that nucleic acids stimulate immune responses via a number of signaling pathways that can be classified as Toll-like receptor (TLR)-dependent or TLR-independent.

TLRs are pattern-recognition receptors of the innate immune system that respond to foreign pathogen-associated molecular products and host-related danger signals.[2] Eleven TLR family members have been identified in mice and ten in humans. The ligand specificities of most TLRs have been identified. A subset of TLRs is specialized for responding to nucleic acids. For example, TLR9 responds to unmethylated CpG motifs characteristic of bacterial and selected viral DNA.[3] dsRNA, a common viral intermediate, has been shown to activate TLR3.[4] Viral-related and synthetic single-stranded (ss) RNAs activate TLR7 and TLR8 on human cells and TLR7 on murine cells (Table 13.1).[5–7] Based on sequence and structural similarities, TLR7, TLR8, and TLR9 form a distinct subfamily. Activation of these receptors and of TLR3[8] depends upon endosomal acidification and leads to production of type I interferons and pro-inflammatory cytokines.

It has recently been discovered that nucleic acids can also induce immune responses via TLR-independent pathways.[9] Two RNA helicases, retinoic acid inducible gene-I (RIG-I)[10] and melanoma differentiation-associated gene 5 (MDA5),[11] are cytoplasmic sensors of dsRNA, while the identity of the cytoplasmic DNA sensor has not yet been established.[12]

dsRNA, in addition to triggering pro-inflammatory signaling cascades that lead to type I interferon production, also directly activates interferon-induced enzymes. These enzymes, including RNA-dependent protein kinase (PKR), RNA-specific

TABLE 13.1
RNA-Responsive Proteins of Innate Immune System

RNA Sensors	Activator RNA	Ref.	Function
Regulators			
TLR3	dsRNA, *in vitro* transcript	4,16	Antiviral and inflammatory
TLR7	ssRNA (oligomer), polyU, *in vitro* transcript	5–7	Antiviral and inflammatory
TLR8	ssRNA (oligomer), *in vitro* transcript	6,7	Antiviral and inflammatory
RIG-I	dsRNA, *in vitro* transcript	10,17	Antiviral and inflammatory
MDA5	dsRNA	11	Antiviral and inflammatory
Effectors			
PKR	dsRNA	18	Inhibition of protein synthesis
	TNF-α mRNA	14	TNF-α production
OAS [RNaseL]	dsRNA	19	Antiviral: degradation of mRNA
ADAR [I-RNase]	dsRNA	20,21	Antiviral: degradation of mRNA
Nalp3 [Caspase-1]	Bacterial RNA	15	IL-1β production

Note: [ ], Enzyme that directly performs function. TLR, Toll-like receptor. RIG-I, retinoic acid inducible gene-I. MDA5, melanoma differentiation-associated gene 5. PKR, RNA-dependent protein kinase. OAS, 2′-5′ oligoadenylate synthetase. RNaseL, latent 2′-5′ oligoadenylate-dependent RNase. ADAR, RNA-specific adenosine deaminase. I-RNase, inosine-dependent RNase. TNF, tumor necrosis factor. Nalp3, NACHT, leucine-rich repeat and pyrin domain-containing protein. Caspase, cysteine-aspartic-acid-protease.

adenosine deaminase (ADAR), and 2'-5'oligoadenylate synthetase (OAS), carry out antiviral functions in association with other enzymes, such as RNaseL and I-RNase.[13] RNA-activated enzymes also participate in the processing of inflammatory cytokines. Splicing of TNF-α mRNA, for example, requires mRNA-mediated PKR activation,[14] while maturation of IL-1β needs functional caspase-1 that can be generated by bacterial RNA-activated Nalp3 (Table 13.1).[15]

Thus "foreign" nucleic acids stimulate immune defenses at multiple levels. They act as *regulators* by inducing inflammatory signaling pathways leading to type I interferon and pro-inflammatory cytokine release. They can also function as *effectors* by activating enzymes that promote or execute antiviral immune functions (Table 13.1). The methylation state of the cytidine in CpG motifs is the structural basis of TLR9 activation by bacterial DNA but not by mammalian DNA.[3] This raises the possibility that in analogy, RNA-responsive TLRs and other RNA sensors with regulatory or effector immune functions may react differently when nucleosides in the activating RNA are modified. RNA has more than 100 different types of nucleoside modifications,[22] some of which are uniquely present in eukaryotic RNA, others in prokaryotic RNA, and most are present in both types. In general, mammalian RNA is more abundantly modified than bacterial RNA.[23] Thus, the extent and quality of nucleoside modifications may alter the immune potency of RNA and may serve as a means to differentiate between host and pathogenic RNA.

NUCLEOSIDE-MODIFIED RNA

All natural RNAs are synthesized from four basic ribonucleotides (ATP, CTP, UTP and GTP); some incorporated nucleosides are then modified post-transcriptionally as part of the RNA maturation process. Most modifications involve methylation of the nucleobase or 2'-*O*-methyl formation on the ribose, but some modifications are more complex and include additions of amino acid (glycine, threonine) or sugar (galactose, mannose) derivatives. Adducts are covalently linked to the carbon or nitrogen atoms of the nucleobase. Nucleoside modifications are considered irreversible. They occur in conserved positions of the polynucleotide chain and in many cases the modifying enzymes are so specialized they can perform the reaction only at one uniquely positioned nucleoside. In general, modifications that do not interfere with Watson-Crick hydrogen bond formation improve the structural stability of the RNA.[24] A comprehensive listing of post-transcriptionally modified nucleosides appears on the RNA Modification Database (<http://medlib.med.utah.edu/RNAmods/>).

Nucleosides in RNA can also be modified via damage by chemicals (e.g., alkylating and oxidative agents) derived from the environment or generated within cells.[25,26] The resulting modifications are different from those generated during RNA maturation. Special enzymes can repair some altered nucleosides in a damaged RNA by reversing alkylation, especially on oxygen atoms of the nucleobase.[27] We will not discuss the impacts of damage-related nucleoside modifications because their immune-stimulatory effects have not been investigated. We will discuss only modifications related to RNA immunity.

Modifications of Ribose

2′-O-methylnucleotide (Nm) — The only known natural ribose modification of RNA is methylation of the 2′-hydroxyl group (Figure 13.1). In eukaryotes, this modification is performed by fibrillarin in association with accessory molecules including a guiding small nucleolar RNA (snoRNA) or Cajal body (sca) RNA that determines nucleoside position within the polynucleotide chain that will undergo 2′-O-methylation.[28] Before discovering its methyltransferase activity, fibrillarin was known for years as a marker recognized by antisera from patients suffering from the autoimmune disease scleroderma.[29] 2′-O-methylnucleosides are present in all types of eukaryotic RNAs including tRNA, rRNA, mRNA, and small nuclear (sn) RNA. 2′-O-methyl modification stabilizes base stacking interactions and increases resistance of RNA to nuclease degradation.[30]

Modifications of Nucleobase

Pseudouridine (Ψ) — Fifty years ago, the first modified nucleoside, pseudouridine, was discovered in RNA.[22] Pseudouridine is generated by 5-ribosyl isomerization of uridine that involves breaking the C–N glycosidic bond, rotating the uridine ring, and forming a C–C glycosidic bond at position 5 (Figure 13.2). In eukaryotes, this reaction is catalyzed by pseudouridine synthase, acting alone[31] or under the guidance of snoRNA/scaRNA and other associated proteins.[32] The presence of Ψ in

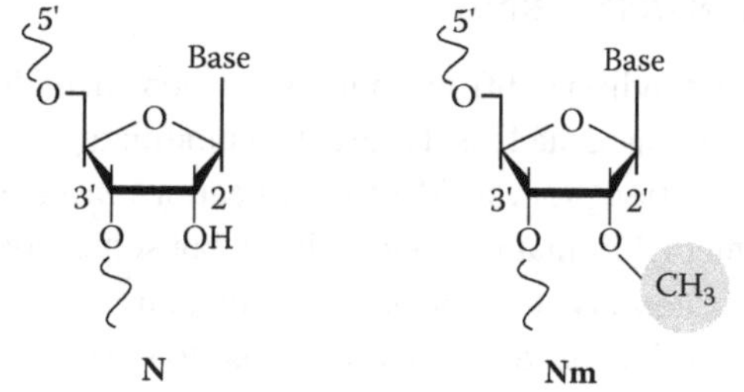

FIGURE 13.1 Structures of a ribonucleoside (N) and a 2′-O-methylnucleoside (Nm).

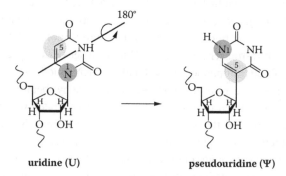

FIGURE 13.2 Uridine isomerization by pseudouridine synthase yields pseudouridine.

RNA promotes base stacking, thereby stabilizing RNA conformations.[33] Pseudouridine is the most abundant modified nucleoside in RNA.

5-Methyluridine (m5U) — Methylation of nucleobases is catalyzed by a number of poorly characterized S-adenosyl-L methionine-dependent methyl transferases. Methylation of uridine at the 5-position of the pyrimidine ring generates m5U, also known as ribothymidine (rT; Figure 13.3).

Homopolymers of m5U form highly ordered structures, suggesting that this nucleoside modification may act to stabilize RNA conformation.[34]

2-Thiouridine (s2U) — Thiolation of uridine at the C-2 position (Figure 13.3) occurs uniquely in tRNAs, and requires cysteine and ATP as substrates.[35] Presence of s2U in RNA strongly stabilizes secondary structures.[36,37]

5-Methylcytidine (m5C) — 5-Methylcytosine is the only modified base that occurs in both DNA and RNA and is also found in organisms from all three domains of life.[38] Introduction of a methyl group at the 5 position of cytidine (Figure 13.4) significantly enhances stacking interactions within the RNA.[39]

N6-Methyladenosine (m6A) — m6A in RNA (Figure 13.5) is generated by MT-A70 methyltransferase (EC 2.1.1.62) that is highly expressed in immune cells (<http://symatlas.gnf.org>). MT-A70 is one of the few nucleoside-modifying enzymes characterized in humans.[40] N6-methyladenosine is the only base-modified nucleoside present in all RNA types including rRNA, tRNA, and snRNA, and also in mRNAs of cellular and viral origins. The methylation in m6A interferes with Watson-Crick base-pairing. Thus, its presence destabilizes double-stranded regions of RNA.[41]

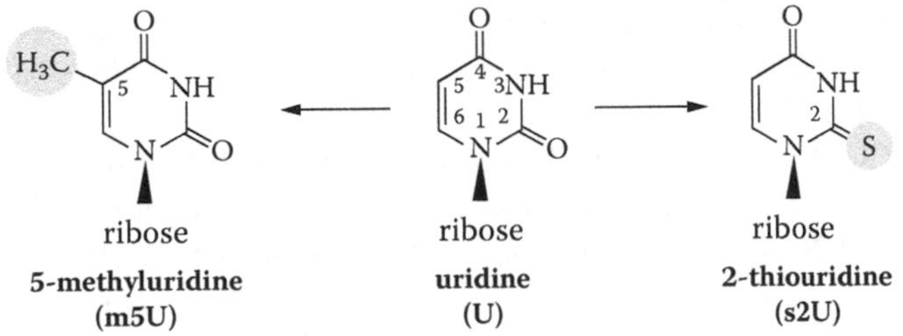

FIGURE 13.3 Structures of uridine and its modified derivatives 5-methyluridine and 2-thiouridine.

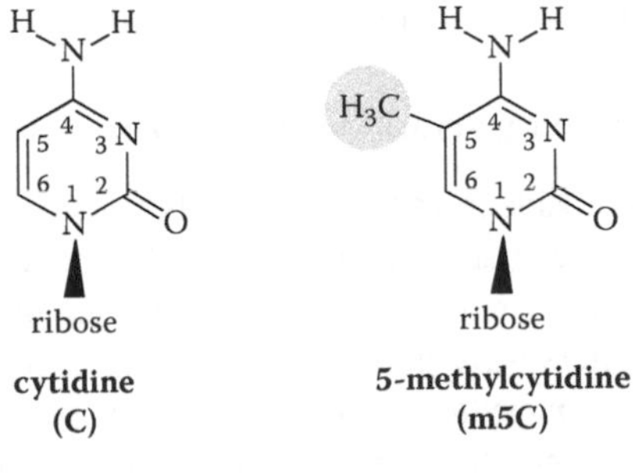

FIGURE 13.4 Structures of cytidine and 5-methylcytidine.

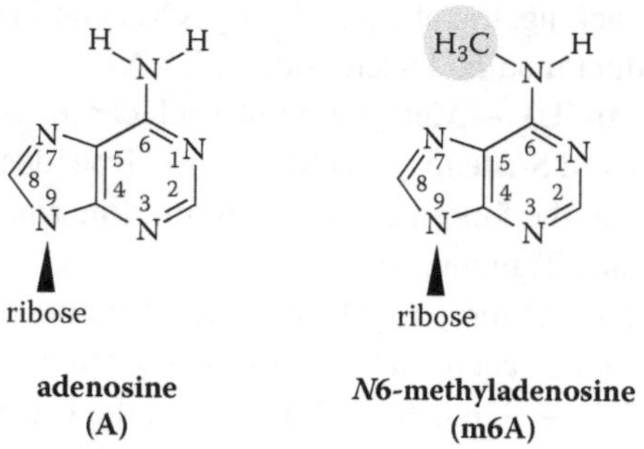

FIGURE 13.5   Structures of adenosine and N6-methyladenosine.

NUCLEOSIDE MODIFICATIONS IN NATURAL RNAS

Nucleoside modifications are most diverse in tRNA and much of our knowledge about modified nucleosides comes from their study. In ribosomal RNA, which is the major constituent of cellular RNA, fewer types of modifications are present (Table 13.2). The modification motifs and their sequence locations tend to be conserved and most of the modifying enzymes remain poorly characterized. The extent and nature of modifications vary and depend on the RNA type and the evolutionary level of the organism from which the RNA is derived. Ribosomal RNA (rRNA) contains significantly more nucleotide modifications when obtained from mammalian cells than bacteria. For example, the pseudouridine (Ψ) content is approximately 10 times greater, and the 2'-O-methylated nucleoside content is approximately 25 times greater in human rRNA compared to bacterial rRNA. Ribosomal RNA derived from mitochondria, a cellular organelle evolved from eubacteria,[42] exhibits only one modification[43] (Table 13.3).

Transfer RNA is the most heavily modified subgroup of RNA in cells. In mammalian tRNAs, up to 25% of the nucleosides can be modified. Fewer modifications occur in prokaryotic tRNAs. Bacterial messenger RNA contains no nucleoside modifications, while mammalian mRNAs contain m5C, m6A, inosine, and many 2'-O-methylated nucleosides, in addition to N7-methylguanosine (m7G) which is part of the 5' terminal cap structure.[23,44] Modified nucleosides (m6A, m5C and Nm) also occur in the internal regions of many viral RNAs including those of influenza,[45] adenovirus,[46] HSV,[47] SV40,[48] and RSV.[49] Surprisingly, modified nucleosides including m6A and m5C occur more frequently in some viral RNAs than in cellular mRNAs.[46]

SYNTHESIZING RNA WITH NUCLEOSIDE MODIFICATION

Attempts to perform nucleoside modification on RNA *in vitro* by emulating the natural mechanisms have not been successful because the process is complex and most of the modifying enzymes are not available. Decades ago, modified nucleoside-containing RNA homopolymers were generated by polymerizing modified nucleotides with polynucleotide phosphorylase.[50] More recently, phage-derived RNA polymerases were used to incorporate modified nucleotides in transcription reactions *in vitro*. Employing this method, m5C-, m6A-, Ψ-, m5U- or s2U-containing RNA could

TABLE 13.2

Distributions of Different Types of Nucleoside Modifications in RNA

		Type of Modification	
RNA	% Total Cellular RNA	Eukaryote	Prokaryote
rRNA	80	23	18
tRNA	15	50	45
mRNA	3 to 5	13	0
snRNA	<2	11	—

Note: snRNA = small nuclear RNA.

TABLE 1.3

Nucleoside-Modified Residues in rRNA

rRNA

		H. sapiens	
Source	*E. coli*	Cytoplasmic	Mitochondrial
Size (nt)	4,446	7,051	1,559
Modification			
2′-O-methyl	4	107	0
Base-methyl	24	10	0
Ψ	10	93	1
Total	38	210	1
%	0.9	3.0	0.06

Note: nt = nucleotide

be generated.[7] Incorporation of other modified nucleotides, such as 2′-O-methylated ones, can be accomplished only through chemical synthesis of short RNA sequences at the present time.

NUCLEOSIDE-MODIFIED DSRNAs AS INDUCERS OF INTERFERON: A SHORT HISTORY

The basic observation that dsRNAs—whether synthetic, viral or phage isolates—induce type I interferon and establish an antiviral state while ssRNAs and mammalian dsDNA produce no such activities was reported 40 years ago.[51-54] The obvious clinical importance of these observations stimulated immediate efforts to determine the structural and sequence-related requirements of dsRNA to exert this biological effect. At the time, since polynucleotide phosphorylase-catalyzed polymerization was the only available method to synthesize RNA, dsRNAs formed between long synthetic homopolymers were studied. Field and colleagues[52] found that polyinosinic:polycytidylic acid [poly(I:C)] was 50-fold more potent an inducer of type I interferon than polyadenylic:polyuridylic acid [poly(A:U)], suggesting that the nucleoside composition of dsRNA was biologically important.

De Clercq and colleagues, however, demonstrated that dsRNA homopolymers formed with 2′-O-methylated nucleotide derivatives [poly(I:Cm) and poly(A:Um)] had reduced or no interferon-inducing activities.[55] Indeed, the greater the degree of 2′-O-methylation, the less potent were dsRNA homopolymers at inducing type I interferon or inhibiting viral replication.[56] This same effect of 2′-O-methylation on activation of two interferon-induced enzymes, 2′-5′ oligoadenylate synthetase and PKR, was also observed.[58] The degree of inhibition was dependent on the degree of methylation, with fully 2′-O-methylated dsRNAs exhibiting no ability to activate either enzyme. Activation of PKR was more sensitive for presence of 2′-O-methyl-nucleosides in dsRNA than the synthetase. On the other hand, methylation of the polypyrimidine strands at C-5 positions of the nucleobases yielding poly(I:m5C) and poly(A:m5U) dsRNA had little influence on induction of antiviral states.[55,57]

These valuable early studies suggest that nucleoside composition is an important determinant of RNA immunogenicity, and that certain post-synthetic nucleoside modifications are capable of suppressing various aspects of this immunogenicity. However, the determination of precise mechanisms suggested by these early studies must be interpreted with caution, because synthetic RNA homopolymers behave differently from natural random RNA sequences in that they can form special tertiary structures (triplexes, tetrads, etc.). Sequence composition and nucleoside modifications may affect those structures, thus only indirectly disrupting interactions with putative dsRNA receptors. Because these early experiments used cultured fibroblasts and rabbits as intact animal models, the target receptors on which the dsRNAs acted to exert induction of interferon and viral inhibition are difficult or impossible to decipher. A more thorough understanding of the precise mechanisms by which RNA and its nucleoside modifications influence immunity requires the ability to study specific models of the immune system and use protocols for manufacturing and manipulating synthetic RNA that more closely simulates natural RNA.

EFFECTS OF NUCLEOSIDE MODIFICATION ON RNA-MEDIATED IMMUNE ACTIVATION

RNA Activation of Dendritic Cells: Not All RNAs Are Equally Potent

Dendritic cells (DCs) are the most powerful antigen-presenting cells of the immune system. Human DCs exposed to RNA transcribed in vitro (therefore containing no post-transcriptional nucleoside modifications) expressed inflammatory cytokines (IL-12, TNF-α, and IFN-α), high levels of MHC class II molecules, and a variety of activation markers (CD80, CD83 and CD86).[59,60] However, naturally produced RNAs have variable abilities to stimulate these DC responses. For example, human DCs secrete high levels of IL-12 when treated with total RNA isolated from bacterial but not from mammalian cells.[61]

Under different circumstances, however, mammalian RNA may also potentiate immune responses in vivo when transfected along with viral particles,[62] or induce IFN-α when delivered to DCs following complexing to cationic lipids.[5] A recent

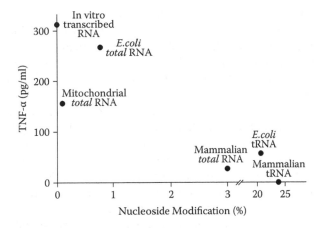

FIGURE 13.6 Production of TNF-α by monocyte-derived dendritic cells incubated with lipofectin-complexed RNA transcribed *in vitro* or isolated from natural sources. TNF-α was measured in the supernatant collected after 8 h incubation.

comparative study[7] demonstrated that RNAs transcribed *in vitro*, isolated from bacterial cells or from mammalian mitochondria (all poorly modified), potently induced human DCs to secrete TNF-α, but that total mammalian RNA and bacterial tRNA (containing more nucleoside modifications) were much less able to do so (Figure 13.6). Mammalian tRNAs (the most modified RNAs) did not induce any detectable production of TNF-α by DCs. Interestingly, DCs can be activated with RNA isolated from *Candida albicans* at the yeast stage but not at the hyphae stage of morphological development.[63] The reason for this is not understood, but an independent study of mRNA from another yeast, *Saccharomyces cerevisiae,* revealed substantial changes in the methylation level of nucleosides, especially formation of m6A, associated with different developmental stages of this organism.[64]

TREATMENT OF TLRs BY MODIFIED NUCLEOSIDE-CONTAINING RNA

It has been demonstrated that *in vitro* transcribed RNA activates human TLR3[16] and murine TLR7,[5] while chemically synthesized oligoribonucleotides (ORNs) stimulate murine TLR7 and human TLR8.[6] Using ORNs, several sequence motifs were identified that enhanced the potentials of RNAs to stimulate TLR7 or TLR8. RNAs containing GUCCUUCAA sequences, and ORNs possessing stretches of oligoG tails, are stimulators of murine TLR7 and human TLR8, respectively.[65,66] Others found that both TLRs can also be stimulated with GU-rich ORNs.[6]

Whether nucleoside modification can impact TLR activation is a question addressed in experiments using long *in vitro* transcripts and short stimulatory ORNs. Sequence compositions become less relevant when long *in vitro* transcripts are used because their random sequences most likely contain at least some of the preferred motifs. Indeed, *in vitro* transcripts have been shown to activate all three RNA-responsive TLRs expressed on HEK293 cells.[7] However, RNAs containing either m5C, m5U, Ψ, s2U, m6A or both m5C and Ψ modified nucleosides did not stimulate TLR7 or TLR8 (Figure 13.7).

Similarly, the presence of m6A and s2U in RNA abrogated cytokine induction by TLR3-transformed cells, while other nucleoside modifications exhibited less suppressive effects (Figure 13.7). The lack of TLR3 activation by m6A-containing RNA was not unexpected, considering that TLR3 binds dsRNA whose structure is destabilized by m6A.[41,67] Using phosphorothioate-linked ORNs, Sugiyama et al. found that the presence of m5C in a CpG motif converts the stimulatory oligomer to nonstimulatory.[68] However, a role for TLR in the impact of m5C modification could not be determined because the unmodified CpG ORNs stimulated only human monocytes, but not HEK293 cells overexpressing RNA-responsive human TLRs.[68]

Considering that ~80% of total cellular RNA is rRNA (Table 13.2) containing a higher level of nucleoside modification in mammalian cells than in bacteria or mitochondria (Table 13.3), studies performed with such natural RNA isolates may determine whether nucleoside modifications suppress the ability of RNA to stimulate TLRs. Indeed, total RNA from mammalian cells did not stimulate any of the RNA-responsive TLRs overexpressed on HEK293 cells (Figure 13.8).[7] However, bacterial total RNA obtained from *Escherichia coli* activated human TLR3, TLR7, and TLR8 (Figure 13.8). Interestingly, mitochondrial RNA, similarly to certain stimulatory ORNs,[6] activated human TLR8 but not TLR7 (Figure 13.8), suggesting subtle differences in RNA sensitivities of these two human TLRs when reconstituted in HEK293 cells.

ACTIVATION OF IMMUNE CELLS BY MODIFIED NUCLEOSIDE-CONTAINING RNA

Several studies have used human monocytes and DCs to investigate whether 2′-O-methylation can alter the immune stimulatory effect of ss or dsORNs. Modifying one or all the uridines to Ums in 21-mer ORNs, for example, is sufficient to abrogate TNF-α induction by DCs[7] or by PBMCs.[69] Replacing all U and/or G nucleosides with their corresponding 2′-O-methylated derivatives diminishes the potential of dsORNs to induce TNF-α and IFN-α in human PBMCs and in mice *in vivo*.[70,71]

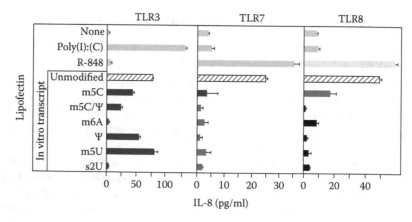

FIGURE 13.7 Production of IL-8 by RNA-treated HEK293 cells expressing human TLR3, TLR7, and TLR8. Cells were exposed to lipofectin-complexed, 1571 nt-long RNA containing the indicated modified nucleosides. IL-8 was measured in the supernatants collected after 8 h incubation.

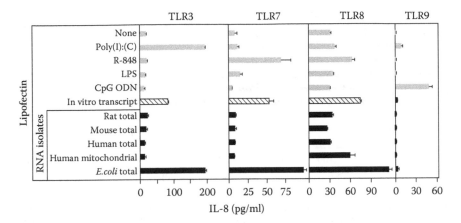

FIGURE 13.8 Production of IL-8 by RNA-treated HEK293 cells expressing human TLR3, TLR7, and TLR8. Cells were exposed to lipofectin-complexed RNA isolates. IL-8 was measured in the supernatants collected after 8 h incubation.

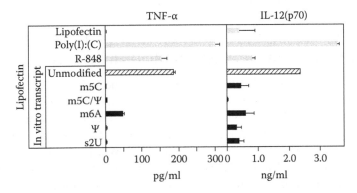

FIGURE 13.9 Monocyte-derived dendritic cells were treated with lipofectin-complexed, 1571 nt-long RNA containing the indicated modified nucleosides. TNF-α and IL-12 were measured in the supernatants after treatment for 8 and 16 h, respectively.

2'-O-methylation is not the only nucleoside modification influencing RNA immunogenicity. Human monocyte-derived DCs exposed to *in vitro* transcripts secreted high levels of TNF-α and IL-12p70, but the introduction of m5C-, m5C/Ψ-, m6A-, Ψ-, or s2U-modified nucleosides ablated cytokine induction by the RNA (Figure 13.9).[7]

Interestingly, in primary human DCs purified directly from peripheral blood, only uridine modification appears to suppress the induction of TNF-α and IFN-α by RNA, while RNA containing m6A and m5C is as stimulatory as unmodified transcripts (Figure 13.10).[7] Additional evidence supports a role for unmodified uridine in the immunogenicity of RNA. Koski et al. demonstrated that the poly(U) homopolymer induced IL-12 secretion in primed DCs.[61] Replacing uridines with adenosine or modifying uridines to 2'-O-methyluridines, abrogated cytokine induction by ORNs,[69] while 2'-O-methylation of cytidines or adenosines did not alter immune activation.[70] Others indicate that even a nucleoside mixture with uridine is sufficient to stimulate human

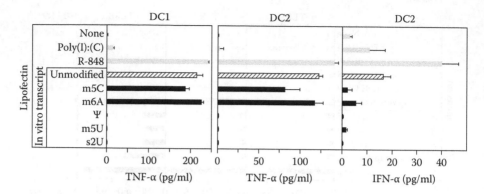

FIGURE 13.10 Primary dendritic cells, DC1 and DC2, were treated with lipofectin-complexed, 1571 nt-long RNA containing the indicated modified nucleosides. TNF-α and IFN-α were measured in the supernatants after treatment for 16 h.

PBMCs to secrete TNF-α. Using DCs from TLR7-null mice, Diebold et al. identified TLR7 as the responding receptor for poly(U) treatment.[5]

POTENTIAL MECHANISMS OF RNA-MEDIATED TLR ACTIVATION AND INHIBITION BY NUCLEOSIDE MODIFICATIONS

The extracellular domains of TLRs are very similar structurally, but highly specific to their respective ligands that can include molecules as diverse as lipids, proteins, and nucleic acids. Current belief is that most TLRs do not directly bind their ligands, but rather utilize accessory proteins. TLR3 and TLR9, however, are exceptions that directly bind their ligands, dsRNA and DNA, respectively.[72,73] Whether TLR7 and TLR8 also directly bind their ssRNA ligands has not been determined. Property changes introduced by nucleoside modifications may alter the way modified RNA interacts with proteins, RNAs, and other molecules. Thus, independently of whether RNA ligands bind TLRs directly or via accessory proteins, modification likely alters this association.

Effects of Nucleoside Modifications on RNA–RNA Interactions

Nucleoside modifications can affect RNA–RNA interactions by altering the ability of RNA to form secondary and tertiary structures. 2′-*O*-methyl, Ψ, m5U, s2U, and m5C modifications increase the stability of double-stranded stems created by base stacking of complementary sequences.[33,39] Pseudouridine introduces rigidity to the RNA chain,[33] while some rare nucleoside modifications, such as acetylated cytidine, are super-stabilizers of RNA structures.[24]

In contrast, methylation of nitrogen or carbon atoms that participate in Watson-Crick base-pairing destabilizes RNA duplexes.[39,41] Therefore the presence of m6A, m1A, and m1G in RNA disrupts double-stranded structures. This may explain the inhibitory effect of m6A modification on dsRNA-dependent TLR3 activation (Figure 13.7). Furthermore, since viral mRNA typically contains significantly more m6A than cellular mRNA,[23] it is tempting to speculate that the accumulation of m6A in viral RNA may serve the virus by enabling it to avoid immune activation via dsRNA sensors (e.g., TLR3, RIG-I, MDA5) and ssRNA sensors (TLR7, TLR8).

EFFECTS OF NUCLEOSIDE MODIFICATIONS ON RNA–PROTEIN INTERACTIONS

Complex modifications of the nucleosides in tRNA provide selective recognition patterns for interacting proteins such as aminoacyl transferase and processing enzymes.[74] The selective introduction of s2U, for example, facilitates RNA interactions with ribosomes.[75] Isomerization of uridine to pseudouridine results in an additional hydrogen bond donor at position N1 of Ψ that is not present on U (Figure 13.2). This hydrogen bond donor enables interactions with other reactive derivatives.[76] Nucleoside modification of RNA can also lead to novel protein–RNA interactions through molecular mimicry. Unique sequences present in the genomes of positive-strand RNA viruses mimic the m5U-Ψ-C loops of tRNA, thus inducing them to undergo a tRNA-like nucleoside modification. The modified nucleosides allow the RNA to form special contacts with proteins that the corresponding unmodified RNA could not allow.[77,78]

STERIC HINDRANCE CAUSED BY NUCLEOSIDE MODIFICATIONS

Nucleoside modifications play important roles in bacterial resistance to antibiotics that inhibit protein synthesis by binding to bacterial ribosomes (e.g., tetracycline, erythromycin, gentamicin).[79] In response to antibiotic exposure, a bacterial methyltransferase modifies selected nucleosides of the bacterial rRNA. The methylated nucleosides then block the access of the antibiotic to the ribosomes by steric hindrance.

RNA MODIFICATION AND AUTOIMMUNE DISEASE

Evidence indicates that nucleic acid-responsive receptors are involved in the pathogenesis of a number of autoimmune diseases.[80] It has recently been shown that increased TLR7 function resulting from receptor gene duplication accelerates the development of autoimmunity in a murine systemic lupus erythematosus model,[81,82] whereas TLR7 deficiency in the same model decreases the severity of disease.[83] TLR3-dependent mechanisms have been implicated in autoimmune liver damage and rheumatoid arthritis.[84,85] Because RNA activates both TLR3 and TLR7, and this activation is prevented by the introduction of selected nucleoside modifications, it is possible that reduced levels of nucleoside modification in natural RNA may contribute to the development of autoimmune disease in genetically susceptible individuals.

Reduced pseudouridine content in RNA, caused by mutation in two pseudouridine synthases (dyskerin and pseudouridine synthase 1), is associated with the human diseases, dyskeratosis congenita and mitochondrial myopathy and sideroblastic anemia. These diseases are not autoimmune; rather they are the consequence of incorrect protein translation due to undermodified rRNA[86] or mitochondrial tRNA.[87] While autoimmune lupus may be induced by drugs that interact with pyrimidine bases and block DNA methylation,[88,89] whether these drugs also interfere with RNA methylation is still unknown, and this particular mechanism for the induction of lupus remains to be investigated.

CONCLUSION

Since its discovery over 40 years ago, RNA-based immunology has been reinvigo-
rated with the observation that multiple TLRs interact with RNA molecules. The
further finding that nucleoside modification alters RNA-mediated TLR signaling
presents a mechanism for long-observed differences in immunogenicity of bacterial, viral, and mammalian RNAs. Further studies analyzing the effects of RNA
modification on other RNA sensors and receptors are ongoing. The involvement of
RNA modification in the pathogenesis of immune disorders and other diseases, and
the implications for designing therapeutics, are still being theorized and will likely
become important considerations in the near future.

REFERENCES

1. Isaacs, A., Cox, R. A., and Rotem, Z. 1963. Foreign nucleic acids as the stimulus to
   make interferon. *Lancet* 282:113.
2. Kaisho, T. and Akira, S. 2006. Toll-like receptor function and signaling. *J. Allergy
   Clin. Immunol.* 117:979.
3. Hemmi, H. et al. 2000. A Toll-like receptor recognizes bacterial DNA. *Nature*
   408:740.
4. Alexopoulou, L. et al. 2001. Recognition of double-stranded RNA and activation of
   NF-κB by Toll-like receptor 3. *Nature* 413:732.
5. Diebold, S. S. et al. 2004. Innate antiviral responses by means of TLR7-mediated recognition of single-stranded RNA. *Science* 303:1529.
6. Heil, F. et al. 2004. Species-specific recognition of single-stranded RNA via Toll-like
   receptor 7 and 8. *Science* 303:1526.
7. Kariko, K. et al. 2005. Suppression of RNA recognition by Toll-like receptors: the
   impact of nucleoside modification and the evolutionary origin of RNA. *Immunity*
   23:165.
8. Johnsen, I. B. et al. 2006. Toll-like receptor 3 associates with c-Src tyrosine kinase on
   endosomes to initiate antiviral signaling. *EMBO J.* 25:3335.
9. Ishii, K. J. and Akira, S. 2005. Innate immune recognition of nucleic acids: beyond toll-
   like receptors. *Int. J. Cancer* 117:517.
10. Yoneyama, M. et al. 2004. The RNA helicase RIG-I has an essential function in dou-
    ble-stranded RNA-induced innate antiviral responses. *Nat. Immunol.* 5:730.
11. Yoneyama, M. et al. 2005. Shared and unique functions of the DExD/H-Box helicases
    RIG-I, MDA5, and LGP2 in antiviral innate immunity. *J. Immunol.* 175:2851.
12. Meylan, E. and Tschopp, J. 2006. Toll-like receptors and RNA helicases: two parallel
    ways to trigger antiviral responses. *Mol. Cell.* 22:561.
13. Samuel, C. E. 2001. Antiviral actions of interferons. *Clin. Microbiol. Rev.* 14:778.
14. Osman, F. et al. 1999. A cis-acting element in the 3'-untranslated region of human
    TNF-α mRNA renders splicing dependent on the activation of protein kinase PKR.
    Genes Dev. 13:3280.
15. Kanneganti, T. D. et al. 2006. Bacterial RNA and small antiviral compounds activate
    caspase-1 through cryopyrin/Nalp3. *Nature* 440:233.
16. Kariko, K. et al. 2004. mRNA is an endogenous ligand for Toll-like receptor 3. *J. Biol.
    Chem.* 279:12542.
17. Samanta, M. et al. 2006. EB virus-encoded RNAs are recognized by RIG-I and acti-
    vate signaling to induce type I IFN. *EMBO J.* 25:4207.
18. Roberts, W. K. et al. 1976. Interferon-mediated protein kinase and low-molecular-
    weight inhibitor of protein synthesis. *Nature* 264:477.

19. Kerr, I. M., Brown, R. E., and Hovanessian, A. G. 1977. Nature of inhibitor of cell-free protein synthesis formed in response to interferon and double-stranded RNA. *Nature* 268:540.

20. Bass, B. L. and Weintraub, H. 1988. An unwinding activity that covalently modifies its double-stranded RNA substrate. *Cell* 55:1089.

21. Scadden, A. D. and Smith, C. W. 1997. A ribonuclease specific for inosine-containing RNA: a potential role in antiviral defence? *EMBO J.* 16:2140.

22. Grosjean, H. 2005. Modification and editing of RNA: historical overview and important facts to remember, in *Fine Tuning of RNA Functions by Modification and Editing*, Grosjean, H., Ed., Springer, Berlin.

23. Bokar, J. A. and Rottman, F. M. 1998. Biosynthesis and functions of modified nucleosides in eukaryotic mRNA, in *Modification and Editing of RNA*, Grosjean, H. and Benne, R., Eds., ASM Press, Washington.

24. Kowalak, J. A. et al. 1994. The role of posttranscriptional modification in stabilization of transfer RNA from hyperthermophiles. *Biochemistry* 33:7869.

25. Singer, B. and Grunberger, D. 1983. *Molecular Biology of Mutagens and Carcinogens*, Plenum Press, New York.

26. Sedgwick, B. 2004. Repairing DNA methylation damage. *Nat. Rev. Mol. Cell Biol.* 5:148.

27. Ougland, R. et al. 2004. AlkB restores the biological function of mRNA and tRNA inactivated by chemical methylation. *Mol. Cell* 16:107.

28. Kiss-Laszlo, Z. et al. 1996. Site-specific ribose methylation of preribosomal RNA: a novel function for small nucleolar RNAs. *Cell* 85:1077.

29. Aris, J. P. and Blobel, G. 1991. cDNA coning and sequencing of human fibrillarin, a conserved nucleolar protein recognized by autoimmune antisera. *Proc. Natl. Acad. Sci. USA* 88:931.

30. Cummins, L. L. et al. 1995. Characterization of fully 2′-modified oligoribonucleotide hetero- and homoduplex hybridization and nuclease sensitivity. *Nucleic Acids Res.* 23:2019.

31. Chen, J. and Patton, J. R. 1999. Cloning and characterization of a mammalian pseudouridine synthase. *RNA* 5:409.

32. Ofengand, J. 2002. Ribosomal RNA pseudouridines and pseudouridine synthases. *FEBS Lett.* 514:17.

33. Charette, M. and Gray, M. W. 2000. Pseudouridine in RNA: what, where, how, and why. *IUBMB Life* 49:341.

34. Szer, W. and Ochoa, S. 1964. Complexing ability and coding properties of synthetic polynucleotides. *J. Mol. Biol.* 12:823.

35. Ikeuchi, Y. et al. 2006. Mechanistic insights into sulfur relay by multiple sulfur mediators Involved in thiouridine biosynthesis at tRNA wobble positions. *Mol. Cell* 21:97.

36. Kumar, R. K. and Davis, D. R. 1997. Synthesis and studies on the effect of 2-thiouridine and 4-thiouridine on sugar conformation and RNA duplex stability. *Nucleic Acids Res.* 25:1272.

37. Nair, T. M., Myszka, D. G., and Davis, D. R. 2000. Surface plasmon resonance kinetic studies of the HIV TAR RNA kissing hairpin complex and its stabilization by 2-thiouridine modification. *Nucleic Acids Res.* 28:1935.

38. Bujnicki, J. M. et al. 2004. Sequence–structure–function studies of tRNA:m5C methyltransferase Trm4p and its relationship to DNA:m5C and RNA:m5U methyltransferases. *Nucleic Acids Res.* 32:2453.

39. Luyten, I. and Herdewijn, P. 1998. Hybridization properties of base-modified oligonucleotides within the double and triple helix motif. *Eur. J. Med. Chem.* 33:515.

40. Bokar, J. A. et al. 1997. Purification and cDNA cloning of the AdoMet-binding subunit of the human mRNA (*N*6-adenosine)-methyltransferase. *RNA* 3:1233.

41. Kierzek, E. and Kierzek, R. 2003. The thermodynamic stability of RNA duplexes and hairpins containing N6-alkyladenosines and 2-methylthio-N6-alkyladenosines. *Nucleic Acids Res.* 31:4472.
42. Margulis, L. and Chapman, M. J. 1998. Endosymbioses: cyclical and permanent in evolution. *Trends Microbiol.* 6:342.
43. Bachellerie, J. P. and Cavaille, J. 1998. Small nucleolar RNAs guide the ribose methylations of eukaryotic rRNAs, in *Modification and Editing of RNA*, Grosjean, H. and Benne, R., Eds., ASM Press, Washington.
44. Desrosiers, R., Friderici, K., and Rottman, F. 1974. Identification of methylated nucleosides in messenger RNA from Novikoff hepatoma cells. *Proc. Natl. Acad. Sci. USA* 71:3971.
45. Krug, R. M., Morgan, M. A., and Shatkin, A. J. 1976. Influenza viral mRNA contains internal N6-methyladenosine and 5'-terminal 7-methylguanosine in cap structures. *J. Virol.* 20:45.
46. Sommer, S. et al. 1976. The methylation of adenovirus-specific nuclear and cytoplasmic RNA. *Nucleic Acids Res.* 3:749.
47. Moss, B. et al. 1977. 5'-Terminal and internal methylated nucleosides in herpes simplex virus type 1 mRNA. *J. Virol.* 23:234.
48. Canaani, D. et al. 1979. Identification and mapping of N6-methyladenosine containing sequences in simian virus 40 RNA. *Nucleic Acids Res.* 6:2879.
49. Kane, S. and Beemon, K. 1987. Inhibition of methylation at two internal N6-methyladenosine sites caused by GAC to GAU mutations. *J. Biol. Chem.* 262:3422.
50. Rottman, F. and Heinlein, K. 1968. Polynucleotides containing 2'-O-methyladenosine. I. Synthesis by polynucleotide phosphorylase. *Biochemistry* 7:2635.
51. Field, A. K. et al. 1967. Inducers of interferon and host resistance. IV. Double-stranded replicative form RNA (MS2-Ff-RNA) from *E. coli* infected with MS2 coliphage. *Proc. Natl. Acad. Sci. USA* 58:2102.
52. Field, A. K. et al. 1967. Inducers of interferon and host resistance. II. Multistranded synthetic polynucleotide complexes. *Proc. Natl. Acad. Sci. USA* 58:1004.
53. Tytell, A. A. et al. 1967. Inducers of interferon and host resistance. III. Double-stranded RNA from reovirus type 3 virions (reo 3-RNA). *Proc. Natl. Acad. Sci. USA* 58:1719.
54. Lampson, G. P. et al. 1967. Inducers of interferon and host resistance. I. Double-stranded RNA from extracts of *Penicillium funiculosum*. *Proc. Natl. Acad. Sci. USA* 58:782.
55. De Clercq, E., Zmudzka, B., and Shugar, D. 1972. Antiviral activity of polynucleotides: Role of the 2'-hydroxyl and a pyrimidine 5-methyl. *FEBS Lett.* 24:137.
56. Greene, J. J. et al. 1978. Interferon induction and its dependence on the primary and secondary structure of poly(inosinic acid):poly(cytidylic acid). *Biochemistry* 17:4214.
57. De Clercq, E., Torrence, P. F., and Witkop, B. 1974. Interferon induction by synthetic polynucleotides: importance of purine N-7 and strandwise rearrangement. *Proc. Natl. Acad. Sci. USA* 71:182.
58. Minks, M. et al. 1980. Activation of 2',5'-oligo(A) polymerase and protein kinase of interferon-treated HeLa cells by 2'-O-methylated poly (inosinic acid):poly(cytidylic acid). *J. Biol. Chem.* 255:6403.
59. Weissman, D. et al. 2000. HIV gag mRNA transfection of dendritic cells (DC) delivers encoded antigen to MHC class I and II molecules, causes DC maturation, and induces a potent human in vitro primary immune response. *J. Immunol.* 165:4710.
60. Ni, H. et al. 2002. Extracellular mRNA induces dendritic cell activation by stimulating tumor necrosis factor-alpha secretion and signaling through a nucleotide receptor. *J. Biol. Chem.* 277:12689.
61. Koski, G. K. et al. 2004. Cutting edge: Innate immune system discriminates between RNA containing bacterial versus eukaryotic structural features that prime for high-level IL-12 secretion by dendritic cells. *J. Immunol.* 172:3989.

62. Riedl, P. et al. 2002. Priming Th1 immunity to viral core particles is facilitated by trace amounts of RNA bound to its arginine-rich domain. *J. Immunol.* 168:4951.
63. Bacci, A. et al. 2002. Dendritic cells pulsed with fungal RNA induce protective immunity to *Candida albicans* in hematopoietic transplantation. *J. Immunol.* 168:2904.
64. Clancy, M. J. et al. 2002. Induction of sporulation in *Saccharomyces cerevisiae* leads to the formation of *N6*-methyladenosine in mRNA: a potential mechanism for the activity of the IME4 gene. *Nucleic Acids Res.* 30:4509.
65. Hornung, V. et al. 2005. Sequence-specific potent induction of IFN-α by short interfering RNA in plasmacytoid dendritic cells through TLR7. *Nat. Med.* 11:263.
66. Peng, G. et al. 2005. Toll-like receptor 8-mediated reversal of CD4+ regulatory T cell function. *Science* 309:1380.
67. Choe, J., Kelker, M. S., and Wilson, I. A. 2005. Crystal structure of human Toll-Like receptor 3 (TLR3) ectodomain. *Science* 309:581.
68. Sugiyama, T. et al. 2005. CpG RNA: Identification of novel single-stranded RNA that stimulates human CD14+CD11c+ monocytes. *J. Immunol.* 174:2273.
69. Sioud, M. 2006. Single-stranded small interfering RNA are more immunostimulatory than their double-stranded counterparts: a central role for 2′-hydroxyl uridines in immune responses. *Eur. J. Immunol.* 36:1222.
70. Judge, A. D. et al. 2006. Design of noninflammatory synthetic siRNA mediating potent gene silencing *in vivo*. *Mol. Therapy* 13:494.
71. Morrissey, D. V. et al. 2005. Potent and persistent in vivo anti-HBV activity of chemically modified siRNAs. *Nat. Biotechnol.* 23:1002.
72. Latz, E. et al. 2004. TLR9 signals after translocating from the ER to CpG DNA in the lysosome. *Nat. Immunol.* 5:190.
73. Bell, J. K. et al. 2005. The molecular structure of the Toll-like receptor 3 ligand-binding domain. *Proc. Natl. Acad. Sci. USA* 102:10976.
74. Engelke, D. R. and Hopper, A. K. 2006. Modified view of tRNA: Stability amid sequence diversity. *Mol. Cell* 21:144.
75. Ashraf, S. S. et al. 1999. Single atom modification (O→S) of tRNA confers ribosome binding. *RNA* 5:188.
76. Helm, M. 2006. Post-transcriptional nucleotide modification and alternative folding of RNA. *Nucleic Acids Res.* 34:721.
77. Baumstark, T. and Ahlquist, P. 2001. Brome mosaic virus RNA3 intergenic replication enhancer folds to mimic a tRNA TΨC-stem loop and is modified *in vivo*. *RNA* 7:1652.
78. Lyons, A. J. and Robertson, H. D. 2003. Detection of tRNA-like structure through RNase P cleavage of viral internal ribosome entry site RNAs near the AUG start triplet. *J. Biol. Chem.* 278:26844.
79. Lapeyre, B. 2005. Conserved ribosomal RNA modification and their putative roles in ribosome biogenesis and translation, in *Fine Tuning of RNA Functions by Modification and Editing*, Grosjean, H., Ed., Springer, Berlin.
80. Colonna, M. 2006. Toll-like receptors and IFN-α: partners in autoimmunity. *J. Clin. Invest.* 116:2319.
81. Subramanian, S. et al. 2006. A Tlr7 translocation accelerates systemic autoimmunity in murine lupus. *Proc. Natl. Acad. Sci. USA* 103:9970.
82. Pisitkun, P. et al. 2006. Autoreactive B cell responses to RNA-related antigens due to TLR7 gene duplication. *Science* 312:1669.
83. Christensen, S. R. et al. 2006. Toll-like receptor 7 and TLR9 dictate autoantibody specificity and have opposing inflammatory and regulatory roles in a murine model of lupus. *Immunity* 25:417.
84. Lang, K. S. et al. 2006. Immunoprivileged status of the liver is controlled by Toll-like receptor 3 signaling. *J. Clin. Invest.* 116:2456.

85. Brentano, F. et al. 2005. RNA released from necrotic synovial fluid cells activates rheumatoid arthritis synovial fibroblasts via toll-like receptor 3. *Arthritis Rheum.* 52:2656.
86. Yoon, A. et al. 2006. Impaired control of IRES-mediated translation in X-linked dyskeratosis congenita. *Science* 312:902.
87. Bykhovskaya, Y. et al. 2004. Missense mutation in pseudouridine synthase 1 (PUS1) causes mitochondrial myopathy and sideroblastic anemia (MLASA). *Am. J. Hum. Genet.* 74:1303.
88. Dubroff, L. M. and Reid, R. J., Jr. 1980. Hydralazine-pyrimidine interactions may explain hydralazine-induced lupus erythematosus. *Science* 208:404.
89. Richardson, B. et al. 1990. Evidence for impaired T cell DNA methylation in systemic lupus erythematosus and rheumatoid arthritis. *Arthritis Rheum.* 33:1665.

14 Activation of Innate Pattern Recognition Pathways by Single-Stranded Ribonucleic Acids

Sandra S. Diebold

ABSTRACT

Innate immune activation triggered by viral single-stranded RNA is mediated by Toll-like receptors 7 and 8, and leads to the induction of type I interferons and other pro-inflammatory cytokines. Interferon production in response to single-stranded RNA ligands is restricted to plasmacytoid dendritic cells and is largely mediated by Toll-like receptor 7, while pro-inflammatory cytokines such as interleukins 6 and 12 are produced by myeloid dendritic cells that express mainly Toll-like receptor 8. In both cases, ligand recognition takes place in a specialized endosomal compartment and is therefore similar to the recognition of bacterial and viral DNA by Toll-like receptor 9. Recognition of single-stranded RNA appears largely sequence-independent and consequently, self RNA can act as an agonist. Recognition of self RNA and DNA plays a role in the progression of autoimmune diseases such as systemic lupus erythematosus. The conditions that allow self RNA to gain access to the endosomal compartments mediating single-stranded RNA recognition, and the extent to which such interactions are involved in initiating autoimmune responses, are currently unclear.

INTRODUCTION

Cells of the innate immune system sense invading pathogens by recognizing evolutionarily conserved molecular structures via germline-encoded pattern recognition receptors (PRRs). The first viral pathogen-associated molecular pattern (PAMP) found to induce immune activation was viral double-stranded RNA (dsRNA).[1] After a few reports of similar immune activation by single-stranded RNA (ssRNA) preparations,[2,3] other studies failed to detect such responses *in vitro* and *in vivo*,[4,5] and the common belief was that ssRNA did not harbor immunostimulatory activity.

However, these early studies employed ssRNA in the form of naked nucleic acid and rapid degradation by RNases was most likely the reason why no immune activation could be detected. In contrast, viral dsRNA is relatively stable and was found to trigger cytoplasmic PRRs such as double-stranded RNA-dependent protein kinase PKR, 2′5′-oligoadenylate synthetase, and the recently identified helicases RIG-I and MDA-5.[6-10] All of these cytoplasmic PRRs are ubiquitously expressed and serve as crucial mediators of interferon responses in infected cells. This enabled a detailed characterization of the immunostimulatory activity of dsRNA long before the pattern recognition hypothesis was postulated.

Only after Janeway's hypothesis was published in 1989[11] was the search for specialized PRRs incited. The pattern recognition hypothesis postulated that cells of the innate immune system express receptors capable of detecting PAMP without the cells having to be infected with the pathogen, distinguishing these receptors from cytoplasmic PRRs such as RIG-I. The PRRs later found to possess these properties all belonged to a single family: the Toll-like receptor (TLR) family.[12] The TLR mediating detection of viral dsRNA was established to be TLR3.[13] It then became obvious that certain ssRNA viruses such as influenza virus and vesicular stomatitis virus activate the immune system via a TLR-mediated TLR3-independent recognition pathway identified as TLR7.[14,15] Previously, the natural ligand for this receptor had been unknown, but small synthetic molecules acting as immune response modifiers (IRMs) had been shown to confer anti-viral activity via a TLR7-mediated mechanism.

Because of the structural similarities of these drugs to nucleosides, genomic viral ssRNA was the most likely candidate for the natural ligand of TLR7. Indeed, purified genomic RNA from influenza virus and synthetic RNA oligonucleotides (ORNs) encoding GU-rich sequences from the U5 region of human immunodeficiency virus (HIV) were shown to trigger TLR7-mediated immune activation.[14,16]

Several studies tried to determine the characteristics of TLR7 ligand binding, but as noted below, controversy still surrounds the RNA motif that triggers TLR7 activation. This is partly due to the fact that ssRNA is recognized by both TLR7 and TLR8. Three issues further complicate the interpretation of the experimental evidence: (1) mouse TLR8 in contrast to human TLR8 is non-functional;[17] (2) different cell types show different expression levels for TLR7 and TLR8; and (3) human TLR7 failed to confer responsiveness to ssRNA ligands in some studies using transfectants.[16,18,19] This chapter provides an overview of our current knowledge about TLR7/8 recognition and the physiological context in which this innate pattern recognition pathway plays a crucial role.

RECOGNITION OF SINGLE-STRANDED RNA BY TLR7 AND TLR8

TLR7- VERSUS TLR8-RESTRICTED RECOGNITION

Since mouse TLR8 is non-functional,[17] all studies exploring ssRNA-mediated immune activation in mouse model systems determined only the characteristics of TLR7 triggering. In contrast, in studies exploring the response of human cells, activation induced by ssRNA was transmitted by TLR7 and TLR8.[17] While the use of cells from TLR-deficient mice in mouse model systems can prove the involvement of a specific TLR, the proof is more difficult in human studies. siRNA technology allows the down-regulation of particular genes in human cells, but the extent to which this affects protein levels is variable. In addition, siRNA molecules have also been shown to trigger TLR7 and TLR8, further complicating the interpretation.[14,19-22]

Another strategy used to determine the involvement of particular human TLRs is the use of HEK293 transfectants, but some studies failed to confer responsiveness to ssRNA ligands by transient expression of human TLR7.[16,18,19] Thus, the characterization of TLR7 ligand recognition in mouse systems has been straightforward; whereas similar studies on human cells are less conclusive.

In the human system, differential expression of TLR7 and TLR8 by different cell types is often used to determine the ligand specificities of these receptors. While human plasmacytoid dendritic cells (PDCs) mainly express TLR7, human myeloid DCs and monocytes express higher levels of TLR8.[23-26] Unlike other DC subsets, PDCs induce type I interferons (IFN-I) and interferon-inducible cytokines in response to TLR7 triggering.[27] In contrast, monocytes that express high levels of TLR8 and only low levels of TLR7[26] are less sensitive to TLR7 agonists and produce pro-inflammatory cytokines such as tumor necrosis factor-α (TNF-α) and interleukin-12 (IL-12) rather than interferon-inducible cytokines.[27] Although these receptors are differentially expressed on different cell types in the steady state, contribution of the other ssRNA-recognizing TLRs cannot be formally excluded in these studies on human cells, taking into account that expression of TLR7 and other TLRs was shown to be inducible by IFN-I.[28,29]

IMMUNE RESPONSE MODIFIERS ACTING AS TLR7 AND TLR8 LIGANDS

Before viral ssRNA was identified as the natural ligand for TLR7 and TLR8, several small molecular drugs had been shown to exert anti-viral activities via a TLR7/8-mediated pathway. The first IRMs triggering TLR7/8 activation cited in the literature included the imidazoquinolins (resiquimod, imiquimod) and the nucleotide analog loxoribine, but many additional IRMs were later described.[17,30-33] These TLR7/8 agonists were shown to exert broad anti-viral[34-37] and anti-tumor activities[38-40] and imiquimod has been approved for treatment of genital warts caused by human papillomavirus infection.[41] While Resiquimod, also known as R848 was demonstrated to activate TLR7 and TLR8; imiquimod (R837) and loxoribine were shown to be TLR7-specific IRMs. More recently, TLR8-specific small molecular compounds were also described.[17,27,30,32]

Comparative analysis of TLR7- and TLR8-specific IRMs was used to dissect the two pathways and their respective roles in the activation of human leukocyte subsets in support of the predominant view that TLR7 mainly triggers PDCs and induces the induction of IFN-I by these cells, whereas TLR8 activation leads to the production of pro-inflammatory cytokines such as IL-6, IL-12 and TNF-α by monocytes and myeloid DCs.[27] However, the interpretation of experiments using TLR7-specific IRM was recently challenged with the discovery that loxoribine can trigger a TLR8-mediated response in the presence of polyT olignucleotides.[42] This surprising finding implies that although some IRMs have preferences for triggering TLR7 instead of TLR8 or vice versa, they may not be pure TLR7 or TLR8 agonists; depending on the experimental set-up, suboptimal activation of the other TLR may contribute to the response.

IMUNOSTIMULATORY ssRNA

The first reports showing that the immunostimulatory activity of ssRNA is mediated via TLR7 already demonstrated that RNA from very different sources triggers this receptor. It became evident that synthetic ORNs encoding HIV-specific gene sequences activate mouse and human DCs via TLR7- and TLR8-dependent pathways, respectively.[16] Similarly, purified genomic influenza RNA was demonstrated to trigger TLR7-mediated activation of mouse PDCs.[14] Other sources of RNA such as in vitro-transcribed RNA, single-stranded ORN designed for siRNA downregulation of murine IL-12 p40, and the polyU RNA homopolymer proved equally potent in triggering this pathway.[14]

The latter finding was surprising because none of the other RNA homopolymers tested (polyA, polyC, polyG, and polyI) were able to trigger TLR7. The variety of RNA preparations exhibiting immunostimulatory activity indicates that it is highly unlikely that the receptor recognizes a specific sequence motif. The RNA homopolymer study indicates that TLR7 exclusively sense uridines but not any other nucleotides. Uridine nucleotides are present in RNA molecules, while they are largely absent from DNA, and therefore may be viewed as a special signature of RNA that is used to discriminate it from DNA.

RESPONSES TO SELF RNA

If TLR7/8 recognizes ssRNA ligands containing sufficient numbers of uridine moieties, it may also recognize self RNA. Many studies have explored this aspect of ssRNA recognition, but whether self RNA triggers these pathways is still controversial. One hypothesis was that viral RNA contains sequence motifs that are less abundant in mammalian RNA, and related studies tried to identify a sequence motif for ssRNA that conferred immunostimulatory activity.[16,20] This hypothesis was questioned by the fact that ORN-encoding sequences specific for mammalian genes were potent inducers of TLR7/8-mediated immune activation.[14,20,22] Furthermore, the discovery that among all tested RNA homopolymers only polyU RNA triggers TLR7 made the existence of a specific motif highly unlikely. Yet, the hypothesis that recognition of ssRNA is sequence-unspecific raised the question of how the immune system can avoid responses to self RNA.

A study exploring molecular differences between microbial and mammalian RNAs offered an explanation by showing that modifications found in mammalian RNA may reduce the immunostimulatory activity of ssRNA.[43] In this study, bacterial RNA, unlike mammalian total RNA, showed good immunostimulatory activity in a gain-of-function experiment using transfected HEK 293 cells expressing human TLR7 and TLR8. The authors concluded from subsequent experiments using modified ORNs that mammalian RNA contains modifications that suppress TLR7/8 stimulation.

Modifications that abrogated the immunostimulatory activities of synthetic ORNs included the introduction of pseudouridines, 2-thiouridines, 5-methylcytidines, 5-methyluridines, 6-methyladenosines, and 2'-O-methyluridines. All these nucleoside modifications are naturally present in mammalian RNA and less abundant in bacterial RNA. However, the extent to which viral RNA may be distinguished from self RNA on the basis of such modifications is unclear. Even though these specific modifications of mammalian RNA may reduce its immunostimulatory activity, purified murine messenger RNA (mRNA) triggered TLR7-mediated immune activation in mouse PDCs.[14] Thus, while total mammalian RNA may be non-immunostimulatory due to inhibitory modifications, the immunostimulatory activity of mammalian mRNA is not sufficiently suppressed to abrogate TLR7 activation.

Mounting evidence indicates that self RNA and self DNA trigger TLR7- and TLR9-mediated immune activation in systemic lupus erythematosus (SLE) patients, and that these pathways play a crucial role in the induction and progression of auto-immunity in these patients. This is further proof that self nucleic acids may trigger these pathways under conditions that facilitate their access to the compartment in which recognition takes place.

ANTAGONISTS

DNA oligonucleotides (ODNs) that act as antagonists for TLR7-mediated immune responses were only recently described. Surprisingly, the inhibitory ODN 2088 originally described as a specific inhibitor of mouse TLR9 was shown to block TLR7-mediated responses of mouse cells via a TLR9-independent mechanism.[44] Another study identified inhibitory ODNs that specifically suppress TLR7- or TLR9-mediated signaling,[45] showing that the DNA motifs that confer inhibitory function are not identical. Interestingly, these inhibitory ODNs suppressed TLR7 signaling in mouse and human cells.

One report compared the abilities of several ODNs to block TLR7 and TLR9 signaling.[46] In another study, a thymidine-rich ODN was used to interfer with activation of human TLR7.[42] Unexpectedly, abolishing TLR7 stimulation with the polyT ODN did not abrogate the immune response to the TLR7 ligand loxoribine, but led to TLR8-mediated immune activation of cells. This result challenges the current view on TLR7- versus TLR8-mediated immune activation, but further studies are necessary to clarify the specificities of other compounds regarding their preferential induction of these two TLRs. Similarly, the ability of antagonists of TLR9 to also block signaling of TLR7 is not well understood and needs further clarification.

All these studies using different ODNs as antagonists for TLR7 signaling indicate that specific DNA motifs can bind to the receptor but fail to induce signaling. However, as for the immunostimulatory siRNA ORN, a clear understanding of the motifs of the TLR7/TLR8 antagonists that confer stimulatory versus inhibitory activity require further clarification. The identification of potent antagonists for TLR7/8 may provide useful therapeutic tools for treating of autoimmune diseases such as SLE for which the involvement of these TLRs in disease progression was shown.

SEQUENCE-SPECIFIC VERSUS SEQUENCE-UNSPECIFIC RECOGNITION

The evidence for sequence-specific ligand recognition by TLR7 is not convincing. The experimental evidence for TLR8 is even less conclusive because mouse TLR8 is non-functional and the contribution of human TLR7 to ssRNA-induced responses cannot be formally excluded. For mouse TLR7, uridine-rich stretches of RNA irrespective of the source of the nucleic acid material are immunostimulatory.[14]

For human TLR7, preferential recognition of uridine-rich RNA must be confirmed in systems in which contributions of human TLR8 can be excluded. Similarly, a detailed characterization of the TLR8-activating nucleic acid ligand awaits further studies that can exclude contributions of human TLR7.

Interestingly, human TLR8 was shown to be triggered by guanosine-rich ODNs.[47] Assuming that the natural ligand of the receptor is RNA rather than DNA, one could hypothesize that human TLR8 recognizes strands of guanosine-rich RNA. The presence of mixed TLR7- and TLR8-mediated responses for many human cells may explain why GU-rich sequences were identified as immunostimulatory components of ssRNA in the human system.[16] Although a lot of experimental evidence suggests sequence-unspecific recognition of ssRNA ligands, much debate surrounds sequence-specific activation by siRNA ORNs.[19-22] This aspect of TLR stimulation clearly requires further clarification.

ACCESS TO TLR7- AND TLR8-CONTAINING ENDOSOMAL COMPARTMENTS

The evidence for sequence-unspecific recognition of RNA and the recognition of self RNA via TLR7 and TLR8 raise the question how the immune system ensures responses to viral ssRNA while simultaneously preventing autoimmune responses to self RNA. Under physiological conditions, viral RNA is present inside virus particles and is thus protected from degradation by RNases. Upon uptake of virus particles in endosomal compartments, these endocytic vesicles are likely to undergo maturation during which the virus particles are digested, allowing the genomic viral ssRNA access to TLR7 and TLR8.

Self RNA, which upon release from necrotic cells is not protected, may degrade too quickly to reach the specialized endosomal compartment where TLR7/8 recognition takes place. In contrast, purified RNA and ORNs present in complexes with cationic lipids, or polycations such as DOTAP and polyethylenimine, are protected from RNase degradation in the extracellular milieu and therefore may be able to trigger these PRRs in experimental set-ups.[14,16,20] Unfortunately, we do not know which receptors at cell surfaces mediate uptake of different cargos such as viruses, RNA–protein or RNA–polycation complexes, and naked nucleic acids such as ODNs, or how these receptors influence the processing of the cargo taken up.

Interestingly, immunoglobulin–nucleic acid complexes present in the blood of SLE patients can trigger TLR7- and TLR9-mediated immunactivation and thus allow access to the endosomal compartment where TLR7/9 recognition takes place. Uptake of such self DNA–immunoglobulin complexes found in SLE patients is mediated via FcγRIIa receptors.[48] How such mechanisms are controlled in healthy individuals remains unclear.

Further evidence supporting the view that the endosomal localization of the TLRs recognizing nucleic acids promotes recognition of viral ligands while at the same time prevents access of self ligand resulted from a study exploring the impact of receptor localization on activation of TLR9.[49] TLR7, TLR8, and TLR9 are evolutionarily closely related and form the TLR9 subfamily.[50] Members of this subfamily are expressed intracellularly and sample the contents of a specialized endosomal compartment.

A mutant TLR9 located at cell surfaces was shown to mediate TLR9-dependent recognition of DNA from apoptotic cells, while wild-type TLR9 located in the endosomal compartment failed to respond to this stimulus.[49] Although this does not indicate how self DNA under physiological conditions is excluded from the TLR9-sensing endosomal compartment, it shows that self-DNA may trigger TLR9-mediated immune activation when it has access to the receptor. A similar scenario is likely for TLR7 and TLR8. Interestingly, mouse splenic CD8α+ DCs that are specialized in taking up material from apoptotic cells[51,52] do not express TLR7,[53] and therefore cannot be activated by self RNA present in apoptotic cells although they are efficiently activated by viral dsRNA present in virus-infected cells.[54]

TLR7- AND TLR8-MEDIATED CELL ACTIVATION

EXPRESSION PATTERNS OF TLR7 AND TLR8

TLR7 and TLR8 are expressed on a variety of cell types in the steady state (Table 14.1). IFN-I induced upon virus infection *in vivo*, however, can lead to increases in the expression levels of TLR7 and can also induce its expression on cells that do not

TABLE 14.1
TLR 7 and TLR8 Expression of Various Human Cell Types

Cell Type	TLR7	TLR8	References
Plasmacytoid DC	+	–	Jarossay et al., 2001; Kadowaki et al., 2001; Hornung et al. 2002; Ito et al., 2002
Myeloid DC	– (+)	+	Jarossay et al., 2001; Kadowaki et al., 2001; Ito et al., 2002
Immature moDC	–	+	Jarossay et al., 2001
Mature moDC	+	+	Peng et al., 2006
IFN-DC	+	+	Mohty et al., 2003
Monocytes	+	+	Jarossay et al., 2001; Kadowaki et al., 2001; Hornung et al. 2002; Ito et al., 2002
Macrophages	+/–*	+	Mietinnen et al., 2001
Eosinophils	+*	–*	Nagase et al., 2003
Neutrophils	+/–	+	Nagase et al., 2003
Blood B cells	+/–*	–	Bekeredjian-Ding et al., 2005
Tonsillar B cells	+	–	Mansson et al., 2006
NK cells	– (+)	– (+)	Gorski et al., 2006; Hart et al., 2005
Mast cells	+	n.d.	Matsushima et al., 2004
Regulatory T cells	–	+	Peng et al., 2006
CD25- T cells	–	+/–	Peng et al., 2006
Keratinocytes	–	–	Mempel et al., 2003
Uterine epithelial cells	+/–	+	Schafer et al., 2004
Cervical epithelial cells	+	–	Andersen et al., 2006

Note: +, Good expression of TLR. +/–, Weak expression of TLR. – No expression of TLR. (+) Lone study demonstrates expression. *, Expression inducible by IFN-I. n.d., Not determined.

express it in the steady state.[28,55] The extent of induction and the cell types on which TLR8 can be induced in response to IFN-I are currently unclear. Among cell types, DC subsets are the most intensively studied regarding their expression pattern. While human pDCs express TLR7 (but no TLR8) in the steady state, human myeloid DCs express TLR8 and little or no TLR7.[23–26] In contrast, among mouse splenic DC subsets, only the CD8α⁺ subset is completely negative for TLR7 expression, while TLR8 is expressed on all four subsets including plasmacytoid, double negative, CD4⁺ DCs, and CD8α⁺ DCs.[53]

Human eosinophils express TLR7 and no TLR8 whereas neutrophils abundantly express TLR8 and to a lesser extent TLR7.[56] NK cells were shown to be negative for both TLRs in one study,[57] and described to express TLR7 and TLR8 in another.[58] In both cases, activation of NK cells by TLR7/8 ligands proved to be indirect. B cells express TLR7,[55] and mast cells express TLR7 but not TLR8,[59] while human regulatory T cells express TLR8 and not TLR7.[47] Expression of these TLRs is different on epithelial cells from different tissues; cervical epithelial cells express only TLR7[60] and uterine epithelial cells show weak expression levels of TLR7 and high expression levels of TLR8.[61] In most studies, expression levels were measured at the RNA level by RT-PCR. While this does not allow conclusions about absolute expression levels of TLR proteins, functional studies usually correlated well with RNA expression data except for NK cells.[58] In human cells, protein expression levels can be studied using TLR7-specific antibodies.[62]

The activation of a variety of different cell types by TLR7/8 ligands is described below and demonstrates that the immune system is likely to be stimulated simultaneously on several levels after infection with ssRNA viruses and *in vivo* application of IRMs. This may explain the induction of severe side effects from systemic applications of IRMs. To date, imiquimod has been approved only for topical application in treating human papillomavirus-induced genital warts[41] and skin malignancies.[63]

ACTIVATION OF PDCs

TLR7 activation induces high levels of IFN-I by PDCs, similarly to TLR9 stimulation with CpG ODNs. PDCs, also known as high or natural interferon producing cells, were thought to be the only cell types capable of producing high levels of IFN-I in response to viral stimuli due to constitutive expression of transciption factor IRF-7.[64–66] Yet, conventional DCs produce similarly high levels of IFN-I in response to cytoplasmic dsRNA,[67] and in response to CpG ODNs manipulated for endosomal retention using cationic lipids.[68] The latter study demonstrated that the difference in endosomal retention of CpG ODNs and the spatiotemporal regulation of MyD88/IRF-7 signaling can account for the different cytokine expression patterns in PDCs and myeloid DCs. Similarly, another study showed that changing the endosomal processing of different classes of CpG ODNs abrogates or promotes IFN-I induction.[69]

TLR7 and TLR8 belong to the TLR9 subfamily and the mechanisms of immune activation induced by these TLRs are very similar. Therefore, one can speculate that similar mechanisms regulate cytokine induction in response to CpG ODNs, ssRNA ligands, and IRMs. Prolonged endosomal retention may lead to TLR7-mediated induction of IFN-I in PDCs, while endosomal retention in myeloid DCs may only be brief, and therefore ssRNA ligands may preferentially induce pro-inflammatory cytokines such as IL-12, IL-6, and TNF-α in this cell type. IRF-7 protein stability is increased in PDCs upon activation by viral stimuli, a phenomenon that could not be observed in other cell types[70] and that may be a result of prolonged MyD88/IRF-7 signaling in PDCs. Based on these observations, we can conclude that PDCs are not the only cells that can produce high levels of IFN-I upon viral infection, but that they are specialized for inducing IFN-I in response to TLR7- and TLR9-mediated activation.

IFN-I produced by PDCs upon TLR7- or TLR9-mediated immune activation can sensitize other cell types for viral recognition by upregulating TLRs such as TLR3 and TLR7.[28,29,55] Upon upregulation of these TLRs, the threshold for activation may be decreased and more cell types may directly become activated by the viral stimuli, thus rendering the immune system more sensitive for viral recognition.

ACTIVATION OF MYELOID DCs

Myeloid DCs express TLR8 rather than TLR7. However, a contribution of TLR7 to ssRNA-mediated responses cannot be completely excluded because one study found reasonable expression of TLR7 mRNA in myeloid DCs,[25] and because TLR7 is inducible by IFN-I. In contrast to PDCs, myeloid DCs do not induce IFN-I in response to TLR7/8 ligands, but induce pro-inflammatory cytokines such as IL-12, IL-6, and TNF-α. The preference for the induction of different cytokines by myeloid DCs versus PDCs in response to the same stimuli may be attributed to different processing of the endosomal compartment, in which recognition takes place as described above.

Monocyte-derived DCs that are more closely related to myeloid DCs than PDCs were initially described as negative for TLR7 and positive for TLR8, similar to myeloid DCs.[23] However, the monocyte-derived DCs examined in this study were immature. In a recent study, mature monocyte-derived DCs were demonstrated to express both TLR7 and TLR8[47] and monocyte-derived DCs generated in the presence of GM-CSF and IFNα were shown to express TLR7 and also induce IFN-α in response to resiquimod.[29]

In summary, it is unclear to what extent myeloid DCs express TLR7 in addition to TLR8 during viral infections, and to what extent both TLRs contribute to the activation of these cells by viral ssRNA under physiological conditions. Regarding monocyte-derived DCs under exploration as cellular vaccines for immunotherapy of cancer, TLR7 expression can be induced using maturation stimuli or IFN-α to exploit the synergy between TLR7/8 and TLR3 activation in response to an IRM such as resiquimod and the viral dsRNA mimic polyI:C.[71,72]

ACTIVATION OF B CELLS

B cells express low levels of TLR7 in the steady state, but TLR7 is dramatically increased upon virus infection or treatment with IFN-I. B cells respond to resiquimod by inducing cytokines such as IL-6 and by upregulating MHC class II expression. In the presence of PDC-produced IFN-I, B cells proliferate in response to TLR7 ligand and secrete immunoglobulin G (IgG).[55] Thus IFN-I from PDCs and stimulation of

TLRs such as TLR7 and TLR9 on B cells can substitute for T cell help, although this seems only to apply for memory B cells.[55,73] This mechanism may explain how autoreactive B cells once induced become independent of T cell help and may be continuously stimulated in the presence of TLR7- and/or TLR9-activating immune complexes.

ACTIVATION OF REGULATORY T CELLS

Natural CD25[+] regulatory T cells were shown to express TLR8 and respond to guanosine-rich ODNs via a TLR8-dependent pathway.[47] Upon triggering of TLR8 on human regulatory T cells, these cells no longer suppress the proliferation of CD25[-] responder T cells. The mechanism by which suppression of responder T cells by regulatory T cells is disabled is currently unclear. Surprisingly TLR8 on these human regulatory T cells was triggered by a variety of ODNs, including CpG, non-CpG, and guanosine-rich ODNs. This questions the specificity of ligand recognition by human TLR8 and results obtained after *in vivo* application of CpG ODNs. Future studies will have to clarify whether effects on regulatory T cell function can contribute to some of the phenomena observed.

ACTIVATION OF GRANULOCYTES

Human eosinophils and neutrophils respond similarly to activation with resiquimod but eosinophils show greater sensitivity for this stimulus.[56] While eosinophils express only TLR7, neutrophils express higher levels of TLR8 than TLR7, so it is likely that the response is mainly mediated via TLR7 in eosinophils and via TLR8 in neutrophils. However the true contribution of each receptor to the response has not been determined. Since both receptors are strongly inducible on eosinophils by IFN-I, contributions of both TLRs may constitute the norm during viral infection. Both granulocyte subtypes respond to resiquimod by increasing CD11b expression on cell surfaces by downregulating L-selectin and generating superoxide.

ACTIVATION OF MAST CELLS

Skin-derived mast cells express TLR7 and induce pro-inflammatory cytokines such as TNF-α and IL-6, and chemokines such as MIP-1α, MIP-2, and RANTES in response to resiquimod.[59] However degranulation of mast cells and induction of IL-13 are not induced by TLR7/8 stimulation.

TLR7- AND TLR8-MEDIATED RECOGNITION OF SINGLE-STRANDED RNA VIRUSES

Under physiological conditions, the main role of TLR7 and TLR8 is to recognize viral genomic ssRNA during viral infections. The involvement of these TLRs in the recognition of a variety of ssRNA viruses has been demonstrated, and the list includes influenza virus, vesicular stomatitis virus, HIV, group B coxsackieviruses, coronaviruses, paramyxoviruses, and parechovirus 1. The experimental evidence for involvement of TLRs in the recognition of ssRNA viruses stems mainly from *in vitro* experiments using mouse or human cells; only a few studies investigated infection models in mice.

Studies of mice or mouse cells cannot address the contribution of TLR8 due to its non-functional status in these models. Therefore, while the activation of mouse TLR7 by influenza virus,[14,15] vesicular stomatitis virus,[15] and coronaviruses[74] was demonstrated, it is unclear whether and to what extent these viruses trigger human TLR7 versus human TLR8. In studies exploring the involvement of human TLR in the recognition of ssRNA viruses, contributions of TLR7 and TLR8 were demonstrated for Sendai virus,[75] parechovirus 1,[76] and group B coxsackieviruses.[77] For HIV, the involvement of TLR7 in the activation of PDCs was determined using inhibitory TLR7/TLR9 ligands, but the contribution of another PRR cannot be excluded.[46].

The induction of TLR7/TLR8 signaling by ssRNA viruses is crucially dependent on endocytosis and endosomal acidification.[14,15,46,75] One study showed that envelope–CD4 interactions are necessary for activation of PDCs, and that the receptors that mediate uptake play a crucial role in virus recognition.[46] Uptake of virus particles via specific cell surface receptors is very different from uptake of oligonucleotides, oligonucleotide–cationic lipid complexes, and IRMs. How the different modes of uptake and endosomal retention affect recognition by TLR7 and TLR8 is currently unclear and awaits further clarification. This is particularly important for therapeutic applications using IRMs or agonistic and antagonistic oligonucleotides for targeting of TLR.

Viruses have evolved a multitude of mechanisms to escape or suppress immune recognition including strategies to block TLR activation. Measles virus and respiratory syncytial virus (RSV) A2 block induction of IFN-I in PDCs, suppress responses to the TLR7/8 ligand resiquimod, and also block TLR9 activation.[78] However, the molecules in the TLR signaling pathway that are targeted by these viruses are still unknown. This example emphasizes that other viruses may have evolved similar or different strategies to suppress TLR activation, and that the failure to demonstrate virus recognition by these TLRs may be due to viral suppression of these pathways, rather than failure of respective TLRs to recognize the genomic nucleic acid of a particular virus.

ROLE OF TLR7- AND TLR8-MEDIATED ACTIVATION IN AUTOIMMUNITY

The immune system has evolved strategies to recognize viral infections by detecting viral nucleic acids present in the form of the viral genome or produced upon virus replication in cellular compartments to which self nucleic acids have no access under physiological conditions. While differences in methylation levels in mammalian and microbial DNAs can skew the responsiveness of DNA-sensing PRRs in favor of detecting microbial ligands rather than self ligands, no clear sequence or structural differences have been found between mammalian and viral RNAs.

Self nucleic acids can trigger these PRRs once they gain access to compartments in which ligand recognition occurs. Where the activation of innate immune responses by endogenous ligands via PRRs is coupled to the initiation of adaptive immune responses against self-antigens, the triggered autoimmune responses can lead to autoimmunity. Our understanding regarding the crucial role of TLRs in such processes has increased dramatically over the last few years. Research focusing on the interactions of endogenous nucleic acids and TLR activation in connection with SLE has furthered our understanding of the mechanisms that lead to autoimmune progression.

SLE is characterized by a break in tolerance to nuclear components accompanied by lymphopenia, a three-fold increase in plasma cells, and the presence of auto-antibodies. The detection of antinuclear antibodies (ANAs) is used for diagnosis of SLE. Anti-dsDNA and anti-RNP antibodies from the sera of SLE patients in the form of immune complexes with DNA and RNA, were shown to induce IFN-I by PDCs.[45,48] Furthermore, it was demonstrated that RNA-associated autoantigen-stimulated immune activation of B cells is TLR7-dependent,[44] and that TLR7-mediated activation promotes disease progression in murine models of lupus.[79,80] Interestingly, female blood leukocytes induce higher levels of IFN-I in response to TLR7 stimulation than male leukocytes, probably due to x-inactivation escape of this gene that is located on chromosome Xp22.2.[81]

Since SLE is a IFN-driven disease,[82] this mechanism may help explain the higher incidence of SLE in women. Despite a prominent role of PDC-produced IFN-I in SLE, it is worth noting that TLR7 and TLR9 are also expressed on B cells, and that nucleic acid-containing immune complexes activate B cells.[44,83] Therefore both PDCs and B cells may be directly activated by nucleic acid ligands, a process that also plays a role in disease progression.

While endogenous nucleic acids have a clear role in TLR activation, what initiates the response remains unclear. Evidence indicates increased apoptosis and defective clearance of apoptotic cells in SLE patients[84] may contribute to disease progression. In addition, circumstantial evidence implicates viral infections in the etiology of SLE.[85] Viral infection may induce cell death, and therefore provide an initial source of self antigen, while at the same time bystander activation of cells may trigger autoimmune responses to these antigens, setting the vicious cycle involving the generation of nucleic acid-containing immune complexes and TLR7/9-mediated activation of PDCs and B cells in motion. Interestingly, self RNA and DNA serve as antigen and adjuvant simultaneously, a circumstance that may facilitate the induction and the progression of this autoimmune disease.

CONCLUSIONS

The recognition of viral ssRNA by TLR7 and TLR8 serves as a mechanism to sense infections by ssRNA viruses. Although these TLRs have evolved to recognize viral ssRNA, activation may be triggered by a variety of RNAs from different sources including synthetic ORNs, endogenous RNA, and IRMs that share structural similarities to nucleosides. Despite expression of these PRRs by a variety of cell types, activation of DCs by ssRNA ligands and IRMs was studied most intensively. The function of direct activation of other cell types via TLR7 and TLR8 during virus infection is less well understood and requires clarification.

Concerning the characteristics of ligand recognition, mouse TLR7 seems to exclusively sense uridine moieties, while recognition of specific nucleotides or motifs is less clear for human TLR7 and human TLR8. Hopefully future studies will reveal which moieties of RNA are crucial for receptor triggering in the human system. Another interesting question is whether mouse TLR8 is really non-functional, or whether it recognizes a yet unidentified ligand.

Future research must clarify what uptake mechanisms are involved in the recognition of ssRNA present in virus particles, and how the different mechanisms involved in uptake of IRMs and synthetic RNA ligands influence TLR activation. Insight into these mechanisms will improve strategies for immune intervention targeting these PRRs by either employing synthetic agonists as adjuvants for vaccination or by using antagonists to suppress immune responses in the context of autoimmunity.

ACKNOWLEDGEMENT

The author's work is funded by a Cancer Research UK Career Development Award.

REFERENCES

1. Tytell, A. A. et al. 1967. Inducers of interferon and host resistance. 3. Double-stranded RNA from reovirus type 3 virions (reo 3-RNA). *Proc Natl Acad Sci USA* 58:1719.
2. Isaacs, A., Cox, R. A., and Rotem, Z. 1963. Foreign nucleic acids as the stimulus to make interferon. *Lancet* 2:113.
3. Jensen, K. E. et al. 1963. Interferon responses of chick embryo fibroblasts to nucleic acids and related compounds. *Nature* 200:433.
4. Lampson, G. P. et al. 1967. Inducers of interferon and host resistance. 1. Double-stranded RNA from extracts of *Penicillium funiculosum*. *Proc Natl Acad Sci USA* 58:782.
5. Field, A. K. et al. 1967. Inducers of interferon and host resistance. 2. Multistranded synthetic polynucleotide complexes. *Proc Natl Acad Sci USA* 58:1004.
6. Williams, B. R. 2001. Signal integration via PKR. *Sci STKE* 2001:RE2.
7. Samuel, C. E. 2001. Antiviral actions of interferons. *Clin Microbiol Rev* 14:778.
8. Yoneyama, M. et al. 2004. The RNA helicase RIG-I has an essential function in double-stranded RNA-induced innate antiviral responses. *Nat Immunol* 5:730.
9. Yoneyama, M. et al. Fujita, T. 2005. Shared and unique functions of the DExD/H-box helicases RIG-I, MDA5, and LGP2 in antiviral innate immunity. *J Immunol* 175:2851.
10. Kato, H. et al. 2006. Differential roles of MDA5 and RIG-I helicases in the recognition of RNA viruses. *Nature* 441:101.
11. Janeway, C. A., Jr. and Medzhitov, R. 1998. Introduction: the role of innate immunity in the adaptive immune response. *Semin Immunol* 10:349.
12. Uematsu, S. and Akira, S. 2006. Toll-like receptors and innate immunity. J Mol Med.
13. Alexopoulou, L. et al. 2001. Recognition of double-stranded RNA and activation of NF-κB by Toll-like receptor 3. *Nature* 413:732.
14. Diebold, S. S. et al. 2004. Innate antiviral responses by means of TLR7-mediated recognition of single-stranded RNA. *Science* 303:1529.
15. Lund, J. M. et al. 2004. Recognition of single-stranded RNA viruses by Toll-like receptor 7. *Proc Natl Acad Sci USA* 101:5598.
16. Heil, F. Et al. 2004. Species-specific recognition of single-stranded RNA via toll-like receptor 7 and 8. *Science* 303:1526.
17. Jurk, M. et al. 2002. Human TLR7 or TLR8 independently confer responsiveness to the antiviral compound R-848. *Nat Immunol* 3:499.
18. Sugiyama, T. et al. 2005. CpG RNA: identification of novel single-stranded RNA that stimulates human CD14+CD11c+ monocytes. *J Immunol* 174:2273.

19. Judge, A. D. et al. 2005. Sequence-dependent stimulation of the mammalian innate immune response by synthetic siRNA. *Nat Biotechnol* 23:457.
20. Hornung, V. et al. 2005. Sequence-specific potent induction of IFN-α by short interfering RNA in plasmacytoid dendritic cells through TLR7. *Nat Med* 11:263.
21. Sioud, M. 2005. Induction of inflammatory cytokines and interferon responses by double-stranded and single-stranded siRNAs is sequence-dependent and requires endosomal localization. *J Mol Biol* 348:1079.
22. Sioud, M. 2006. Single-stranded small interfering RNA are more immunostimulatory than their double-stranded counterparts: a central role for 2'-hydroxyl uridines in immune responses. *Eur J Immunol* 36:1222.
23. Jarrossay, D. et al. 2001. Specialization and complementarity in microbial molecule recognition by human myeloid and plasmacytoid dendritic cells. *Eur J Immunol* 31:3388.
24. Kadowaki, N. et al. 2001. Subsets of human dendritic cell precursors express different toll-like receptors and respond to different microbial antigens. *J Exp Med* 194:863.
25. Ito, T. et al. 2002. Interferon-alpha and interleukin-12 are induced differentially by Toll-like receptor 7 ligands in human blood dendritic cell subsets. *J Exp Med* 195:1507.
26. Hornung, V. et al. 2002. Quantitative expression of toll-like receptor 1-10 mRNA in cellular subsets of human peripheral blood mononuclear cells and sensitivity to CpG oligodeoxynucleotides. *J Immunol* 168:4531.
27. Gorden, K. B. et al. 2005. Synthetic TLR agonists reveal functional differences between human TLR7 and TLR8. *J Immunol* 174:1259.
28. Miettinen, M. et al. 2001. IFNs activate toll-like receptor gene expression in viral infections. *Genes Immun* 2:349.
29. Mohty, M. et al. 2003. IFN-alpha skews monocyte differentiation into Toll-like receptor 7-expressing dendritic cells with potent functional activities. *J Immunol* 171:3385.
30. Hemmi, H. et ak, 2002. Small anti-viral compounds activate immune cells via the TLR7 MyD88-dependent signaling pathway. *Nat Immunol* 3:196.
31. Gibson, S. J. et al. 2002. Plasmacytoid dendritic cells produce cytokines and mature in response to the TLR7 agonists, imiquimod and resiquimod. *Cell Immunol* 218:74.
32. Heil, F. et al. 2003. The Toll-like receptor 7 (TLR7)-specific stimulus loxoribine uncovers a strong relationship within the TLR7, 8 and 9 subfamily. *Eur J Immunol* 33:2987.
33. Lee, J. et al. 2003. Molecular basis for the immunostimulatory activity of guanine nucleoside analogs: activation of Toll-like receptor 7. *Proc Natl Acad Sci USA* 100:6646.
34. Bernstein, D. I., Miller, R. L., and Harrison, C. J. 1993. Adjuvant effects of imiquimod on a herpes simplex virus type 2 glycoprotein vaccine in guinea pigs. *J Infect Dis* 167:731.
35. Tomai, M. A. et al. 1995. Immunomodulating and antiviral activities of the imidazoquinoline S-28463. *Antiviral Res* 28:253.
36. Miller, R. L. et al. 1999. Treatment of primary herpes simplex virus infection in guinea pigs by imiquimod. *Antiviral Res* 44:31.
37. Spruance, S. L. et al. T. C. 2001. Application of a topical immune response modifier, resiquimod gel, to modify the recurrence rate of recurrent genital herpes: a pilot study. *J Infect Dis* 184:196.
38. Palamara, F. et al. 2004. Identification and characterization of pDC-like cells in normal mouse skin and melanomas treated with imiquimod. *J Immunol* 173:3051.
39. Barnetson, R. S. et al. 2004. Imiquimod induced regression of clinically diagnosed superficial basal cell carcinoma is associated with early infiltration by CD4 T cells and dendritic cells. *Clin Exp Dermatol* 29:639.
40. Wolf, I. H. et al. 2005. Treatment of lentigo maligna (melanoma in situ) with the immune response modifier imiquimod. *Arch Dermatol* 141:510.

41. von Krogh, G. et al. 2000. European course on HPV associated pathology: guidelines for primary care physicians for the diagnosis and management of anogenital warts. *Sex Transm Infect* 76:162.
42. Jurk, M. et al. 2006. Modulating responsiveness of human TLR7 and 8 to small molecule ligands with T-rich phosphorothiate oligodeoxynucleotides. *Eur J Immunol* 36:1815.
43. Kariko, K. et al. 2005. Suppression of RNA recognition by Toll-like receptors: the impact of nucleoside modification and the evolutionary origin of RNA. *Immunity* 23:165.
44. Lau, C. M. et al. 2005. RNA-associated autoantigens activate B cells by combined B cell antigen receptor/Toll-like receptor 7 engagement. *J Exp Med* 202:1171.
45. Barrat, F. J. et al. 2005. Nucleic acids of mammalian origin can act as endogenous ligands for Toll-like receptors and may promote systemic lupus erythematosus. *J Exp Med* 202:1131.
46. Beignon, A. S. et al. 2005. Endocytosis of HIV-1 activates plasmacytoid dendritic cells via Toll-like receptor-viral RNA interactions. *J Clin Invest* 115:3265.
47. Peng, G. et al. 2005. Toll-like receptor 8-mediated reversal of CD4+ regulatory T cell function. *Science* 309:1380.
48. Bave, U. et al. 2003. Fc gamma RIIa is expressed on natural IFN-alpha-producing cells (plasmacytoid dendritic cells) and is required for the IFN-alpha production induced by apoptotic cells combined with lupus IgG. *J Immunol* 171:3296.
49. Barton, G. M., Kagan, J. C., and Medzhitov, R. 2006. Intracellular localization of Toll-like receptor 9 prevents recognition of self DNA but facilitates access to viral DNA. *Nat Immunol* 7:49.
50. Wagner, H. 2004. The immunobiology of the TLR9 subfamily. Trends Immunol 25:381.
51. Iyoda, T. et al. 2002. The CD8+ dendritic cell subset selectively endocytoses dying cells in culture and in *vivo*. *J Exp Med* 195:1289.
52. Schulz, O. and Reis e Sousa, C. 2002. Cross-presentation of cell-associated antigens by CD8alpha+ dendritic cells is attributable to their ability to internalize dead cells. *Immunology* 107:183.
53. Edwards, A. D. et al. 2003. Toll-like receptor expression in murine DC subsets: lack of TLR7 expression by CD8 alpha+ DC correlates with unresponsiveness to imidazoquinolines. *Eur J Immunol* 33:827.
54. Schulz, O. et al. 2005. Toll-like receptor 3 promotes cross-priming to virus-infected cells. *Nature* 433:887.
55. Bekeredjian-Ding, I. B. et al. 2005. Plasmacytoid dendritic cells control TLR7 sensitivity of naive B cells via type I IFN. *J Immunol* 174:4043.
56. Nagase, H. et al. 2003. Expression and function of Toll-like receptors in eosinophils: activation by Toll-like receptor 7 ligand. *J Immunol* 171:3977.
57. Gorski, K. S. et al. 2006. Distinct indirect pathways govern human NK-cell activation by TLR-7 and TLR-8 agonists. *Int Immunol* 18:1115.
58. Hart, O. M. et al. 2005. TLR7/8-mediated activation of human NK cells results in accessory cell-dependent IFN-gamma production. *J Immunol* 175:1636.
59. Matsushima, H. et al. 2004. TLR3-, TLR7-, and TLR9-mediated production of proinflammatory cytokines and chemokines from murine connective tissue type skin-derived mast cells but not from bone marrow-derived mast cells. *J Immunol* 173:531.
60. Andersen, J. M., Al-Khairy, D., and Ingalls, R. R. 2006. Innate immunity at the mucosal surface: role of toll-like receptor 3 and toll-like receptor 9 in cervical epithelial cell responses to microbial pathogens. *Biol Reprod* 74:824.

61. Schaefer, T. M. et al. 2004. Toll-like receptor (TLR) expression and TLR-mediated cytokine/chemokine production by human uterine epithelial cells. *Immunology* 112:428.

62. Mansson, A. et al. 2006. A distinct Toll-like receptor repertoire in human tonsillar B cells, directly activated by PamCSK, R-837 and CpG-2006 stimulation. *Immunology* 118:539.

63. Sauder, D. N. 2005. The emerging role of immunotherapy in the treatment of non-melanoma skin cancers. *Nat Clin Pract Oncol* 2:326.

64. Kerkmann, M. et al. 2003. Activation with CpG-A and CpG-B oligonucleotides reveals two distinct regulatory pathways of type I IFN synthesis in human plasmacytoid dendritic cells. *J Immunol* 170:4465.

65. Izaguirre, A. et al. 2003. Comparative analysis of IRF and IFN-alpha expression in human plasmacytoid and monocyte-derived dendritic cells. *J Leukoc Biol* 74:1125.

66. Coccia, E. M. et al. 2004. Viral infection and Toll-like receptor agonists induce a differential expression of type I and lambda interferons in human plasmacytoid and monocyte-derived dendritic cells. *Eur J Immunol* 34:796.

67. Diebold, S. S. et al. 2003. Viral infection switches non-plasmacytoid dendritic cells into high interferon producers. *Nature* 424:324.

68. Honda, K. et al. 2005. Spatiotemporal regulation of MyD88-IRF-7 signalling for robust type-I interferon induction. *Nature* 434:1035.

69. Guiducci, C. et al. 2006. Properties regulating the nature of the plasmacytoid dendritic cell response to Toll-like receptor 9 activation. *J Exp Med* 203:1999.

70. Prakash, A. and Levy, D. E. 2006. Regulation of IRF7 through cell type-specific protein stability. *Biochem Biophys Res Commun* 342:50.

71. Napolitani, G. et al. 2005. Selected Toll-like receptor agonist combinations synergistically trigger a T helper type 1-polarizing program in dendritic cells. *Nat Immunol* 6:769.

72. Warger, T. et al. 2006. Synergistic activation of dendritic cells by combined Toll-like receptor ligation induces superior CTL responses *in vivo*. *Blood* 108:544.

73. Poeck, H. et al. 2004. Plasmacytoid dendritic cells, antigen, and CpG-C license human B cells for plasma cell differentiation and immunoglobulin production in the absence of T-cell help. *Blood* 103:3058.

74. Cervantes-Barragan, L. et al. 2006. Control of coronavirus infection through plasmacytoid dendritic cell-derived type I interferon. *Blood.*

75. Melchjorsen, J. et al. 2005. Activation of innate defense against a paramyxovirus is mediated by RIG-I and TLR7 and TLR8 in a cell type-specific manner. *J Virol* 79:12944.

76. Triantafilou, K. et al. 2005. TLR8 and TLR7 are involved in the host's immune response to human parechovirus 1. *Eur J Immunol* 35:2416.

77. Triantafilou, K. et al. 2005. Human cardiac inflammatory responses triggered by Coxsackie B viruses are mainly Toll-like receptor (TLR) 8-dependent. *Cell Microbiol* 7:1117.

78. Schlender, J. et al. 2005. Inhibition of toll-like receptor 7- and 9-mediated alpha/beta interferon production in human plasmacytoid dendritic cells by respiratory syncytial virus and measles virus. *J Virol* 79:5507.

79. Subramanian, S. et al. 2006. A Tlr7 translocation accelerates systemic autoimmunity in murine lupus. *Proc Natl Acad Sci USA* 103:9970.

80. Christensen, S. R. et al. 2006. Toll-like receptor 7 and TLR9 dictate autoantibody specificity and have opposing inflammatory and regulatory roles in a murine model of lupus. *Immunity* 25:417.

81. Berghofer, B. et al. 2006. TLR7 ligands induce higher IFN-alpha production in females. *J Immunol* 177:2088.

82. Banchereau, J., Pascual, V., and Palucka, A. K. 2004. Autoimmunity through cytokine-induced dendritic cell activation. *Immunity* 20:539.
83. Leadbetter, E. A. et al. 2002. Chromatin-IgG complexes activate B cells by dual engagement of IgM and Toll-like receptors. *Nature* 416:603.
84. Kaplan, M. J. 2004. Apoptosis in systemic lupus erythematosus. *Clin Immunol* 112:210.
85. Denman, A. M. 2000. Systemic lupus erythematosus: is a viral aetiology a credible hypothesis? *J Infect* 40:229.

15 RNA Interference in Scope of Immune System

Andrea Ablasser, Gunther Hartmann, and Veit Hornung

ABSTRACT

Recent years have seen the rapid rise of RNA interference as a powerful tool to silence gene expression at the post-transcriptional level. The most widely-employed technique to initiate RNA interference is the delivery of small interfering RNAs into the cytosols of target cells. Next to knock-down studies *in vitro*, high hopes surround the development of gene-specific drugs to prevent or cure human disease, although a number of studies have raised serious concerns about the specificity and safety of siRNA applications *in vitro* and *in vivo*. One major unanticipated and unwanted effect is the activation of the immune system upon siRNA application. In addition to potential dose limiting toxicities *in vivo*, siRNA-mediated activation of the immune system can produce a pleiotropy of secondary responses, thereby masking specific effects. This chapter reviews the problem of non-specific recognition of siRNAs by the immune system, along with our current understanding of the underlying mechanisms. We tackle the question of how RNAi can coexist with characterized RNA recognition pathways of the innate immune system, and highlight ways in which immune recognition can be bypassed when employing RNAi. Extending our knowledge on these intriguing topics is an inevitable prerequisite before this touted technology can be introduced as an innovative clinical therapeutic tool.

INTRODUCTION

Interferon (IFN) was first described in 1957 by Isaacs and Lindemann and its history as an antiviral agent has been ongoing since then.[1-3] Because IFN-mediated potent resistance to viral infection, it was then known as a factor that conferred virus interference. Extensive research efforts during recent decades provided a large body of knowledge on the mechanistic level of interferon: its sources, its induction, and its versatile effects as a paramount cytokine, bridging innate and adaptive immunity.

It appears that the immune system must be able to sense the presence of pathogens before it mounts an interference response. Many organisms drive highly specific immune responses when exposed to foreign genetic material to intercept these infectious agents before they are able to propagate. However, in light of their genetic instability and resulting capability of adaptation, reliably and specifically detecting viruses of microorganisms is an intricate challenge.

In order to face this task with a limited number of germline-encoded receptors, the vertebrate immune system evolved the strategy of recognizing highly conserved molecular patterns, specific and unique for entire subsets of pathogens.[4] In the 1960s, long before this concept was born, Maurice Hilleman's group made a fundamental discovery when they identified dsRNA as a molecular pattern that evoked the production of large amounts of interferon.[5,6] Almost 30 years later, Charles Janeway postulated the presence of so-called pattern recognition receptors (PRRs) on theoretical grounds and subsequently proved their existence experimentally.[7] The first group of PRRs characterized is the family of Toll-like receptors (TLRs)[8,9] that harbor specificity for a variety of pathogen-associated molecular patterns (PAMPs), including the two entities of nucleic acids: DNA and RNA.[10–12] More recently, a family of cytoplasmic RNA recognition receptors has been added into the repertoire of the cellular antiviral defense machinery.[13]

Compared to the sophisticated vertebrate PRR systems, plants and invertebrates harness another tactic to control viral infection, and dsRNA plays a key role also within this defense system.[14] The strategy underlying this antiviral mechanism was discovered only a few years ago and termed RNA interference (RNAi).[15,16] While trying to suppress genes in the nematode *C. elegans* by the use of antisense RNA complementary to the target mRNA, Fire et al. made the surprising observation that small amounts of the corresponding dsRNA strongly enhanced the silencing effect.[15] RNAi was found to represent a universal highly-conserved gene regulatory mechanism shared throughout different species including plants,[17] fungi,[18] Drosophila,[19] and mammals.[20,21]

Although dsRNA was clearly the initial trigger of RNAi, the cellular processes that followed were not known. Extensive studies of the mechanisms behind this phenomenon revealed a highly-conserved pathway responsible for regulating a large number of genes. Figure 15.1 illustrates the core concept of RNAi. In the cytoplasm, long dsRNA of exogenous or endogenous origin is cleaved by an RNase-III-type enzyme called DICER.[22] Processing by DICER results in short dsRNA molecules of 21 to 28 base pairs containing characteristic 2-nucleotide overhangs at the 3′ ends of both RNA strands. This short dsRNA product is subsequently passed onto a multi-protein complex called RNA-induced silencing complex (RISC). Within this complex, Argonaut 2 cleaves the passenger strand of the small dsRNA molecule, which is a function of the thermodynamic properties of the duplex. RISC then associates with the cognate mRNA transcript, an event leading to degradation of the target mRNA. The net result of this process is the silencing of a gene at the post-transcriptional level.

In 2001, Tuschl and colleagues published a pioneering study that sparked the propagation of RNAi.[23] They demonstrated that DICER processing may be bypassed when short synthetic dsRNA mimicking endogenous DICER products is introduced into cells. Also known as small interfering RNAs (siRNAs), these molecules retain the function of silencing genes in a sequence-specific manner.

Importantly, this study revealed that synthetic siRNA was short enough to circumvent cell-autonomous antiviral pathways recognizing long dsRNA. The observed

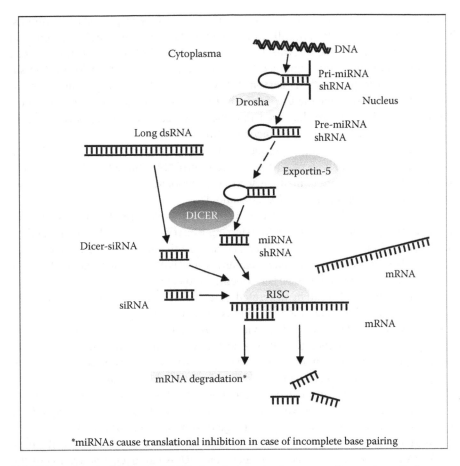

FIGURE 15.1 Mechanisms of RNA interference in the mammalian system. Endogenous genes or artificially introduced plasmids are transcribed into pre-miRNAs or shRNAs that subsequently undergo processing by the RNase III enzyme Drosha in the nucleus. The resulting double-stranded hairpin RNA interacts with exportin-5 (Exp5) that mediates translocation of pri-miRNAs into the cytoplasm. Upon processing by the RNase III enzyme DICER, the resulting short RNA duplex is passed onto a multi-protein RNA-induced silencing complex (RISC). Activated RISC associates with the cognate mRNA transcript and regulates gene expression. Perfect sequence complementarity leads to mRNA degradation, whereas incomplete sequence match leads to translational repression, leaving the mRNA transcript intact. Long dsRNA can also be directly processed by DICER, resulting in short siRNA duplexes that interact with RISC. Alternatively, synthetic siRNAs mimicking endogenous DICER products are incorporated into RISC and subsequently guide sequence-specific RNA degradation.

complete shutdown in protein synthesis had in fact hampered the use of long dsRNA to initiate RNAi in somatic mammalian cells. Unlike classical genetic knock-out approaches, RNAi is simple, cost- and time-effective, and thus rapidly evolved as a method of choice to study gene functions in cells and whole organisms. In addition to the great possibilities for target validation studies, siRNA technology holds great promise to silence disease-related genes in therapeutic settings.

Based on the advanced PRR system and powerful IFN responses of mammals, it seems unlikely that RNAi still plays an important role in antiviral defense in this class of animals. The discovery of micro RNAs (miRNAs) orchestrating gene expression at the post-transcriptional level added a physiological RNAi mechanism to mammalian responses.[14] Primary miRNAs (pri-miRNAs) are genome-encoded natural hairpin RNAs that mature into miRNAs while being exported into the cytoplasm (Figure 15.1) where they feed into the "classical" RNAi pathway by being processed by DICER, followed by incorporation into RISC.

Depending on the degree of sequence homology, these non-coding RNAs can work in two modes. In case of a perfect match, miRNAs guide the cleavage of the bound mRNA; an imperfect match leads to translational arrest of the mRNA, however, without disrupting the mRNA molecule.[14]

Based on the understanding of the mechanisms underlying RNAi, a variety of methods have proven successful in transferring RNAi into mammalian cells. Endogenous transcription of self-complementary short hairpin RNA (shRNA) from an imported DNA plasmid mimics endogenous pri-miRNA that follows along the natural miRNA route to mature into an active degradation template (Figure 15.1). Alternatively, siRNA molecules can be generated in test tubes and subsequently transferred into cells to enter the route of RNAi, bypassing endogenous processing. One option is the enzymatic generation of RNA via *in vitro* transcription, achieved by the use of T7 phage RNA polymerase that produces RNA from a DNA template containing a T7 promoter upstream of the region coding for the RNA transcript of interest. Using this technique, siRNA can be generated via two short separate strands, or via a long dsRNA precursor that can be digested by recombinant DICER or RNase III to yield siRNA molecules. However, the most common method for composing siRNA is chemical synthesis that slows a great range of additional modifications.[24]

More recent studies continue to fuel concerns about the specificity and safety of siRNA applications. Evidence indicates that siRNA can trigger activation of the innate immune system. Initial studies focused on so-called nonspecific effects mediated by activation of the IFN pathway and induction of IFN-inducible genes in cell lines via the application of RNAi.[25-27] These events can mask specific gene silencing effects initiated by the RNAi pathway, and therefore potentially complicate the interpretation of data. Additional drawbacks emerged when *in vivo* siRNA delivery was shown capable of invoking severe systemic immune responses.[28-30]

NONSPECIFIC siRNA EFFECTS

Recent studies curbed the initial enthusiasm about the use of siRNA because unwanted effects were reported. Cross-hybridization of siRNA and mRNAs closely related to the target of interest propagates the unintentional silencing of the respective gene products. These so-called "off-target effects" result from loose requirements for base pair complementarity, as observed in miRNA-mediated regulation, and demand careful siRNA design to circumvent this obstacle. siRNA can trigger intrinsic responses accompanied by the induction of type I IFN that can result in a global shutdown of protein synthesis. In contrast to the off-target effects, these phenomena are called "nonspecific effects." Bear in mind that the mechanism behind off-target effects is indeed the genuine RNAi machinery, whereas nonspecific effects are due to recognition pathways independent of this system.

Upregulation of a variety of IFN-stimulated genes occurred after transfection of siRNA into tumor cell lines (T98G and RCC1), and this was controlled by the intrinsic IFN feedback loop leading to the activation of the Jak-Stat-pathway. Both chemically and enzymatically, synthesized siRNAs elicited respective nonspecific effects, whereas the single-stranded counterparts were not active.[25] Similar but somewhat contrasting observations were reported by Bridge and colleagues. While synthetic siRNA failed to exert an effect on the IFN responses in cell lines (HLFs and HeLa cells), this group reported a potent stimulation of the IFN-inducible 2'5'-oligoadenylate synthetase gene by the endogenous expression of shRNA.[26] Primarily responsible for the shRNA-mediated IFN-response was the U6 promoter sequence harboring an AA dinucleotide near the transcription start site.[31]

Kim and colleagues described IFN induction in multiple cell lines (HEK-293, HeLa, and CV1) by siRNA synthesized using T7 phage RNA polymerase (T7RNAP) but not by chemically-designed siRNAs. The critical difference in these preparations may be attributed to the 5' end: unmodified T7RNAP-derived RNA transcripts harbor a triphosphate moiety at the 5' end, whereas synthetic RNA is typically unphosphorylated at the 5' end. Dephosphorylation of the 5' ends of *in vitro* transcribed RNA rendered this type of RNA completely inert in terms of immunostimulation.[27]

Findings by Kariko et al.[32] further support the notion of siRNA triggering nonspecific immune stimulation. Both chemically and enzymatically generated siRNA molecules elicited profound cytokine production in cell lines (HEK 293, HaCaT), primary macrophages, and dendritic cells concomitant with global protein suppression.

Despite confusion about the pathways responsible for triggering IFN, initial reports proposed involvement of the cytoplasmic sensor double-stranded RNA-activated protein kinase (PKR) or TLR3,[25-32] However, conflicting data made it difficult to generalize about the source of siRNA capable of evoking nonspecific responses. It is noteworthy that cell lines routinely used in laboratories often display dysfunctional IFN induction or response pathways, and thus greatly vary in their abilities to respond to siRNAs. In view of these observations, cell lines may not be adequate models for clarifying the mechanisms underlying siRNA recognition and monitoring nonspecific siRNA effects prior to the *in vivo* setting. These early experiments indicated that introduction of siRNA into cells is prone to elicit effects beyond sequence-specific gene regulation by initiating nonspecific translational inhibition via type I IFNs.

siRNA UNDER TOLL SURVEILLANCE

The non-specific effects mediated by long dsRNA were conventionally attributed to the cytoplasmic dsRNA recognition protein known as PKR.[21,33] The molecular mediators of pattern recognition are, however, not mutually exclusive; they share overlapping specificities. In addition to cytoplasmic RNA receptors, cells of the innate immune system are equipped with a broad repertoire of transmembrane glycoproteins known as Toll-like receptors (TLRs).[4] TLRs 3, 7, 8, and 9 share the distinctive ability to recognize nucleic acids in the form of RNA[11,13,34] or DNA.[10] The expression profiles among different cell populations are not identical.[35]

TLR3 recognizes dsRNA in a sequence-independent manner, and unlike other RNA-detecting TLRs, it is not exclusively expressed on immune cells.[34,36] Nucleic acids in the form of DNA exert potent immunostimulatory effects by triggering TLR9 in plasmacytoid dendritic cells (PDCs) and B cells.[10,37,38] It is well established that this TLR functions through species-specific recognition of CpG motifs, a common element of bacterial or viral genomes that is suppressed within mammalian DNA. Based on sequence and underlying secondary structure, different classes of CpG-ODNs preferentially initiate distinct immune responses in different cell types. Therefore one can subdivide CpG-ODNs into three distinct classes: CpG-A, CpG-B, and CpG-C ODNs.[38–40] TLR7 and TLR8 also belong to the TLR subfamily whose members act cooperatively to recognize nucleic acids.

Recent advances revealed a crucial role for TLR7 and TLR8 in sensing RNA.[11,12] It was established earlier that imidazoquinolines elicit activation of the innate immune system through these two TLRs.[41] Major representatives of this class of ligands are imiquimod and resiquimod. Imiquimod is already in clinical use as an antiviral for treating papillomavirus infection.[42] Although reminiscent in their chemical structures of the adenosine and guanosine purine bases, these compounds do not represent viral products in nature. The physiological substrate for both receptors was identified to be single-stranded RNA, initially derived from the genome of influenza virus and HIV. In a first attempt to unravel the immunostimulatory sequence within RNA strands, poly-U RNA and RNA rich in guanosine and uridine appeared very potent in triggering IFN responses in PDCs. However, understanding sequence-specific recognition by TLR7 and TLR8 in terms of guanosine- and uridine-rich RNA sequences may be an oversimplification and may not cover all aspects of preferred sequence characteristics as illustrated later in this chapter.

The production of type I IFN is a universally shared feature among diverse cell types, although PDCs are by far the major sources of INF-α in the human immune system.[43] To meet this requirement, PDCs express TLR7 and TLR9 within endosomal compartments. Upon TLR stimulation, subsequent signaling events are initiated, finally culminating in the production of IFN-α. In an effort to silence TLR9 in PDCs by means of RNAi, we made the surprising observation that transfection of siRNA boosted the production of IFN-α in PDCs.[28] Single-stranded components of the siRNA duplex promoted IFN-α production in the same range as full siRNA duplexes.[28,44] The latter observation raised the suspicion that remaining, nonannealed ssRNA molecules may be responsible for the siRNA-mediated immunostimulatory effect.

To address this question, siRNA was pretreated with RNase, selectively degrading ssRNA. Identical amounts of IFN-α were still triggered in PDCs, and thus substantiated the hypothesis of a direct influence of siRNA on its respective receptor. Despite significant differences in the levels of IFN-α produced by individual siRNAs, a remarkable IFN signal could also be elicited by siRNAs targeting genes not expressed in PDCs. Equally interesting, both sense or antisense strands could elicit activation.

In conclusion, siRNA sequences with no direct mRNA targets induced IFN-α production and both the sense and antisense strands may be active. These results led to the hypothesis that IFN production works independently of RNAi activity but still represents an active immunological response toward siRNA. This assumption was confirmed by the observation that siRNA molecules active in terms of gene silencing could be rendered completely non-immunostimulatory by the introduction of base-modified nucleotides. The short dsRNA duplexes were found to require minimal lengths of about 19 bases, while shorter immunostimulatory motifs could be flanked by non-stimulatory sequences.[28] Detection of siRNA leading to IFN-α production required endosomal maturation.[29] Our group found that TLR7 was the receptor responsible for recognition (Figure 15.2). Moreover, we and other groups observed that systemic application of siRNA encapsulated with cationic lipids resulted in a systemic inflammatory response *in vivo* and severe signs of morbidity.[28–30] Intravenous injections of siRNA into mice markedly enhanced systemic levels of IFN-α and

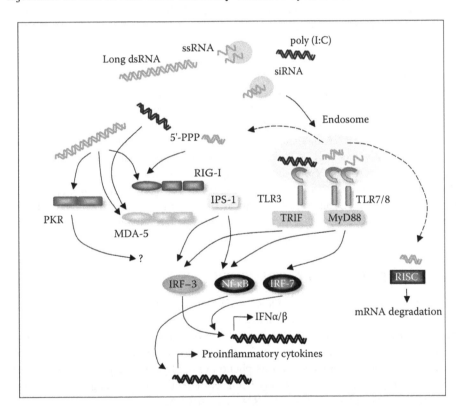

FIGURE 15.2 RNA recognition pathways. 5′ triphosphate RNA of single- or double-stranded configuration represents a ligand for RIG-I, whereas poly(I:C) is recognized by MDA-5. IPS-1 (also known as MAVS, Cardif, or VISA) is recruited as the first adaptor molecule linking RNA recognition through these two helicases with the translocation of transcription factors NF-κb and IRFs into the nucleus. The precise role of PKR, the exact nature of its ligand, and the subsequent signaling cascade are not fully understood, but long dsRNA appears to channel into this pathway. Beyond the cytosolic recognition machinery, professional APCs express TLRs 3, 7, and 8 that recognize long dsRNA, ssRNA, and siRNA, respectively. Activation of TLRs recruits TIR-domain-containing adaptor molecules and subsequently initiates transcription of type I interferons and pro-inflammatory cytokines.

caused activation of immune effector cells. Unlike PDCs, CD4+ T cells, CD8+ T cells, and MDCs do not express TLR7, and thus their activation was due to secondary effects. In fact, a complete loss of IFN production and secondary immune cell activation were seen in TLR7-deficient mice upon siRNA injection.[28] Further clues about systemic siRNA-mediated toxicities appeared when Ian MacLachlan's group reported transient thrombocytopenia, leukopenia, body weight loss, hunched posture, and piloerection induced by siRNA.[29,30,45]

Note that transfection of chemically synthesized siRNA is recognized by a highly conserved immunological mechanism. Consecutively, this divorces the application of this type of siRNA depending on the presence or absence of TLR7. When delivered into cells lacking TLR7, as is commonly the case in cell culture experiments, chemically synthesized RNA is predicted to be secure and thus specific in terms of RNA interference. However the same molecule administered *in vivo*—an immune-competent surrounding where TLR7 is present—will activate immune cells leading to deleterious systemic responses.

The ability of TLR3 to interact with dsRNA makes it a logical candidate receptor for siRNA recognition. Initial studies saw TLR3 as a mediator of siRNA-induced IFN-β production in HEK 293 cells overexpressing TLR3.[32] However, the fact that TLR3 recognition is thought to occur independently of dsRNA sequence[34] contrasts with recent reports of sequence-specific siRNA immune effects *in vitro* and *in vivo*.[28–30] Indeed, siRNA-mediated IFN induction in murine bone marrow cells was completely lost when TLR7-deficient cells were analyzed. In line with this observation, bone marrow cells from mice lacking TRIF, a necessary adaptor protein of TLR3, were not impaired in the recognition of siRNA (Hornung et al. unpublished data). Recent publications reveal that most of the IFN-α–inducing activity previously ascribed to TLR3 is due to the cytoplasmic helicase melanoma differentiation associated protein 5 (MDA5; Figure 15.2).[46,47]

IMPACT OF STRUCTURE VERSUS SEQUENCE

While progress has been made in siRNA design for gene silencing activity through prediction algorithms,[48] the exact nature of the components within the RNA molecules that account for immune activation remain poorly understood. For avoiding or intentionally introducing immunostimulatory activity into ssRNA or siRNA molecules, it is crucial to understand this phenomenon. Specific sequences and certain structural requirements have recently been reported to impact recognition and subsequent immune responses toward ssRNA and siRNA.

The events that trigger the immune system via ssRNA-mediated TLR7 activation were initially described to depend on the underlying sequence.[11,12] Consistently, our panel of siRNAs showed significant differences in levels of IFN induction, and further reports substantiated the observation that TLR7-mediated recognition of ssRNA or siRNA is strongly influenced by the base composition.[28,29,49]

In searches for sequence requirements, we were able to show that the mere contents of guanosine or uridine bases were not sufficient to render RNA stimulatory, as had been suggested; the identical number of GU dinucleotides displayed differential activity and ssRNAs with few or no GU residues were also immunologically potent.[49]

Instead, the internal 9-mer motif 5'-GUCCUUCAA-3' could be identified as a strong trigger by our group. Others have also noted that the presence of 5'-UGUGU-3' in siRNA duplexes constitutes another putative stimulatory motif (Figure 15.3).[29] These motifs, however, mirror only representatives of particularly active base arrangements, while additional sequences or motifs are likely to be added to this repertoire.

Introduction of adenosines lowers stimulatory properties, while uridines seem to drive immunorecognition.[49] Flanking regions or positional shifts within the RNA strand may also influence immunostimulatory properties. In addition, CpG-RNA motifs flanked by poly-G tails were recently reported to induce the upregulation of co-stimulatory molecules and cytokines within monocytes.[50] A second point in considering sequence-specific effects caused by ssRNAs and siRNAs concerns the differences between TLR7 and TLR8. Despite interacting with the identical class of ligands and their close evolutionary relationship,[51] accurate discrimination of both

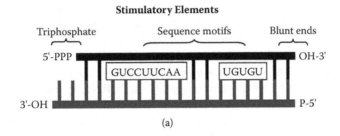

(a)

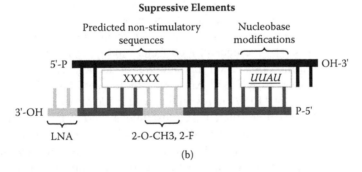

(b)

FIGURE 15.3 Immunomodulatory components within are siRNA duplex. (a) Primary siRNA duplexes may contain intrinsic structural and sequence-dependent immunostimulatory elements. SiRNA synthesis by means of T7 RNA polymerase leads to a triphosphate moiety at the 5' end that is recognized via the RIG-I cytosolic RNA-receptor. Blunt-end siRNA duplexes in the range of 27 nt have also been reported to activate RIG-I. On the other hand, TLR7- and TLR8-dependent recognition involves characteristic sequence motifs. (b) Strategies aimed to avoid immunostimulation while retaining silencing activity. Chemical modifications that inhibit immunorecognition include locked nucleic acids (LNAs) and backbone modifications such as 2-*O*-methyl and 2-fluoro. Algorithms that select for low immunostimulatory activity of a specific RNA strand represent another potential approach. Nucleobase modifications may dampen immunostimulation via siRNA, yet their role in maintaining the functionality of the siRNA duplex in terms of silencing activity remains to be assessed.

receptors is essential. Sequence preference by TLR7 may be distinct from that of TLR8. Notably, some RNA sequences preferentially promote activation of one type of immune effector cell and remain unrecognized in others.[12,29,50] It is noteworthy that a G-to-A replacement within an immunostimulatory RNA sequence abrogated IFN-α induction by PDCs, whereas TNF-α, IL-6, and IL-12 in PBMCs were not affected by this substitution.[12] Beyond a universal stimulatory RNA sequence, both the type of TLR triggered and the cell subset stimulated critically impacted induced activities. As with different cell populations, murine TLR7 has been reported to favor distinct sequence motifs compared to its human counterpart.[12] This finding is analogous to the species-specific preference for flanking regions of CpG motifs by murine and human TLR9.[10,52] However, in contrast to the human system, murine TLR8 has been reported to be dysfunctional and thus seems not to contribute to RNA recognition processes.[12]

Another characteristic property of RNA determination of immunostimulatory activity lies within its single- or doubled-stranded configuration. The finding that a stimulatory motif can confer activity to both ssRNA and dsRNA points to a model in which immunostimulatory sequences are equally recognized in the context of either RNA conformation. The striking observation that monocytes, unlike PDCs, seem highly responsive only when stimulated with ssRNA and not the cognate dsRNA[49] (Hornung et al., unpublished data) suggests that ssRNA molecules are critical for recognition by monocytes or TLR8 .

Although it is clear that TLR recognition relies on sequence motifs, a unifying model combining cell type-specific responses and the contribution of RNA configuration with receptor specificity remains to be found. The recently described crystal structure of TLR3 postulates direct interactions of RNA with TLR domains,[53] as shown for CpG and TLR9.[54,55] Aside from the main actors, the presence of auxiliary proteins or co-receptors may facilitate uptake and recognition in some cell types and limit it in others. The presence of a co-factor modifying RNA-responsiveness is implied based on the observation that addition of autologous plasma enhances inflammatory responses to RNAs.[29] Answering these questions calls for binding studies with recombinant TLR7/8 to reveal the exact mode of interaction.

The profound effects that single base substitutions exert in terms of immunostimulation imply that TLR-dependent RNA recognition strongly depends on base sequence. This contrasts with cytoplasmic RNA recognition, whereby structural motifs such as 5′ triphosphate groups are in fact critical components (Figure 15.3). Thus, both structure and sequence are crucial components that determine the way siRNA triggers non-specific immune responses.

RECOGNITION OF TRIPHOSPHATE RNA VIA RIG-I

The observation by Kim et al. that *in vitro*-transcribed siRNA induced potent IFN-α responses in various cell lines pointed to another interesting aspect of RNA recognition by the innate immune system.[27] Addressing the impacts of long ssRNA molecules on activities of human monocytes, our group used *in vitro* transcription to generate ssRNA because chemical synthesis is impractical for generating ssRNAs larger than 100 nucleotides. Transfection of long ssRNA molecules into monocytes showed high potency in terms of IFN induction and short synthetic ssRNA was completely inactive.

This activity, however, could not be attributed to certain motifs or secondary structures of the *in vitro*-transcribed RNA molecule, but was critically dependent on the 5' moiety of the RNA transcript. Indeed, viral polymerases classically employed for *in vitro* transcription reactions leave 5' triphosphate groups on the RNA transcripts, whereas synthetic RNA is typically generated to contain 5' hydroxyl group. Consistent with this notion, dephosphorylation of *in vitro*-transcribed RNA completely abolished the IFN-α-inducing activity of *in vitro*-transcribed RNA.

While endosomal maturation and TLR3 and TLR7 were not required for recognition, retinoic acid-inducible gene I (RIG-I) was the critical receptor required for recognition of 5' triphosphate RNA, whereas the closely related RNA helicase MDA5 was not involved. RIG-I ligand activity of 5' triphosphate RNA was independent of dsRNA formation; single- and double-stranded 5' triphosphate RNA molecules were equally active. Both exogenous 5' triphosphate RNA transfected into cells and endogenously formed 5' triphosphate RNA activated RIG-I. Genomic RNA prepared from a negative strand RNA virus and RNA prepared from virus-infected cells, but not RNA from non-infected cells, triggered potent IFN-α responses in a 5' phosphorlylation-dependent manner.

Binding studies of RIG-I and 5' triphosphate RNA revealed a direct molecular interaction. All these results showed that 5' triphosphate RNA present in viruses and in virus-infected cells but otherwise absent in eukaryotic cells serves as a molecular signature for detecting viral infections via RIG-I.[56] Based on the broad expression of RIG-I in primary somatic cells and even tumor cell lines, *in vitro*-transcribed siRNA constitutes a universal trigger of type I IFN responses. As a consequence, extreme caution is required in drawing conclusions from siRNA studies carried out with *in vitro*-transcribed RNA molecules.

DISCRIMINATION OF SELF AND FOREIGN NUCLEIC ACIDS

The pattern recognition strategy of the innate immune system rests on highly conserved molecular patterns unique to microbes and absent in the host. Nucleic acids are essential constituents of every living organism, and thus our immune system. Recognizing foreign RNA demands a mechanism that allows discrimination of self and non-self RNAs. This raises the important questions of what molecular determinants are associated with pathogen-derived RNA and how host RNA can escape recognition by endogenous RNA receptors.

According to their ligand specificities, individual TLRs differ in their subcellular localizations. TLRs that specialize in detection of viral and bacterial nucleic acids are expressed in intracellular organelles. The intracellular expression of TLR3 is controlled by its cytoplasmic linker region, and the transmembrane domain regulates expression for TLR7 and TLR9.[36,57-60] Such compartmentalization may serve dual functions. First, the low pH microenvironment of endosomes facilitates the uncoating of viral capsids and the release of TLR ligands, namely RNA or DNA, for adequate recognition. Second, TLR localization within the endosomal niche may have evolved to scan for viral or bacterial RNA and aid in determining sources of nucleic acids by providing access to viral RNA while avoiding self-RNA encounters.

Under normal conditions, host RNA is abundantly found in cytoplasm or within nuclei and circumvents the endosomal pathway. Notably, a number of viruses have been reported to enter host cells via the endosomal route.[61] The latter hypothesis is supported by the observation that host RNA is not generally ignored by TLRs but can regain the ability to stimulate immune effector cells under pathological circumstances when endogenous RNA breaks this barrier and enters endosomes.[62] Additional support of the concept of compartmentalization of PRRs that renders host RNA non-stimulatory came from John Rossi's group. They showed that lipid-delivered, exogenous siRNA induced IFN in CD34+ progenitor-derived hematopoietic cells, whereas the same sequences endogenously expressed via shRNAs were non-stimulatory.[63]

Although sequestration of TLRs certainly increases the probability that immune responses are exquisitely mounted against foreign RNA, further reports hint at additional signatures rendering RNA immunogenic. It has long been appreciated that bacterial DNA contains markedly more unmethylated CpG dinucleotides—well-known TLR9 ligands—than vertebrate DNA in which this specific nucleotide combination is underrepresented.[52] Consequently, certain nucleotide combinations may also be suppressed within the mammalian RNA spectrum. Data indicate that UG sequences occur more frequently in viral RNA, but precise information on nucleotide suppression in mammalian RNA is scarce.[64]

Recent evidence suggests that certain nucleoside modifications within RNA occur more frequently in vertebrates than in lower organisms, adding an additional facet of self versus non-self recognition.[65] RNA derived from prokaryotes elicits a stronger activation than mammalian RNA, correlating with lower degrees of nucleotide modifications in prokaryotic RNAs.[66] On the other hand, endogenous eukaryotic RNA is subject to numerous modifications.[67] Besides 5′ capping, eukaryotic RNA undergoes several other post-transcriptional maturation steps, including modifications of various nucleobases of the RNA transcript and methylation of the backbone ribose at the 2′ hydroxyl position. As to the immunosuppressive effects of nucleobase modifications, TLR7 and 8 appear particularly sensitive to alterations of uridine, with substitution of this base by pseudouridine, 5-methyluridine or 2-thiouridine, resulting in a substantial decrease in activation. Interestingly, TLR3 differs from TLR7 and TLR8 in its susceptibility to 2′ methylated RNAs. The activation profiles in primary DCs were not identical to those of TLR-expressing cell lines, suggesting that additional mechanisms of RNA recognition were involved in these cell types.[66]

In addition to hampering TLR recognition, nucleoside modifications also strongly impact the activity of 5′ triphosphate RNA to trigger via RIG-I.[56] Introducing pseudouridine or 2-thiouridine into 5′ triphosphate RNA almost completely abolished the IFN-α-inducing activity of *in vitro*-transcribed RNA oligonucleotides. In addition, methylation of the ribose at the 2′ hydroxyl position also strongly decreased IFN-α-inducing activity of 5′ triphosphate RNA. While 5′ capping of *in vitro*-transcribed RNA appeared not to alter recognition via the TLR system, RIG-I mediated activity was reduced proportionally to the number of RNA transcripts capped. These results indicate that post-transcriptional modifications commonly found in mature eukaryotic RNA species suppress the immunostimulatory activities of RNA oligonucleotides via the TLR-system and RIG-I, thereby providing additional molecular structures that allow the distinction of self and non-self RNAs.

Evidence indicates that short synthetic dsRNAs such as siRNA do not elicit type I IFN responses in non-immune cells and immune cells devoid of TLR7 and TLR8.[21,27,28] A recent study suggests that the two nucleotides overhanging the 3′ ends of DICER cleavage products are essential for the lack of immunorecognition of short dsRNA. The same study proposed that synthetic blunt-end short dsRNA is recognized via RIG-I. The conclusion that RIG-I is the receptor for such dsRNA is based on experiments using RIG-I overexpression and anti-RIG-I siRNA (short dsRNA with two nucleotide 3′ overhangs) on top of stimulation with blunt-end short dsRNA stimulation. RIG-I-deficient cells were not examined.[68]

An emerging model suggests that the innate defense machinery has multiple stages of control that act cooperatively and thereby allow an ostensibly paradoxical situation, namely detecting RNA as a danger signal (Figure 15.2). The system incorporates (1) sequestration of RNA sensors in niches, where under normal circumstances self RNA is bypassed, and (2) structure- and sequence-based disparities between endogenous and exogenous RNA products. The exact characterization of possible immunostimulatory components should facilitate the development of siRNA to circumvent such determinants—a major goal in advancing this technology to clinical application.

BYPASSING IMMUNOSTIMULATION OF siRNAS

Based on the strong potential for siRNA-based drugs, enormous research effort focused on strategies to enhance the effectiveness of *in vivo* application of siRNA. Multiple parameters must be considered to achieve effective and prolonged silencing activity: stability of the RNA molecule, biodistribution, pharmacokinetics, biocompatibility, and lack of toxicity.[69] Toxicities associated with administration of siRNA *in vivo* may be partially explained by its immunostimulatory component[28–30,45] and represent fundamental barriers to translating the method into a clinically practicable setting. Immunostimulation and gene silencing are two independent functions of siRNA, and modulation of one function does not necessarily mutually influence the other.[25,28,29] Present research focuses on evading immune recognition.

The finding that the siRNA sequence is the central culprit of immunostimulation points to the need to consider the primary RNA sequence as an important criterion for safe therapeutic use. It is most likely that stimulatory sequences occur at high frequency in randomly designed siRNAs and therefore additional algorithms that select for low immunostimulatory sequences are required. In a comprehensive analysis we created a database of stimulatory IFN-inducing RNA sequences. Using this RNA library, we then devised a mathematical algorithm that allowed the prediction of the putative stimulatory value of any RNA strand (Hornung et al., unpublished data). Complementing available siRNA predicting platforms with this algorithm may provide a valuable platform for effective and safe siRNA design.

Chemical RNA modifications intended to enhance the pharmacokinetics of RNA therapeutics such as antisense ORNs and ribozymes[70–72] may also serve to limit siRNA-caused immunostimulation. It is advantageous to classify chemical modifications into (1) backbone modifications compromising the phosphodiester bond and the ribose sugar, and (2) nucleobase modifications altering the base chemistry. While backbone modifications generally do not directly affect Watson-Crick base pairing to the target mRNA but can increase stability toward nucleases, they are appreciated for enhancing siRNA efficiency.[73] These modifications include phosphorothioate (PT) duplexes containing phosphor-to-sulfur substitutions of phosphodiester bonds that were found to be less susceptible to nuclease degradation. This may be advantageous for *in vivo* delivery even though PT does not significantly promote gene suppression or affect immunostimulatory properties of siRNA in complex with liposomes.[74]

Other groups studied the impacts of different substituents at the 2′ position of the ribose on stimulatory properties of RNA.[30,49,50] The introduction of 2′-fluoro, 2′-deoxy, and 2′-*O*-methyl groups in ssRNA abrogated TNF-α in human PBMCs.[49] Most of these derivatives did not interfere with siRNA activity, at least when the antisense strand was not affected. MacLachlan's group confirmed that as few as two 2′-*O*-methyl-modified uridines rendered siRNA non-stimulatory.[30] Interestingly, this inhibitory effect was also operational in *trans* immunosuppression was seen for an active siRNA duplex when a stimulatory ssRNA strand was left unmodified. Although completely 2′-*O*-methyl-modified duplexes lack their silencing ability, the introduction of 2′-*O*-methyl substitutes to a lesser extent was well tolerated and induced efficient gene inhibition.[21,75–77]

The locked nucleic acid (LNA) modification is another tool for modulating silencing activity. Containing a methylene linkage between the 2′ oxygen and the 4′ carbon of the ribose ring, LNA strongly enhances binding affinity[78] and thus increases the melting temperature of hybridized RNA. When positioned into the wing region, LNAs do not compete with mRNA suppression. Introduction of LNAs into both ends of the sense strand maintains the full silencing activity of the siRNA duplex.[28,78] However, in contrast to the above described 2′-*O*-methyl modification, LNA introduction into the ssRNA of an siRNA duplex does not have an inhibitory effect in *trans* on a stimulatory motif within the same strand or the complementary strand.[28] Altogether, siRNA molecules that combine high degrees of backbone modifications (>90%) appear to promote effective RNAi-dependent silencing of HBV RNA *in vivo*.[45]

Nucleobase modifications offer other opportunities to exert suppressive effects on the production of pro-inflammatory cytokines. Methylation of cytosines in mammalian DNA strongly suppresses immunostimulatory properties of host DNA and more recently it has been shown that methylation of cytidines rendered RNA non-active in human monocytes.[10,50] Several base modifications, especially on uridines, were shown to alter TLR ligand activity.[66] Although these parameters may also be applied in designing non-stimulatory siRNAs, they may be less reliable because they directly interfere with complementary base pairing. In this respect, the presence of 6-methyladenosine within RNA markedly destabilizes RNA duplexes.[79] However, further investigation is needed to validate the impacts of base modifications on mRNA degradation via RNAi.[78]

Conclusions to be drawn from these experiments are not straightforward. A variety of reasons may explain differential effects of chemical modifications. While some modifications may confer a broad inhibition of the immunostimulatory activity (2′-*O*-methyl), certain modifications may be relevant only in particular sequential contexts or meticulously defined positions (LNAs). It is not yet possible to predict how particular alterations in chemical structures of siRNA molecules impact their gene regulatory functions or stimulatory potencies. This underscores the need to test all potentially harmful siRNAs in immunocompetent milieus before they advance to *in vivo* applications.

Another technological obstacle related to immunostimulation lies in delivery vehicles that shield RNA from nucleases and mediate uptake into the target cells. Because the RNAi machinery is located within the cytoplasm, exogenous siRNA must cross cell membranes to reach this compartment to initiate the gene-silencing process. Two major obstacles hamper the effective delivery of siRNA into cells or tissues. First, RNA is highly susceptible to degradation by ubiquitous enzymes in the blood and other tissue fluids. Second, siRNA molecules are not readily taken up into cells. This demands stabilization and modification of siRNA to increase the efficiency of delivery.

Strategies to overcome both problems typically involve use of macromolecular carriers that bind to siRNA. One widespread technique is the use of polycationic lipids or peptides to complex siRNA molecules. Cationic lipids, however, do not directly fuse with plasma membranes; instead they are incorporated into the cellular endocytosis pathway[80] where TLRs 3, 7, 8, and 9 are located. The mechanism of TLR9 activation has been studied in detail and most likely this pathway is attributable to TLR7 and 8 as well. Non-activated TLR9 resides in the endoplasmic reticulum and is recruited to endosomes upon stimulation with CpG where the encounter with DNA subsequently triggers downstream signaling.[54,57] Therefore, if RNA traffics through the endosomal compartments of innate immune cells, it is predicted to induce TLR stimulation. Thus, RNA bonds to cationic polymers do not circumvent natural endocytosis and, on the contrary, may enhance their stimulatory effects through this route. Of interest, the peptide-driven delivery preferentially activates myeloid immune cells, whereas cationic lipids appear to be stronger stimuli for PDCs.[29] Administration of naked siRNA via high pressure tail vein injection bypassed this delivery route in mice and led to target-specific mRNA degradation.[81] This critical treatment allows rapid injection of huge volumes over a short period and not rarely produces lethal outcomes. Alternatively, *in vivo* delivery without the need for transfection agents can be achieved by 2′-F-modified siRNA, rendering siRNA resistant to RNase A.[82]

Another remarkable improvement in siRNA administration was achieved by covalent conjugation of a cholesterol to modified siRNAs (two PT linkages and two 2′-O-methyl groups at the 3′ ends of the siRNA molecule).[83] The covalent linkages of bulky molecules at the 3′ ends of the sense strands do not interfere with gene silencing.[78] Successful immune evasion was most likely achieved through the chemical modifications used and through cholesterol modification.

One novel approach took delivery one step further and aimed to target specific cell types *in vivo*. Song et al. engineered a fusion protein that contained a positively charged protamine domain and a heavy chain antibody fragment F(ab). The positively charged peptide portion non-covalently tethered siRNA and the antibody fragment served to carry siRNA specifically to cells of interest, thus bypassing nontargeted immune cells.[84]

As highlighted above, current efforts focus on methods of circumventing immunostimulation while retaining full silencing activity. Unraveling the precise stimulatory triggers within siRNA and improving our understanding of self versus non-self recognition are critical issues for designing a highly specific and safe siRNA that meets sequence, chemical modification, and delivery system requirements.

CONCLUSION

Several types of molecules have fueled our expectations for selectively impeding protein expression and thereby acting as miracle drugs to cure a variety of diseases. The era of antisense ODNs brought Fomivirsen[85] into clinical use to treat CMV retinitis, and second generation ODNs have been approved for clinical trials. An even more promising spin-off of this technology has been the development of immunostimulatory ODNs with CpG-DNA as their most clinically advanced representative. The advent of RNAi brought another actor onto the antisense stage. This development holds great promise, because it exploits a natural and ubiquitous regulatory mechanism. The power of siRNA to silence virtually any gene of interest underscores the potential of RNAi.[86]

Its remarkable ability of targeting adverse conditions such as chronic viral infection and cancer has been shown in mouse models *in vivo*, and further areas of siRNA intervention are undergoing investigation. Despite these initially promising results and high expectations, siRNA-based drugs must overcome several hurdles before they achieve approval for clinical applications. Based on their similarity to viral particles in terms of composition and route of entry, siRNAs activate the immune system and thus launch severe systemic immune responses. Studying the physiological role of RNA recognition and its players has led to advancements in this area but further studies are needed to fully explore and manipulate these immunostimulatory properties and open additional avenues for innovative strategies in immunotherapy: Gene-specific silencing concomitant with potent immune activation may provide a valuable synergy in the treatment of cancers and chronic infections.

REFERENCES

1. Lindenmann, J., Burke, D. C., and Isaacs, A. 1957. Studies on the production, mode of action and properties of interferon. *Br J Exp Pathol* 38:551.
2. Isaacs, A. and Lindenmann, J. 1957. Virus interference. I. The interferon. *Proc R Soc Lond B Biol Sci* 147:258.
3. Isaacs, A., Lindenmann, J., and Valentine, R. C. 1957. Virus interference. II. Some properties of interferon. *Proc R Soc Lond B Biol Sci* 147:268.
4. Takeda, K. and Akira, S. 2005. Toll-like receptors in innate immunity. *Int Immunol* 17:1.
5. Lampson, G. P. et al. 1967. Inducers of interferon and host resistance. I. Double-stranded RNA from extracts of *Penicillium funiculosum*. *Proc Natl Acad Sci USA* 58:782.
6. Tytell, A. A. et al. 1967. Inducers of interferon and host resistance. 3. Double-stranded RNA from reovirus type 3 virions (reo 3-RNA). *Proc Natl Acad Sci USA* 58:1719.
7. Janeway, C. A., Jr. 1989. Approaching the asymptote? Evolution and revolution in immunology. Cold Spring Harbor Symposia on Quantitative Biology 54 Pt 1:1.

8. Lemaitre, B. et al. 1996. The dorsoventral regulatory gene cassette spatzle/Toll/cactus controls the potent antifungal response in Drosophila adults. *Cell* 86:973.
9. Medzhitov, R., Preston-Hurlburt, P., and Janeway, C. A., Jr. 1997. A human homologue of the Drosophila Toll protein signals activation of adaptive immunity. *Nature* 388:394.
10. Hemmi, H. et al. 2000. A Toll-like receptor recognizes bacterial DNA. *Nature* 408:740.
11. Diebold, S. S. et al. 2004. Innate antiviral responses by means of TLR7-mediated recognition of single-stranded RNA. *Science* 303:1529.
12. Heil, F. et al. 2004. Species-specific recognition of single-stranded RNA via toll-like receptor 7 and 8. *Science* 303:1526.
13. Yoneyama, M. et al. 2004. The RNA helicase RIG-I has an essential function in double-stranded RNA-induced innate antiviral responses. *Nat Immunol* 5:730.
14. Cullen, B. R. 2006. Is RNA interference involved in intrinsic antiviral immunity in mammals? *Nat Immunol* 7:563.
15. Fire, A. et al. 1998. Potent and specific genetic interference by double-stranded RNA in *Caenorhabditis elegans. Nature* 391:806.
16. Hamilton, A. J. and Baulcombe, D. C. 1999. A species of small antisense RNA in post-transcriptional gene silencing in plants. *Science* 286:950.
17. Napoli, C., Lemieux, C., and Jorgensen, R. 1990. Introduction of a chimeric chalcone synthase gene into petunia results if reversible co-suppression of homologous genes. *Plant Cell* 2:279.
18. Romano, N. and Macino, G. 1992. Quelling: transient inactivation of gene expression in *Neurospora crassa* by transformation with homologous sequences. *Mol Microbiol* 6:3343.
19. Aravin, A. A. et al. 2001. Double-stranded RNA-mediated silencing of genomic tandem repeats and transposable elements in the *D. melanogaster* germline. *Curr Biol* 11:1017.
20. Wianny, F. and Zernicka-Goetz, M. 2000. Specific interference with gene function by double-stranded RNA in early mouse development. *Nat Cell Biol* 2:70.
21. Elbashir, S. M. et al. 2001. Duplexes of 21-nucleotide RNAs mediate RNA interference in cultured mammalian cells. *Nature* 411:494.
22. Bernstein, E. et al. 2001. Role for a bidentate ribonuclease in the initiation step of RNA interference. *Nature* 409:363.
23. Elbashir, S. M. et al. 2001. Functional anatomy of siRNAs for mediating efficient RNAi in *Drosophila melanogaster* embryo lysate. *EMBO J* 20:6877.
24. Amarzguioui, M., Rossi, J. J., and Kim, D. 2005. Approaches for chemically synthesized siRNA and vector-mediated RNAi. *FEBS Lett* 579:5974.
25. Sledz, C. A. et al. 2003. Activation of the interferon system by short-interfering RNAs. *Nat Cell Biol* 5:834.
26. Bridge, A. J. et al. 2003. Induction of an interferon response by RNAi vectors in mammalian cells. *Nat Genet* 34:263.
27. Kim, D. H. et al. 2004. Interferon induction by siRNAs and ssRNAs synthesized by phage polymerase. *Nat Biotechnol* 22:321.
28. Hornung, V. et al. 2005. Sequence-specific potent induction of IFN-alpha by short interfering RNA in plasmacytoid dendritic cells through TLR7. *Nat Med* 11:263.
29. Judge, A. D. et al. 2005. Sequence-dependent stimulation of the mammalian innate immune response by synthetic siRNA. *Nat Biotechnol* 23:457.
30. Judge, A. D. et al. 2006. Design of noninflammatory synthetic siRNA mediating potent gene silencing *in vivo. Mol Ther* 13:494.
31. Pebernard, S. and Iggo, R. D. 2004. Determinants of interferon-stimulated gene induction by RNAi vectors. *Differentiation* 72:103.

32. Kariko, K. et al. 2004. Exogenous siRNA mediates sequence-independent gene suppression by signaling through toll-like receptor 3. *Cells Tissues Organs* 177:132.
33. Williams, B. R. 2001. Signal integration via PKR. *Science* STKE [electronic resource] 2001:RE2.
34. Alexopoulou, L. et al. 2001. Recognition of double-stranded RNA and activation of NF-κB by Toll-like receptor 3. *Nature* 413:732.
35. Hornung, V. et al. 2002. Quantitative expression of toll-like receptor 1-10 mRNA in cellular subsets of human peripheral blood mononuclear cells and sensitivity to CpG oligodeoxynucleotides. *J Immunol* 168:4531.
36. Matsumoto, M. et al. 2003. Subcellular localization of Toll-like receptor 3 in human dendritic cells. *J Immunol* 171:3154.
37. Krieg, A. M. et al. 1995. CpG motifs in bacterial DNA trigger direct B-cell activation. *Nature* 374:546.
38. Hartmann, G. and Krieg, A. M. 2000. Mechanism and function of a newly identified CpG DNA motif in human primary B cells. *J Immunol* 164:944.
39. Krug, A. et al. 2001. Identification of CpG oligonucleotide sequences with high induction of IFN-α/β in plasmacytoid dendritic cells. *Eur J Immunol* 31:2154.
40. Hartmann, G. et al. 2003. Rational design of new CpG oligonucleotides that combine B cell activation with high IFN-alpha induction in plasmacytoid dendritic cells. *Eur J Immunol* 33:1633.
41. Jurk, M. et al. 2002. Human TLR7 or TLR8 independently confer responsiveness to the antiviral compound R-848. *Nat Immunol* 3:499.
42. Slade, H. B. 1998. Cytokine induction and modifying the immune response to human papilloma virus with imiquimod. *Eur J Dermatol* 8:13.
43. Colonna, M., Trinchieri, G., and Liu, Y. J. 2004. Plasmacytoid dendritic cells in immunity. *Nat Immunol* 5:1219.
44. Sioud, M. 2005. Induction of inflammatory cytokines and interferon responses by double-stranded and single-stranded siRNAs is sequence-dependent and requires endosomal localization. *J Mol Biol* 348:1079.
45. Morrissey, D. V. et al. 2005. Potent and persistent *in vivo* anti-HBV activity of chemically modified siRNAs. *Nat Biotechnol* 23:1002.
46. Kato, H. et al. 2006. Differential roles of MDA5 and RIG-I helicases in the recognition of RNA viruses. *Nature* 441:101.
47. Gitlin, L. et al. 2006. Essential role of mda-5 in type I IFN responses to polyriboinosinic:polyribocytidylic acid and encephalomyocarditis picornavirus. *Proc Natl Acad Sci USA* 103:8459.
48. Reynolds, A. et al. 2004. Rational siRNA design for RNA interference. *Nat Biotechnol* 22:326.
49. Sioud, M. 2006. Single-stranded small interfering RNA are more immunostimulatory than their double-stranded counterparts: a central role for 2′-hydroxyl uridines in immune responses. *Eur J Immunol* 36:1222.
50. Sugiyama, T. et al. 2005. CpG RNA: identification of novel single-stranded RNA that stimulates human CD14+CD11c+ monocytes. *J Immunol* 174:2273.
51. Roach, J. C. et al. 2005. The evolution of vertebrate Toll-like receptors. *Proc Natl Acad Sci USA* 102:9577.
52. Krieg, A. M. 2002. CpG motifs in bacterial DNA and their immune effects. *Annu Rev Immunol* 20:709.
53. Choe, J., Kelker, M. S., and Wilson, I. A. 2005. Crystal structure of human toll-like receptor 3 (TLR3) ectodomain. *Science* 309:581.
54. Latz, E. et al. 2004. TLR9 signals after translocating from the ER to CpG DNA in the lysosome. *Nat Immunol* 5:190.

55. Rutz, M. et al. 2004. Toll-like receptor 9 binds single-stranded CpG-DNA in a sequence- and pH-dependent manner. *Eur J Immunol* 34:2541.
56. Hornung, V. et al. 2006. 5'-triphosphate RNA is the ligand for RIG-I. *Science*.
57. Leifer, C. A. et al. 2004. TLR9 is localized in the endoplasmic reticulum prior to stimulation. *J Immunol* 173:1179.
58. Funami, K. et al. 2004. The cytoplasmic 'linker region' in Toll-like receptor 3 controls receptor localization and signaling. *Int Immunol* 16:1143.
59. Nishiya, T. et al. 2005. TLR3 and TLR7 are targeted to the same intracellular compartments by distinct regulatory elements. *J Biol Chem* 280:37107.
60. Barton, G. M., Kagan, J. C., and Medzhitov, R. 2006. Intracellular localization of Toll-like receptor 9 prevents recognition of self DNA but facilitates access to viral DNA. *Nat Immunol* 7:49.
61. Smith, A. E. and Helenius, A. 2004. How viruses enter animal cells. *Science* 304:237.
62. Lau, C. M. et al. 2005. RNA-associated autoantigens activate B cells by combined B cell antigen receptor/Toll-like receptor 7 engagement. *J Exp Med* 202:1171.
63. Robbins, M. A. et al. 2006. Stable expression of shRNAs in human CD34+ progenitor cells can avoid induction of interferon responses to siRNAs *in vitro*. *Nat Biotechnol* 24:566.
64. Karlin, S., Doerfler, W., and Cardon, L. R. 1994. Why is CpG suppressed in the genomes of virtually all small eukaryotic viruses but not in those of large eukaryotic viruses? *J Virol* 68:2889.
65. Cavaille, J. and Bachellerie, J. P. 1998. SnoRNA-guided ribose methylation of rRNA: structural features of the guide RNA duplex influencing the extent of the reaction. *Nucleic Acids Res* 26:1576.
66. Kariko, K. et al. 2005. Suppression of RNA recognition by Toll-like receptors: the impact of nucleoside modification and the evolutionary origin of RNA. *Immunity* 23:165.
67. Rozenski, J., Crain, P. F., and McCloskey, J. A. 1999. RNA Modification Database: Update. *Nucleic Acids Res* 27:196.
68. Marques, J. T. et al. 2006. A structural basis for discriminating between self and non-self double-stranded RNAs in mammalian cells. *Nat Biotechnol* 24:559.
69. Dorsett, Y. and Tuschl, T. 2004. siRNAs: applications in functional genomics and potential as therapeutics. *Nat Rev Drug Discov* 3:318.
70. Pieken, W. A. et al. 1991. Kinetic characterization of ribonuclease-resistant 2'-modified hammerhead ribozymes. *Science* 253:314.
71. Levin, A. A. 1999. A review of the issues in the pharmacokinetics and toxicology of phosphorothioate antisense oligonucleotides. *Biochim Biophys Acta* 1489:69.
72. Beigelman, L. et al. 1995. Chemical modification of hammerhead ribozymes: catalytic activity and nuclease resistance. *J Biol Chem* 270:25702.
73. Amarzguioui, M. et al. 2003. Tolerance for mutations and chemical modifications in a siRNA. *Nucleic Acids Res* 31:589.
74. Braasch, D. A. et al. 2003. RNA interference in mammalian cells by chemically-modified RNA. *Biochemistry* 42:7967.
75. Chiu, Y. L. and Rana, T. M. 2003. siRNA function in RNAi: a chemical modification analysis. *RNA* 9:1034.
76. Czauderna, F. et al. 2003. Structural variations and stabilising modifications of synthetic siRNAs in mammalian cells. *Nucleic Acids Res* 31:2705.
77. Jackson, A. L. et al. 2006. Position-specific chemical modification of siRNAs reduces "off-target" transcript silencing. *RNA* 12:1197.
78. Manoharan, M. 2004. RNA interference and chemically modified small interfering RNAs. *Curr Opin Chem Biol* 8:570.

79. Kierzek, E. and Kierzek, R. 2003. The thermodynamic stability of RNA duplexes and hairpins containing N6-alkyladenosines and 2-methylthio-N6-alkyladenosines. *Nucleic Acids Res* 31:4472.

80. Almofti, M. R. et al. 2003. Cationic liposome-mediated gene delivery: biophysical study and mechanism of internalization. *Arch Biochem Biophys* 410:246.

81. Heidel, J. D. et al. 2004. Lack of interferon response in animals to naked siRNAs. *Nat Biotechnol* 22:1579.

82. Capodici, J., Kariko, K., and Weissman, D. 2002. Inhibition of HIV-1 infection by small interfering RNA-mediated RNA interference. *J Immunol* 169:5196.

83. Soutschek, J. et al. 2004. Therapeutic silencing of an endogenous gene by systemic administration of modified siRNAs. *Nature* 432:173.

84. Song, E. et al. 2005. Antibody mediated *in vivo* delivery of small interfering RNAs via cell-surface receptors. *Nat Biotechnol* 23:709.

85. Sereni, D. et al. 1999. Pharmacokinetics and tolerability of intravenous trecovirsen (GEM 91), an antisense phosphorothioate oligonucleotide, in HIV-positive subjects. *J Clin Pharmacol* 39:47.

86. Howard, K. 2003. Unlocking the money-making potential of RNAi. *Nat Biotechnol* 21:1441.

16 Recognition of RNA and Synthetic Compounds by TLR7 and TLR8

Svetlana Hamm and Stefan Bauer

ABSTRACT

The mammalian immune system senses pathogens through pattern recognition receptors and responds with activation. The Toll-like receptors (TLRs) expressed on antigen-presenting cells (APCs) such as macrophages and dendritic cells (DCs) play a critical role in this process. Their signaling activates these cells and leads to an innate immune response with subsequent initiation of an adaptive immune response. TLRs 7 and 8 that recognize single-stranded RNA and nucleoside analogs have been shown to lead to cellular activation and cytokine production, influencing immune responses against viruses and bacteria. The stimulation of these TLRs will be exploited for adjuvant therapy and immune response-directing anti-tumor and allergy treatments.

INTRODUCTION

The immune systems of vertebrates can be broadly categorized as adaptive and innate. Adaptive immunity relies on clonally distributed T and B cells that confer specific and memory responses against pathogens. The innate immune system has developed germline-encoded pattern recognition receptors (PRRs) that promote rapid responses to microbial pathogens during the invading phase. These receptors recognize conserved pathogen-associated molecular patterns (PAMPs) that are not present in the host and are usually important for pathogenicity and/or survival of the pathogens.[1] Sensing of these patterns by innate immune cells activates and directs the emanating responses against pathogens.[2]

Toll receptors are type I transmembrane proteins that are evolutionarily conserved in insects and vertebrates.[3] Toll was first identified as an essential molecule for dorsoventral patterning of Drosophila embryos, and subsequently as a key molecule for antifungal immune responses in adult animals.[4,5] A homologous family designated Toll-like receptors (TLRs) exists in vertebrates,[3] and 13 members (TLRs 1 through 13) have been reported as fundamental for recognizing PAMPs[6,7] from different pathogenic origins such as bacteria, viruses, fungi, and protozoan parasites.[8]

The sources and qualities of TLR ligands have been reviewed extensively.[8–10] This chapter focuses on ligands for TLR7 and TLR8, their expression and signaling.

LIGANDS OF TLR7 AND TLR8

TLR7 and TLR8 located on chromosome X are closely related and show significant homology with TLR9.[11,12] Recent studies revealed that TLRs 7, 8, and 9 are expressed in endosomal and lysosomal compartments and recognize nucleic acid and nucleoside analogs[13,14] (Table 16.1 and Table 16.2). Initially, imidazoquinoline derivatives including imiquimod (R-837, S-26308) and resiquimod (R-848) were identified as TLR7 ligands in mice.[15] These derivatives are small synthetic compounds that induce type I interferon (IFN) and IL-12 production in immune cells.

The antiviral activity of imiquimod was first discovered in guinea pigs infected with herpes simplex virus.[16] Based on effective antiviral and antitumor activities in various animal models, imiquimod and resiquimod were approved for treatment of genital warts caused by human papillomavirus. The more potent R-848 is anticipated

TABLE 16.1
Synthetic TLR7 and TLR8 Ligands

Compound	Structure	TLR activity	Compound	Structure	TLR activity
Imiquimod(R-837, S-26308) 1-(2-methyl propyl)-1H-imidazo[4,5-c]quinolin 4-amine		mTLR7, hTLR7 (15, 21)	Resiquimod (R-848, S-28463) 4-amino-2-ethoxymethyl-a, a-dimethyl-1H-imidazo[4, 5-c]quinoline-1-ethanol		mTLR7, hTLR7, hTLR8 (15, 20)
3M-001 N-[4-(4-amino-2-ethyl-1H-imidazo[4,5-c]quinolin-1-yl)butyl]metha nesulfonamide		hTLR7 (22)	3M-002 2-propyl-thiazolo[4, 5-c]quinolin-4-amine		hTLR8, mTLR8 (if poly (dT)₁₇ added)(22, 29), and this review
3M-003 4-amino-2-(ethoxymethyl)-α,α-dimethyl-6,7,8,9-tetrahydro-1H-imidazo[4,5-c]quinoline-1-ethanol		hTLR7, hTLR8 (22)	SM360320(9-benzyl-8-hydroxy-2-(2-methoxyethoxy) adenine)		hTLR7 (28)
Bropirimine (2-amin-5-bromo-6-phenyl-4(3H)-pyrimidinone)		TLR7(23)	Loxoribine (7-allyl-8-oxoguanosine)		mTLR7, hTLR7 (13, 21) and hTLR8 (if poly (dT)₁₇ added) (30)

TABLE 16.2
RNA Sequences Stimulating TLR7 and TLR8

Name	Sequence	TLR Activity/Ref.
RNA40	GsCsCsCsGsUsCsUsGsUsUsGsUsGsUsGsAsCsUsC	h/mTLR7, hTLR8 (32)
RNA42	AsCsCsCsAsUsCsUsAsUsUsAsUsAsUsAsAsCsUsC	hTLR8 (32)
Poly U	UUUUUUUUUUUUUUUUUUUUUUUUUU (variable in length)	mTLR7, hTLR7 (31)
9.2	AGCUUAACCUGUCCUUCAA	m/hTLR7 (37)
RNA27	GUCCGGGCAGGUCUACUUUTT	Not determined (39)

Note: s = Phosphorothioate linkage.

for treatment of recurring genital herpes and hepatitis C infections.[16–19] Imiquimod is a TLR7-specific ligand found in humans and mice; R-848 activates human and murine TLR7 and human TLR8.[15,20,21]

Further variants of R-848 are specific for TLR7 (3M-001, N-[4-(4-amino-2-ethyl-1H-imidazo[4,5-c]quinolin-1-yl)butyl]methanesulfonamide) or TLR8 (3M-002, 2-propylthiazolo[4,5-c]quinolin-4-amine), or activate both receptors (3M-003, 4-amino-2-(ethoxymethyl)-α,α -dimethyl-6,7,8,9-tetrahydro-1H-imidazo[4,5-c]quinoline-1-ethanol hydrate).[22] Other base analogs such as adenosine, guanosine, and pyrimidine derivatives have been shown to activate TLR7 and/or TLR8. Accordingly, loxoribine (7-allyl-8-oxoguanosine), a guanosine analog, and bropirimine (2-amin-5-bromo-6-phenyl-4(3H)-pyrimidinone), a pyrimidine analog, activate human and murine TLR7.[13,21,23]

Loxoribine showed antiviral and antitumor activity in murine animal models because it activates NK cells and B cells and induces various cytokines.[24,25] Bropirimine is an orally active immunomodulator that induces IFN-α and is used for immunotherapy of carcinoma *in situ* of the bladder and upper urinary tract.[26] The synthetic 8-hydroxy adenine analog SM360320 (9-benzyl-8-hydroxy-2-(2-methoxyethoxy)adenine)[27,28] activates human and murine TLR7 and, in combination with inosine monophosphate dehydrogenase inhibitor, reduces hepatitis C virus (HCV) levels in liver tissue.[28] The synthetic nucleoside analogs serve as universal compounds for activating TLR7 and/or TLR8 and modulate immune responses in fighting viral infections and tumors.

The structural similarities of these synthetic compounds fueled a search for the natural ligand of TLR7 and TLR8 which is single-stranded RNA (ssRNA).[31–33] TLR7 senses synthetic RNA oligonucleotides derived from the U5 region of HIV, poly U RNA, and RNA from influenza virus, Newcastle disease virus (NDV), and vesicular stomatitis virus (VSV).[31–33] Human TLR8 recognizes RNA rich in guanosine and uridine and is involved in the recognition of coxsackie B virus and human parechovirus 1.[32,34,35] The sequence dependency of RNA recognition by TLR7 and TLR8 is not well defined, although uridine is the most critical nucleoside for TLR7/8 activation[32,36] (Table 16.2). Interestingly, guanosine and uridine-rich RNA oligonucleotides mediate TLR7-dependent IFN-α production, whereas adenosines and uridine-rich RNA oligonucleotides induce pro-inflammatory cytokines such as TNF-α and IL-12 in a TLR8-dependent manner.[32] siRNA has been identified as a stimulus for TLR7 and TLR8.[37–39] It is conceivable that both receptors recognize short double-stranded RNA in addition to ssRNA, although we hypothesize that the siRNA duplex may dissociate in endosomal compartments with consecutive stimulation of TLR7 and TLR8 by ssRNA molecules.

The recognition of ssRNA by TLR7 and TLR8 is inhibited by incorporation of modified nucleosides such as 5-methylcytosine (m5C), N6-methyladenosine (m6A), 5-methyluridine (m5U), 2-thiolated uridine (s2U), or pseudouridine into RNA.[40,41] These modifications are common in endogenous RNA. Since DCs exposed to such modified RNA express significantly fewer cytokines and activation markers than those treated with unmodified RNA, it is believed that the innate immune system utilizes these modifications to discriminate between endogenous and stimulatory bacterial or viral RNA.[41]

EXPRESSION AND FUNCTIONALITY OF TLR7 AND TLR8

TLR7 and TLR8 are expressed in a variety of different cell types of the immune system such as dendritic cells, B cells, monocytes, NK cells, and certain T cells (Table 16.3). TLR7 and TLR8 are functionally distinct in human DCs and monocytes. TLR7-mediated viral recognition is prominent in a DC subtype known as the plasmacytoid dendritic cell (pDC) or natural interferon-α producing cell (NIPC). This cell type is characterized by its ability to secrete high amounts of IFN-α in response to viral infection.[42–45] Other cell types such as conventional DCs and fibroblasts also produce type I IFN upon infection with single-stranded RNA viruses in a TLR7-independent manner.[46] This observation suggests that alternative pathways sense viral infection. These TLR-independent receptors in non-pDCs that confer recognition of viral RNA and subsequent type I IFN production are the RIG-I RNA helicase and MDA-5.[47–49]

TABLE 16.3
TLR7 and TLR8 Expression in Immune Cells

Cell Type	TLR7	TLR8	Ref.
Murine CD11c+ CD8-	+	+	(15,56,57)
Murine CD11c+ CD8+	(+)	+	(56)
Murine plasmacytoid DCs	+	+	(56)
Murine B cells	+	?	(15)
Human blood DCs	+	+	(57)
Human MDDCs	+/–	+	(58)
Human plasmacytoid DCs	+	–	(59)
Human B cells	+	–	(59)
Human NK cells	+/–	+/–	(59-61)
Human monocytes	+	+	(58,59)
Human eosinophils	+	n.d.	(62)
Human regulatory T cells	n.d.	+	(63)
Human T cells	–	–	(59)

Note: n.d. = Not determined.

TLR8 functions in monocytes and myeloid DCs and is involved in the production of pro-inflammatory cytokines such as TNF-α. TLR8 agonists therefore may drive strong cell-mediated immune responses or Th1-like immune responses, whereas TLR7 agonists may be important in driving humoral responses.

The role of TLR7 in pathogen-driven immune responses in B cells is not well analyzed, although it has been demonstrated that TLR triggering via TLR2 and TLR9 ligands directly influences B cell activation and antibody production in humans and mice.[50,51] In contrast, a role for B cells-expressed TLR7 in autoimmune disease has recently emerged. TLR7 recognizes endogenous RNA and sustains B cell-driven autoimmunity in systemic lupus erythematosus (SLE).[52–55]

The expression of TLR7 and TLR8 on human NK cells is controversial.[60,61] Hart et al. described TLR7 and TLR8 expression on NK cells and their direct activation via these receptors. Gorski et al. noted that TLR7 and TLR8 were not expressed on mRNA or protein levels,[60,61] Both agreed that indirect mechanisms driven by TLR7 and TLR8 activation of innate immune cells such as IL-12 or IL-18 activated NK cells in PBMC mixtures to produce IFN-γ. Eosinophils constitutively express TLR1, TLR4, TLR7, TLR9, and TLR10 mRNAs, but only the resiquimod TLR7 and TLR8 ligand regulated adhesion molecule (CD11b and L-selectin) expression and prolonged survival, indicating that both receptors are important PRRs for eosinophil function.[62]

Wang et al. observed a direct effect of TLR activation on regulatory T cells (Tregs). They demonstrated that synthetic and natural ligands for human TLR8 (RNA40 and polyG) may reverse Treg function. This effect was independent of DCs but required functional TLR8–MyD88–IRAK4 signaling in Tregs. This Treg-suppressive function of TLR8 was underscored using *in vivo* adoptive transfer experiments. Transferring TLR8 ligand-stimulated Tregs into tumor-bearing mice dramatically enhanced anti-tumor immunity.[63]

Information about the expression of TLR7 and TLR8 on non-immune cells is scarce. Recent work demonstrated that TLR8 is dynamically expressed during mouse brain development and localizes to neurons and axons. Agonist stimulation of TLR8 causes inhibition of neurite outgrowth and induces apoptosis. These findings reveal a novel function—distinct from the classical role of TLRs in immunity—for TLR8 in the mammalian nervous system.[64]

FUNCTION OF MURINE TLR8

Although the functionality of human TLR8 has been demonstrated, a role for murine TLR8 in the recognition of PAMPs has not been demonstrated. Reports from genetic complementation assays using gene-deficient mice indicate that the resiquimod (R-848) antiviral imidazoquinoline and guanosine–uridine rich RNA activate immune cells via human and murine TLR7 as well as human TLR8.[15,20,32] With human TLR8 acting as a sensor for R-848, it is surprising that TLR7-deficient mice would fail to respond to R-848 if murine TLR8 had a ligand repertoire similar to human TLR8.[15]

Accordingly, genetic complementation assays in HEK293 cells revealed that only human TLR8 conferred responsiveness to R-848; murine TLR8 was mute (Figure 16.1A). This observation may be explained by one of the following hypotheses.

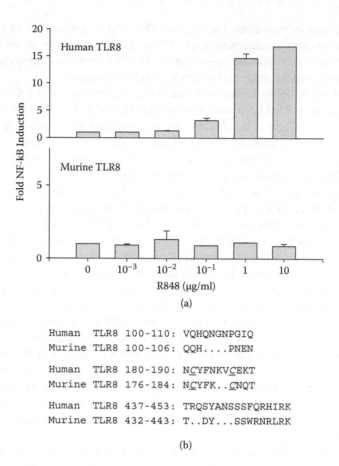

FIGURE 16.1   (A) HEK293 cells were transfected with human or murine TLR8 expression plasmid and stimulated with increasing amounts of R848. (B) Partial alignment of murine and human TLR8 protein sequences.

First, similar to species specificity in ligand recognition reported for TLR4 and TLR9,[65,66] murine TLR8 may be unable to sense R-848 while recognizing another unidentified ligand. Second, TLR8 may utilize an unidentified tightly regulated cofactor for efficient ligand recognition. Third, murine TLR8 may be non-functional. Supporting the latter view, sequence comparison of human and murine TLR8 revealed three spots of amino acid deletions in murine TLR8 that may account for its mute phenotype (Figure 16.1B).

Two deletions are located in the N terminal part of the protein. The third is found in a non-structured (hinge) region that is not well conserved among TLRs 7, 8, and 9. Interestingly, the two amino acid deletions at position 181 and 182 of murine TLR8 (verified in C57BL/6, BALB/c, and C3H/HeN mouse strains, unpublished observation) alter the spacing of two cysteines that are highly conserved among TLRs 7, 8, and 9. Although these cysteines are not part of the CXXC domain involved in R-848- and RNA-driven signaling,[67,68] an additional role in ligand recognition is probable (unpublished observation). However, filling the deletions at amino acid positions 103 and 181 independently or in combination with corresponding amino acids from human TLR8 did not restore murine TLR8 activity. It is therefore possible that the deletion in the hinge region critically influences the activity, and future site-directed mutagenesis should address this possibility.

A recent report indicated that the addition of a homopolymer oligo $(dT)_{17}$ containing a phosphorothioate backbone enhanced R-848-driven NF-κB activation and cytokine production via human TLR8.[30] The mechanism is unknown, but it was hypothesized that poly(dT) may act as an allosteric activator. Accordingly, we tested whether oligo$(dT)_{17}$ added to murine TLR8-transfected HEK293 cells would restore recognition of the R-848 TLR7/8 ligand, the 3M-002 human TLR8-specific ligand, or the hTLR8 guanosine–uridine-rich RNA40 ligand.

HEK293 transfected with human TLR8 showed an increase in NF-κB activation when stimulated with the RNA, synthetic ligands, and poly$(dT)_{17}$. Transfection of murine TLR8 conferred a weak activity toward the 3M-002/ poly$(dT)_{17}$ combination, although R-848 and RNA40 were not recognized (Figure 16.2). These data suggest that murine TLR8 is functional under certain conditions and with selected ligands. TLR8 is dependent on additional co-factors that are presumably mimicked by phosphorothioate-modified poly$(dT)_{17}$.

The functionality of mTLR8 was also reported by Gorden et al.[29] Transfection of mTLR8 conferred recognition of 3M-002 and TLR7-deficient immune cells responded to 3M-002 in a Myd88-dependent manner. These data suggest that murine TLR8 may be functional in immune cells although its natural ligand remains unclear.

TLR7 AND TLR8 SIGNALING

In general, the engagement of TLRs by their cognate ligands leads to the activation of a signaling cascade. Three major pathways are activated. The first culminates in the activation of the NF-κB transcription factor that acts as a master switch for inflammation. The second leads to activation of the MAP kinases known as p38 and Jun amino-terminal kinase (JNK) that also participate in increased transcription. The third pathway leads to type I IFN production via IFN regulatory factors (IRFs).[69]

Upon ligand binding, the TLRs presumably dimerize, undergo conformational change, and recruit adapter molecules to their intracellular domain (the Toll/IL1-receptor-like or TIR domain) shared by the TLR and IL1-receptors.[70,71] Four adapter molecules have been identified: (1) MyD88 (myeloid differentiation protein 88), (2) TIRAP (TIR-associated protein)/MAL (MyD88-adaptor like), (3) TRIF (TIR domain-containing adaptor protein-inducing IFNβ/TICAM1 [TIR domain containing molecule 1]) and, (4) TRIF-related adaptor molecule (TRAM).[69]

TLR7 and TLR8 solely utilize MyD88 to induce cytokines[15,32] (Figure 16.3). Recruitment of MyD88 is followed by engagement of IL-1 receptor-associated kinase 4 (IRAK-4) and IL-1 receptor-associated kinase 1 (IRAK-1) which is phosphorylated by IRAK-4 and associates with TNF receptor associated factor-6 (TRAF-6) and/or TNF receptor associated factor-3 (TRAF-3)[72–74] (Figure 16.3). TRAF3 is recruited

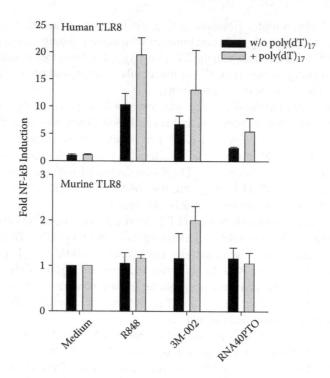

FIGURE 16.2 HEK293 cells were transfected with human or murine TLR8 expression plasmid and stimulated with 10 μM R-848, 10 μM 3M-002, or 10 μM RNA40 complexed to cationic DOTAP lipid with or without 10 μM poly(dT)$_{17}$.

along with TRAF6 and is essential for the induction of type I IFN but it is dispensable for expression of pro-inflammatory cytokines.[75,76] Interferon regulatory factors (IRFs) are critical components of immune activation and type I interferon regulation. IRF-3 and IRF-7 regulate the expression of IFN-inducible genes by binding to interferon stimulated response element (ISRE).[77,78] IRF-7 is the key component in IFN-α induction via TLR7 and requires MyD88, IRAK-4, IRAK-1, and TRAF-3.[79,80] In addition, IRF5 is activated by the TLR7 and TLR8 signaling pathway that utilizes MyD88, IRAK-1, and TRAF-6.[81,82]

It is unclear why TLR7 and TLR8 induce a different cytokine pattern that is mainly characterized by IFN-α for TLR7 and pro-inflammatory cytokines for TLR8. This may be due to cell type-specific expression of TLR7, TLR8, and IRF-5/IRF-7. The cytoplasmic domains of TLR7 and TLR8 may be different in such a way that only TLR7 specifically interacts with MyD88 and IRF-7. Alternatively, TLR7 and TLR8 may be expressed in different compartments such as early and late endosome, with only the early endosome allowing the initiation of IRF-7 mediated IFN-α production (Figure 16.3).

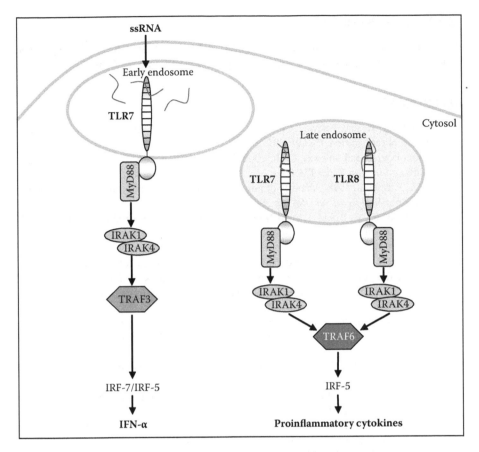

FIGURE 16.3   TLR7 and TLR8 signalling. Activation of TLR7 and TLR8 induces secretion of pro-inflammatory cytokines and type I IFN. Expression of pro-inflammatory cytokines is induced via IRAK1, IRAK4, TRAF-6, and TRAF-3. Type I IFN production is controlled by TRAF3 and is also dependent on IRF-7 and IRF-5 transcription factors.

CONCLUSION

Recent TLR research has revealed how the innate immune system senses invading pathogens and initiates the adaptive immune response. Current research on TLR7 and TLR8 focuses on their direct functioning on cells of the adaptive immune system and their role in autoimmunity. Future research will exploit this knowledge to efficiently manipulate immune responses or treat autoimmune diseases aided by synthetic antagonists and the development of small molecules with antagonistic functions for TLR7 and TLR8.

ACKNOWLEDGMENTS

We would like to thank Dr. Sabine Amslinger for help with the structures of the synthetic TLR7 and TLR8 ligands. This work was supported by Transregio-SFB22 (S.B.).

REFERENCES

1 Medzhitov, R. and Janeway, C. A., Jr. 1997. Innate immunity: virtues of a nonclonal system of recognition. *Cell* 91: 295.

2 Medzhitov, R. and Janeway, C., Jr. 2000. Innate immune recognition: mechanisms and pathways. *Immunol Rev* 173: 89.

3 Rock, F. L. et al. 1998. A family of human receptors structurally related to Drosophila Toll. *Proc Natl Acad Sci USA* 95: 588.

4 Anderson, K. V., Bokla, L., and Nusslein-Volhard, C. 1985. Establishment of dorsal-ventral polarity in the Drosophila embryo: the induction of polarity by the Toll gene product. *Cell* 42: 791.

5 Lemaitre, B. et al. 1996. The dorsoventral regulatory gene cassette spatzle/Toll/cactus controls the potent antifungal response in Drosophila adults. *Cell* 86: 973.

6 Takeda, K. and Akira, S. 2005. Toll-like receptors in innate immunity. *Int Immunol* 17: 1.

7 Tabeta, K. et al. 2004. Toll-like receptors 9 and 3 as essential components of innate immune defense against mouse cytomegalovirus infection. *Proc Natl Acad Sci USA* 101: 3516.

8 Akira, S., Uematsu, S., and Takeuchi, O. 2006. Pathogen recognition and innate immunity. *Cell* 124: 783.

9 Pasare, C. and Medzhitov, R. 2005. Toll-like receptors: linking innate and adaptive immunity. *Adv Exp Med Biol* 560: 11.

10 Beutler, B. et al. 2003. How we detect microbes and respond to them: the Toll-like receptors and their transducers. *J Leukoc Biol* 74: 479.

11 Chuang, T. H. and Ulevitch, R. J. 2000. Cloning and characterization of a subfamily of human toll-like receptors: hTLR7, hTLR8 and hTLR9. *Eur Cytokine Netw* 11: 372.

12 Du, X. et al. 2000. Three novel mammalian toll-like receptors: gene structure, expression, and evolution. *Eur Cytokine Netw* 11: 362.

13 Heil, F. et al. 2003. Toll-like receptor 7 (TLR7)-specific stimulus loxoribine uncovers a strong relationship within the TLR7, 8 and 9 subfamily. *Eur J Immunol* 33: 2987.

14 Latz, E. et al. 2004. TLR9 signals after translocating from the ER to CpG DNA in the lysosome. *Nat Immunol* 5: 190.

15 Hemmi, H. et al. 2002. Small anti-viral compounds activate immune cells via the TLR7 MyD88-dependent signaling pathway. *Nat Immunol* 3: 196.

16 Harrison, C. J. et al. 1988. Modification of immunological responses and clinical disease during topical R-837 treatment of genital HSV-2 infection. *Antiviral Res* 10: 209.

17 Tomai, M. A. et al. 1995. Immunomodulating and antiviral activities of the imidazoquinoline S-28463. *Antiviral Res* 28: 253.

18 Testerman, T. L. et al. 1995. Cytokine induction by the immunomodulators imiquimod and S-27609. *J Leukoc Biol* 58: 365.

19 Spruance, S. L. et al. T. C. 2001. Application of a topical immune response modifier, resiquimod gel, to modify the recurrence rate of recurrent genital herpes: a pilot study. *J Infect Dis* 184: 196.

20 Jurk, M. et al. 2002. Human TLR7 or TLR8 independently confer responsiveness to the antiviral compound R-848. *Nat Immunol* 3: 499.

21 Lee, J. et al. 2003. Molecular basis for the immunostimulatory activity of guanine nucleoside analogs: activation of Toll-like receptor 7. *Proc Natl Acad Sci USA* 100: 6646.
22 Gorden, K. B. et al. 2005. Synthetic TLR agonists reveal functional differences between human TLR7 and TLR8. *J Immunol* 174: 1259.
23 Akira, S. and Hemmi, H. 2003. Recognition of pathogen-associated molecular patterns by TLR family. *Immunol Lett* 85: 85.
24 Smee, D. F. et al. 1990. Roles of interferon and natural killer cells in the antiviral activity of 7-thia-8-oxoguanosine against Semliki Forest virus infections in mice. *Antiviral Res* 13: 91.
25 Pope, B. L. et al. 1994. 7-Allyl-8-oxoguanosine (loxoribine) inhibits the metastasis of B16 melanoma cells and has adjuvant activity in mice immunized with a B16 tumor vaccine. *Cancer Immunol Immunother* 38: 83.
26 Sarosdy, M. F. 1997. A review of clinical studies of bropirimine immunotherapy of carcinoma in situ of the bladder and upper urinary tract. *Eur Urol* 31 Suppl 1: 20.
27 Kurimoto, A. et al. 2004. Synthesis and evaluation of 2-substituted 8-hydroxyadenines as potent interferon inducers with improved oral bioavailabilities. *Bioorg Med Chem* 12: 1091.
28 Lee, J. et al. 2006. Activation of anti-hepatitis C virus responses via Toll-like receptor 7. *Proc Natl Acad Sci USA* 103: 1828.
29 Gorden, K. K. et al. 2006. Cutting edge: activation of murine TLR8 by a combination of imidazoquinoline immune response modifiers and polyT oligodeoxynucleotides. *J Immunol* 177: 6584.
30 Jurk, M. et al. 2006. Modulating responsiveness of human TLR7 and 8 to small molecule ligands with T-rich phosphorothiate oligodeoxynucleotides. *Eur J Immunol* 36: 1815.
31 Diebold, S. S. et al. 2004. Innate antiviral responses by means of TLR7-mediated recognition of single-stranded RNA. *Science* 303: 1529.
32 Heil, F. et al. 2004. Species-specific recognition of single-stranded RNA via toll-like receptor 7 and 8. *Science* 303: 1526.
33 Lund, J. M. et al. 2004. Recognition of single-stranded RNA viruses by Toll-like receptor 7. *Proc Natl Acad Sci USA* 101: 5598.
34 Triantafilou, K. et al. 2005. TLR8 and TLR7 are involved in the host's immune response to human parechovirus 1. *Eur J Immunol* 35: 2416.
35 Triantafilou, K. et al. 2005. Human cardiac inflammatory responses triggered by coxsackie B viruses are mainly Toll-like receptor (TLR) 8-dependent. *Cell Microbiol* 7: 1117.
36 Sioud, M. 2006. Single-stranded small interfering RNA are more immunostimulatory than their double-stranded counterparts: a central role for 2'-hydroxyl uridines in immune responses. *Eur J Immunol* 36: 1222.
37 Hornung, V. et al. 2005. Sequence-specific potent induction of IFN-α by short interfering RNA in plasmacytoid dendritic cells through TLR7. *Nat Med* 11: 263.
38 Judge, A. D. et al. 2005. Sequence-dependent stimulation of the mammalian innate immune response by synthetic siRNA. *Nat Biotechnol* 23: 457.
39 Sioud, M. 2005. Induction of inflammatory cytokines and interferon responses by double-stranded and single-stranded siRNAs is sequence-dependent and requires endosomal localization. *J Mol Biol* 348: 1079.
40 Maden, B. E. 1990. The numerous modified nucleotides in eukaryotic ribosomal RNA. *Prog Nucleic Acid Res Mol Biol* 39: 241.
41 Kariko, K. et al. 2005. Suppression of RNA recognition by Toll-like receptors: the impact of nucleoside modification and the evolutionary origin of RNA. *Immunity* 23: 165.

42 Hochrein, H. et al. 2004. Herpes simplex virus type-1 induces IFN-α production via Toll-like receptor 9-dependent and -independent pathways. *Proc Natl Acad Sci USA* 101: 11416.

43 Krug, A. et al. 2004. TLR9-dependent recognition of MCMV by IPC and DC generates coordinated cytokine responses that activate antiviral NK cell function. *Immunity* 21: 107.

44 Lund, J. et al. 2003. Toll-like receptor 9-mediated recognition of herpes simplex virus-2 by plasmacytoid dendritic cells. *J Exp Med* 198: 513.

45 Siegal, F. P. et al. 1999. The nature of the principal type 1 interferon-producing cells in human blood. *Science* 284: 1835.

46 Kato, H. et al. 2005. Cell type-specific involvement of RIG-I in antiviral response. *Immunity* 23: 19.

47 Gitlin, L. et al. 2006. Essential role of MDA-5 in type I IFN responses to polyriboinos inic:polyribocytidylic acid and encephalomyocarditis picornavirus. *Proc Natl Acad Sci USA* 103: 8459.

48 Kato, H. et al. 2006. Differential roles of MDA5 and RIG-I helicases in the recognition of RNA viruses. *Nature* 441: 101.

49 Yoneyama, M. et al. 2004. The RNA helicase RIG-I has an essential function in double-stranded RNA-induced innate antiviral responses. *Nat Immunol* 5: 730.

50 Ruprecht, C. R. and Lanzavecchia, A. 2006. Toll-like receptor stimulation as a third signal required for activation of human naive B cells. *Eur J Immunol* 36: 810.

51 Pasare, C. and Medzhitov, R. 2005. Control of B-cell responses by Toll-like receptors. *Nature* 438: 364.

52 Lau, C. M. et al. 2005. RNA-associated autoantigens activate B cells by combined B cell antigen receptor/Toll-like receptor 7 engagement. *J Exp Med* 202: 1171.

53 Vollmer, J. et al. 2005. Immune stimulation mediated by autoantigen binding sites within small nuclear RNAs involves Toll-like receptors 7 and 8. *J Exp Med* 202: 1575.

54 Savarese, E. et al. 2006. U1 small nuclear ribonucleoprotein immune complexes induce type I interferon in plasmacytoid dendritic cells through TLR7. *Blood* 107: 3229.

55 Subramanian, S. et al. 2006. A Tlr7 translocation accelerates systemic autoimmunity in murine lupus. *Proc Natl Acad Sci USA* 103: 9970.

56 Edwards, A. D. et al. 2003. Toll-like receptor expression in murine DC subsets: lack of TLR7 expression by CD8 alpha+ DC correlates with unresponsiveness to imidazoquinolines. *Eur J Immunol* 33: 827.

57 Doxsee, C. L. et al. 2003. The immune response modifier and Toll-like receptor 7 agonist S-27609 selectively induces IL-12 and TNF-α production in CD11c+CD11b+CD8-dendritic cells. *J Immunol* 171: 1156.

58 Ito, T. et al. 2002. Interferon-α and interleukin-12 are induced differentially by Toll-like receptor 7 ligands in human blood dendritic cell subsets. *J Exp Med* 195: 1507.

59 Hornung, V. et al. 2002. Quantitative expression of toll-like receptor 1-10 mRNA in cellular subsets of human peripheral blood mononuclear cells and sensitivity to CpG oligodeoxynucleotides. *J Immunol* 168: 4531.

60 Gorski, K. S. et al. 2006. Distinct indirect pathways govern human NK-cell activation by TLR-7 and TLR-8 agonists. *Int Immunol* 18: 1115.

61 Hart, O. M. et al. 2005. TLR7/8-mediated activation of human NK cells results in accessory cell-dependent IFN-γ production. *J Immunol* 175: 1636.

62 Nagase, H. et al. 2003. Expression and function of Toll-like receptors in eosinophils: activation by Toll-like receptor 7 ligand. *J Immunol* 171: 3977.

63 Peng, G. et al. 2005. Toll-like receptor 8-mediated reversal of CD4+ regulatory T cell function. *Science* 309: 1380.

64 Ma, Y. et al. 2006. Toll-like receptor 8 functions as a negative regulator of neurite outgrowth and inducer of neuronal apoptosis. *J Cell Biol* 175: 209.

65   Bauer, S. et al. 2001. Human TLR9 confers responsiveness to bacterial DNA via spe-
     cies-specific CpG motif recognition. *Proc Natl Acad Sci USA* 98: 9237.
66   Poltorak, A. et al. 2000. Physical contact between lipopolysaccharide and toll-like
     receptor 4 revealed by genetic complementation. *Proc Natl Acad Sci USA* 97: 2163.
67   Gibbard, R. J., Morley, P. J., and Gay, N. J. 2006. Conserved features in the extracellular
     domain of human toll-like receptor 8 are essential for pH-dependent signaling. *J Biol
     Chem* 281: 27503.
68   Bauer, S. and Wagner, H. 2002. Bacterial CpG-DNA licenses TLR9. *Curr Top Micro-
     biol Immunol* 270: 145.
69   O'Neill, L. A. 2006. How Toll-like receptors signal: what we know and what we don't
     know. *Curr Opin Immunol* 18: 3.
70   O'Neill, L. A. and Greene, C. 1998. Signal transduction pathways activated by the IL-1
     receptor family: ancient signaling machinery in mammals, insects, and plants. *J Leu-
     koc Biol* 63: 650.
71   Medzhitov, R. et al. 1997. A human homologue of the Drosophila Toll protein signals
     activation of adaptive immunity. *Nature* 388: 394.
72   Li, S. et al. 2002. IRAK-4: a novel member of the IRAK family with the properties of
     an IRAK-kinase. *Proc Natl Acad Sci USA* 99: 5567.
73   Muzio, M. et al. 1998. The human toll signaling pathway: divergence of nuclear factor
     kappaB and JNK/SAPK activation upstream of tumor necrosis factor receptor-associ-
     ated factor 6 (TRAF6). *J Exp Med* 187: 2097.
74   Cao, Z. et al. 1996. TRAF6 is a signal transducer for interleukin-1. *Nature* 383: 443.
75   Hacker, H. et al. 2006. Specificity in Toll-like receptor signalling through distinct
     effector functions of TRAF3 and TRAF6. *Nature* 439: 204.
76   Oganesyan, G. et al. 2006. Critical role of TRAF3 in the Toll-like receptor-dependent
     and -independent antiviral response. *Nature* 439: 208.
77   Honda, K. et al. 2005. IRF-7 is the master regulator of type-I interferon-dependent
     immune responses. *Nature* 434: 772.
78   Sakaguchi, S. et al. 2003. Essential role of IRF-3 in lipopolysaccharide-induced
     interferon-β gene expression and endotoxin shock. *Biochem Biophys Res Commun* 306:
     860.
79   Honda, K. et al. 2004. Role of a transductional-transcriptional processor complex
     involving MyD88 and IRF-7 in Toll-like receptor signaling. *Proc Natl Acad Sci USA*
     101: 15416.
80   Kawai, T. et al. 2004. Interferon-α induction through Toll-like receptors involves a
     direct interaction of IRF7 with MyD88 and TRAF6. *Nat Immunol* 5: 1061.
81   Schoenemeyer, A. et al. 2005. The interferon regulatory factor, IRF5, is a central medi-
     ator of toll-like receptor 7 signaling. *J Biol Chem* 280: 17005.
82   Takaoka, A. et al. 2005. Integral role of IRF-5 in the gene induction programme acti-
     vated by Toll-like receptors. *Nature* 434: 243.

17 Helicases at Frontline of RNA Virus Recognition

Leonid Gitlin and Marco Colonna

ABSTRACT

Interferon induction is a critical decision made by cells. As a signal of virus invasion, interferon must be induced in a timely fashion if the host is to survive the infection. Interferon response must also be limited to the duration of infection to avoid potential autoimmune effects. Therefore, it is not surprising that only a limited number of interferon-inducing signals have been identified to date, chief among them, double-stranded RNA. While its interferon-inducing effects have been known for a long time, the protein sensors that mediate recognition of double-stranded RNA remained obscure. Recent discovery of the RIG-I and MDA5 RNA helicases has begun to lift the veil on this long-standing mystery and promises to guide us to better treatment of viral infections and better management of autoimmune disorders involving interferon. This chapter discusses our progress in understanding interferon induction by double-stranded RNA.

INTRODUCTION

Since its discovery in 1957,[1] interferon was recognized as a central antiviral messenger. Most, if not all, nucleated mammalian cells are capable of producing type I interferon upon infection, and some cell types can detect virus presence and produce interferon while not being infected. Type I interferons (IFNs) are now known to constitute a family of molecules including over a dozen IFN-α members (14 in mice) and a single IFN-β.[2] They all share the same receptor and activate largely the same signaling pathway. Downstream of this pathway lie hundreds of genes whose expression prepares the cell to combat viral infection, shifting it into a so-called antiviral state; the adaptive immune system may also be mobilized.[3,4] This impressive antiviral armada is commanded by type I inteferons. It follows that signals leading to IFN production must be reliable indicators of viral infection, and that the recognition of such signals must be carefully regulated.

Only a handful of such signals have been identified. Molecules shared by various pathogens and not found in the host are called pathogen-associated molecular patterns (PAMPs).[5] While many such molecules exist in free-living pathogens, viruses are by definition harder to distinguish from their host. Indeed the only established PAMPs associated with viruses are nucleic acids. Chief among them is double-stranded RNA (dsRNA) that is not thought to be present in mammalian cells in significant amounts.[6] RNA viruses, on the other hand, must replicate through a double-stranded intermediate, providing a convenient Achilles' heel for virus recognition machinery.

dsRNA was the first and remains among the major IFN-inducing molecules found; first described as an interferon inducer in 1967,[7-10] it remains the inducer of choice in laboratories. Usually, polyriboinosinic:polyribocytidylic acid (polyI:C), a mimic of true dsRNA, is used for this purpose because it is particularly effective.[9] Cytoplasmic transfection of dsRNA or simply its extracellular application (leading to its endosomal localization) induces IFN. In contrast, the other known viral PAMPs seem to be recognized specifically within certain cellular compartments (Figure 17.1). This may be because the other patterns are double-stranded DNA (provided it is cytoplasmic) and single-stranded RNA (ssRNA) or unmethylated CpG DNA (provided it is endosomal).[11-15]

One would expect to find viral, but not host, genomic material in endosomes because some viruses go through endosomal uptake and acidification in their life cycles.[16] Furthermore, only some cell types are capable of endosomal recognition since it is accomplished by Toll-like receptors (TLRs)—transmembrane proteins expressed in certain cells such as dendritic cells (DCs) and macrophages. Specifically, TLR7 (recognizing ssRNA[11,12,14]) and TLR9 (recognizing CpG DNA[13]) are highly expressed in plasmacytoid dendritic cells (pDCs).[17,18] Because DCs and macrophages are specialized immune cells, they constantly monitor the extracellular environment and dying host cells for foreign material. This environmental sampling provides another route to the endosomes. As mentioned earlier, the endosomes are also equipped with their own dsRNA recognition system, TLR3.

In contrast to TLRs, the systems sensing viral genetic material in the cytoplasm (dsRNA and cytoplasmic DNA) must be nearly ubiquitous; indeed most cells should, and are equipped to signal viral infection. It is only now, 40 years after the recognition

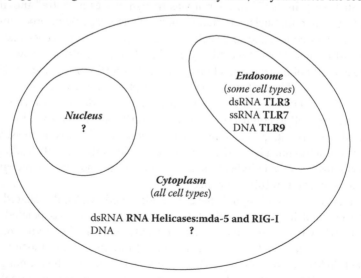

FIGURE 17.1 Intracellular topography of PAMP recognition. The nucleic acid-based PAMPs are listed next to their sensors and the intracellular compartments they occupy. It is not known whether pattern recognition receptors exist in the mammalian cell nucleus.

that dsRNA induces interferon, we are beginning to understand the sensors of this viral pattern. This chapter will summarize our current, and still incomplete understanding of the dsRNA-induced pathways, focusing on the antiviral RNA helicases designated RIG-I and MDA5. These helicases likely account for a significant fraction of interferon induced following RNA virus infection.

INTERFERON β AND α PROMOTERS AND THEIR REGULATION

All virus sensors have interferon promoters as their ultimate targets. With the exception of pDCs, the interferon cascade begins at interferon β (see below). Our understanding of the IFN-β promoter has advanced steadily over the past 20 years. The promoter contains four positively regulated domains (PRD I through PRD IV) that bind factors arising from three signaling pathways: interferon regulatory factors (IRFs), nuclear factor-κB (NF-κB), and AP-1 (ATF2 and c-Jun heterodimer, activated by MAP kinases) as shown in Figure 17.2. It has been thought that all of these pathways have to be activated in order to induce the promoter.[19] Generally speaking, NF-κB is activated by a variety of inducers such as pro-inflammatory cytokines, phorbol esters, viruses, and bacterial PAMPs that are loosely termed immune stress inducers. In the same manner, one could call IRF3 and IRF7, the two IRF proteins relevant for IFN-β induction, virus signaling factors.

A case has been made for IFN-β induction through IRF-dependent pathways without activation of MAP kinases or the classical NF-κB pathway.[20] Indeed, dsDNA seems to act through IRFs without involvement of NF-κB or MAP kinases.[15] This was not foreseen and may be an interesting window into non-dsRNA inducers of this promoter.

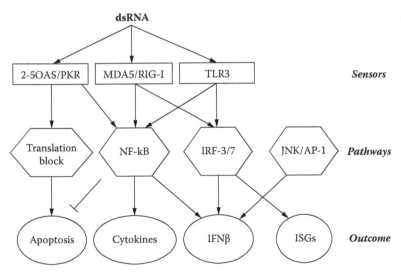

FIGURE 17.2 Bird's eye view of dsRNA sensing pathways. Major signaling pathways and their outcomes are shown, with the emphasis on their interrelationships. Signaling paths known to lead to IFN-β transcription are denoted by bold arrows. Activation of JNK and AP-1 may be due to PKR/2-5OAS and/or to TLR3 sensors (see text).

Another interesting concept is constant low-level interferon presence and signaling in normal uninfected tissues. While this activity was first observed over 20 years ago,[21,22] a rationale for it has now been proposed.[23] Interferon signaling may keep the threshold for IFN induction lower and in this way prepare cells for the moment when they become infected (see IRF Pathways below). Until recently, it was difficult to know which factors were behind this background interferon level, but the discovery of the interferon inducers may provide the answers.

IRF PATHWAYS

The first IRF discovered by virtue of its binding to the IFN-β promoter and stimulating transcription was IRF1.[24,25] It is not currently considered a major contributor to IFN-β gene induction.[19,26] However, IRF1 knockout embryonic fibroblasts are impaired in IFN response to polyI:C, although the response can be fully restored by IFN pre-treatment.[27] There seemed to be no difference in replication of encephalomyocarditis virus (EMCV), vesicular stomatitis virus (VSV), and herpes simplex virus (HSV) between the knockout and wildtype cells. However, pre-treatment with IFN was not as effective in IRF1$^{-/-}$ cells against EMCV as in the wild-type cells, resulting in higher EMCV titers of knockout cells.[28]

It later became clear from biochemical and overexpression data that IRF3 and IRF7 are major players in the induction of the IFN-β gene; additionally, adenovirus protein E1A proved capable of disrupting IRF3-mediated IFN induction.[29–33] IRF3 and IRF7 could induce IFN-β and IFN-α4 transcription, while only IRF7 efficiently induced other IFN-αs.[34,35] Based on the expression patterns of IRF3 (constitutive and unaffected by IFN) and IRF7 (low in untreated cells and highly induced by IFN-α/β) it was proposed that interferon induction occurs in two phases.[36]

In the first phase, viral infection induces IFN-β and IFN-α4 through IRF3, NF-κB, and AP-1. IFN, in a feedback loop, then acts in both autocrine and paracrine fashion to induce IRF7. Once IRF7 is expressed, further viral replication stimulates transcription of a variety of IFN-α subtypes. It has been suggested that this two-tier system represents an amplification loop whereby two IFN genes are activated initially and most or all of them are activated later.[23,26]

Another (and not necessarily mutually exclusive) way to rationalize this arrangement is that the initial "decision" at the IFN-β promoter is made by integration of several signaling pathways (see above). However, once IRF7 is expressed, it is sufficient to activate only the IRF signaling pathway to achieve IFN gene transcription; in effect, this appears to lower the threshold for IFN induction. It should be noted that, despite a lower threshold, viral replication would still be required for IRF7 phosphorylation and IFN transcription.

One would expect that this lowered threshold would not be sustained much beyond viral infection. Indeed, IRF7 is very unstable, with a half-life of 0.5 to 1 hour.[37] Therefore, disappearance of virus from tissues would quickly result in a disappearance of IFN-α/β induced by viral replication. Within hours, IRF7 degradation would return the threshold back to normal. It is likely that a deregulation of this loop in a way that would shift the normal cellular state toward the lower threshold characterized by IRF7 presence may lead to autoimmune disorders.

One group of cells in the mammalian organism constitutively expresses IRF7. The pDCs are ready to produce high amounts of IFN-α rapidly due to the high levels of IRF7.[38–40] While the two-tier IFN induction model holds for fibroblasts and apparently other cell types,[39,40] it is clearly not true in these immune cells. It is now believed that a MyD88–IRF7 interaction is responsible for TLR7 and TLR9 signaling to the IFN-β promoter in pDCs.[41,42] Interestingly, the positive feedback loop buoying IRF7 expression levels appears to be required for the functioning of pDCs. While IRF3 is dispensable for pDC response to TLR7 and TLR9 ligands, IRF7 and IFN-α/β receptor (IFNAR) are not.[26] This demonstrates that the IFN feedback loop, whose signal is transduced through IFN-αR, is indispensable for the pDC function, and the upregulation of IRF7 may be its major target. An interesting future question relates to the mechanism of keeping IFN production and signaling active because it differs in pDCs and other cells.

TLR3, a dsRNA sensor, also targets the IFN-β promoter,[43] but signals differently from TLRs 7 and 9. Even in cells with low IRF7 levels, TLR3 can induce IFN-β transcription through its association with TRIF (TIR domain-containing adaptor-inducing IFN, also called TICAM1).[44–47] This TLR–IFN-β axis has yet another arm: TLR4 can associate with TRIF in the presence of an adaptor called TRAM.[48,49] Thus TLR4 ligand LPS can also induce IFN-β. The significance of bacterial PAMP signaling to IFN-β is not yet clear; on the other hand, our list of TLR4 ligands may not be exhaustive.

Because of the importance of IRF3 and IRF7 in activating both intracellular and TLR-dependent IFN-β transcription, their relative contributions to the pathway have been subject to debate. Recent descriptions of IRF3[37] and IRF7[26] knockouts clarified these issues. Somewhat unexpectedly in light of early predictions,[34] IRF7 proved indispensable for all IFN responses, both at the cytosolic pathway and the TLR pathways, for nucleic acid inducers and for viruses such as VSV, HSV, and EMCV.

IRF3 deficiency produces significant but less severe effects on the cytoplasmic pathway; it seems particularly important for the activation of IFN-β through TLR4.[50] These findings indicate that IRF7 is present in IFN-primed and in unstimulated cells, although probably at very low levels. The IRF7/3 heterodimers or IRF7 homodimers may fulfill the IFN-β induction task previously assigned to IRF3 homodimers during the first (IFN-β) stage of the two-stage IFN induction. However, it is worth remembering that IRF3 was purified in a complex with CBP/p300 transcriptional co-activators as an inducer of dsRNA-stimulated genes.[51] It is therefore possible that while IFN-β depends on IRF7, some (or most) other dsRNA-stimulated genes may depend mostly on IRF3.

NF-κB AND AP-1 PATHWAYS

It has been proposed that dsRNA sets two competing programs in motion. It stimulates a pro-apoptotic translational shutoff by sensors such as PKR (see below) and also activates NF-κB to promote cell survival.[52] NF-κB signaling leads to activation of a variety of genes including pro-inflammatory cytokines. It is possible that dsRNA-induced cytokines are activated to a large extent by NF-κB-responsive elements in their promoters. Of the several NF-κB subunits, only p65 and p50 are thought to be relevant to IFN-β promoter induction.[53] Normally NF-κB (p65/p50 heterodimer) exists in the cytoplasm in a complex with IκB, its inhibitory subunit. Activation of NF-κB is achieved by an IKK (IκB kinase) complex consisting of IKKα, IKKβ, and NEMO. IKKβ is the kinase that phosphorylates IκB, thus targeting it for degradation and releasing NF-κB.[54] NF-κB translocates to the nucleus where it induces transcription of its target genes, some of them anti-apoptotic and promoting cell proliferation (IAP-1, XIAP, cyclin D).[55]

Three different MAP kinases can be activated in response to dsRNA: JNK, p38 MAP kinase, and ERK.[56] Little is known about the role of ERK in dsRNA-induced gene transcription, if such a role exists. Interestingly, while JNK is likely to be the kinase responsible for activation of AP-1 through phosphorylating c-Jun, its relevance to dsRNA pathways was discovered fairly late.[54] Moreover, it appeared likely that the MAP kinase pathways have different upstream inducers than the NF-κB pathway.[54] Further study of the MAP kinases added p38 to the list of relevant activated proteins.[57] It is distinguished from JNK by virtue of activity in RNAse L-deficient cells. It seems that p38 cannot substitute for JNK in inducing IFN-β. JNK2 knockout fibroblasts revealed profound defects in IFN-β transcription in response to dsRNA or virus.[54] It is possible that JNK1 or a related kinase is responsible for residual IFN-β gene activation.

dsRNA SIGNALING

It is interesting to note that dsRNA, the model IFN-β inducer, has been used largely due to its ability to induce all three pathways leading to IFN-β transcription (IRF, NF-κB, and AP-1; see Figure 17.2). Theoretically, if only such strong types of inducers existed for the interferon promoter, the induction pathway would have no need for three divisions. It is tempting to speculate about a variety of "weaker" interferon inducers that will activate one or two of the pathways so that the IFN-β promoter would fire only in case of synergy among two or three such signals. While this is currently a speculation, we note that commonly-used viral inducers are consistently less sensitive to genetic perturbations of the IFN induction pathway components than polyI:C treatment. For example, PKR and IRF-1 knockout cells[27,58] were reported to have defects in response to polyI:C but not Newcastle disease virus (see below). Furthermore, alternative ways of inducing IFN-β, bypassing MAP kinases and p50/p65 heterodimers of NF-κB, argue for a variety of possibilities to access the promoter.[15,20]

Much of our knowledge about IFN-β induction comes from experiments using dsRNA (usually polyI:C) treatment. The potency of dsRNA as an IFN inducer is such that a single molecule per mammalian cell was observed to yield interferon output.[59] Such sensitivity may be required of the host to respond to viruses. While dsRNA is the required intermediate for RNA virus replication, the extent of the double-stranded duplex formed during the replication process is still a matter of debate.[60] Moreover, viral genetic material is often tightly associated with membranes or viral proteins,[61,62] likely as a viral strategy to protect it from surveillance by the host.

However, it is difficult to envision a perfectly foolproof mechanism whereby no defective genome incapable of replication would ever be produced, and even if replicating RNA were absolutely invisible to the host, its dead-end intermediates could become a convenient handle to grip the viral dsRNA PAMP. In this scenario, the ability to detect very low amounts of dsRNA could prove important. It is even possible to imagine a host machinery designed to actively search for dsRNA within proteinaceous or membranous structures.

At the other end of dsRNA signaling is not only IFN, but a variety of other genes whose expression is altered by dsRNA. A distinction must be made between dsRNA-induced genes and genes induced by IFN protein binding to its receptor. Distinguishing them is not trivial because dsRNA induces IFN. Therefore, the first requirement for parsing out the two induction pathways and their outcomes is comparing the gene induction by dsRNA versus induction by IFN in cells deficient in type I IFN response. Such dsRNA-dependent induction has been observed.[63] In fact, a transcription factor complex termed DRAF1[51] containing IRF3 and transcriptional co-activators CBP/p300 has been suggested to activate dsRNA-responsive genes. In contrast, IFNAR ligation at cell surfaces leads to phosphorylation of STAT1 and STAT2 that associate with IRF9 (also known as p48) to form the ISGF3 transcription complex. Interestingly, the ISRE (IFN-stimulated response element) binding these two distinct complexes may contain the same physical DNA sequence.[64]

Which genes are induced by dsRNA, and how are they related to the genes induced by IFN treatment? Unfortunately, comprehensive microarray studies of the entire genome are lacking, but an important step to answer this question has been undertaken with a 4600-gene array.[65] The authors used the GRE glioma cell line that lacks IFNα and β genes, and found that treatment with 100 µg/ml polyI:C upregulated 175 genes and downregulated 95. These genes could be divided into several groups based on kinetics of their expression and profiles of responses to certain cytokines. Indeed, a large group of downregulated genes was a surprise because little effort had been made to describe them. The most significant conclusion from these experiments is not the identity of the genes, but the multitude of expression programs apparently induced by dsRNA. Importantly, for all the signal transduction pathways mediated by dsRNA, we know next to nothing about the transcriptional repression program turned on by the dsRNA. This presents interesting opportunities for future study.

The next step on this path may be even more comprehensive. Making use of the more complete gene arrays available today, it should be possible to test several cell populations from IFN-αR knockout mice to obtain an accurate picture of the dsRNA action across different cell types. Furthermore, our current knowledge of dsRNA sensors can help us understand the relative contributions of these sensors to the transcriptional program. The next sections review the expansion of our knowledge about candidate interferon-inducing proteins.

THE AGE OF PKR

Clearly, for anti-RNA virus surveillance to work, there should be a host-encoded activity binding dsRNA regardless of its sequence. dsRNA is an A-form helix molecule.[66] We know of at least three different, probably evolutionarily unrelated, mammalian protein structures (dsRNA-binding motif in PKR, the 2'-5' oligoadenylate synthetase (2-5OAS) family, and TLR3) that recognize double-stranded RNA. Indeed, the activities of PKR and the 2-5OAS family as inhibitors of the translation system[67–69] were discovered early.

PKR consists of two double-stranded RNA binding motifs (dsRBMs) at its N terminus and a classical kinase domain at the C terminus. In the absence of dsRNA inducer, the kinase is dormant, but dsRNA binding is thought to result in kinase dimerization and autophosphorylation.[70] The phosphorylated dimer is then capable of docking onto and activating eIF2α, its major substrate.[71] Its phosphorylation blocks translation initiation by eIF2. The existence of other PKR substrates has not been conclusively proven.

PKR would seem to be ideally suited for a signal transduction role; the 2-5A/RNAse L pathway, once untangled, also presented signaling potential with an intriguingly novel second messenger 2-5A, a 2'-5' linked oligoadenylate. Indeed, PKR was long thought to be a major sensor of dsRNA and the factor inducing IFNβ[72] based on its ability to activate NF-κB[73,74] and the fact that its pharmacological inhibitor 2-aminopurine (2-AP) blocks IFN-β induction.[75] However, doubts have been raised about the specificity of 2-AP[76] and mice deficient in PKR are capable of IFNβ induction by polyI:C.[58] Embryonic fibroblasts from PKR knockout mice were deficient in NF-κB activation and IFN-α/β induction, but the signaling was restored in primed with type I or type II (γ) interferon. Further research[54,77] appeared to confirm the finding that PKR is critical for NF-κB and IFN-β induction in embryonic fibroblasts. However, a subsequent paper[52] described a precisely opposite result: polyI:C-mediated induction of NF-κB signaling and IFN-β transcription were intact in PKR knockout cells. Two different PKR-deficient fibroblasts were used in this work.[58,78] Therefore, the question of PKR inducing NF-κB is a matter of controversy. Finally, IRF pathway induction is not known to be mediated by PKR.

Despite two attempts at a PKR knockout, the gene has still not been dealt a knockout blow.[79] The first knockout[58] still expresses a low level of an interferon-inducible catalytically active kinase that cannot be activated by dsRNA due to the absence of dsRNA-binding domains (dsRBMs). The second knockout[78] expresses a protein whose kinase is catalytically dead. It binds dsRNA and is likely to retain NF-κB activating capacity because such capacity seems to reside in its N terminal portion[80] and may not be dependent on kinase activity.[54,77,81–83] PKR is thought to possess NF-κB activating capacity because PKR binds and activates IKKβ. One recent suggestion is that PKR links to NF-κB activation through TRAF2 and TRAF5, two proteins of the TNF receptor-associated factor (TRAF) family,[84] much to the same effect. Once free from the IκB interaction, NF-κB translocates to the nucleus where transcription of NF-κB-responsive genes commences.

Since overexpressed kinase-inactive PKR and IKKα can be co-immunoprecipitated even in the absence of transfected polyI:C,[77,82] the mechanism and regulation of the activation of IKK complex by PKR, if any, remain unclear. It is possible that under physiologic non-overexpression conditions, autophosphorylation of PKR is a prerequisite for its activation of the IKK complex.

An interesting hypothesis has been advanced recently. Based on time course experiments with a cell line that stably and inducibly expressed PKR, and with cells infected by VSV and measles viruses, it was proposed (see Figure 17.2) that PKR has two activities: anti-apoptotic NF-κB signaling followed by pro-apoptotic eIF-2α phosphorylation.[85] This intriguing possibility may provide a way to reconcile the two opposing observations mentioned above whereby PKR was or was not required for the activation of NF-κB signaling in embryonic fibroblasts. The transience of PKR-dependent NF-κB signaling and the dependence of its kinetics on culture conditions and passage history of the fibroblasts may account for the discrepancy.

A further candidate signaling pathway may involve the 2-5 oligoadenylate (2-5A) synthetases.[86] Their principal role has generally been considered to be translational shutoff, much like the function of PKR. The classical view of the pathway holds that the large family of 2-5OAS enzymes bind dsRNA in cells and produce an unusual second messenger, 2-5 oligoadenylate that contains three or more adenylates linked 2′-5′ on their riboses. This messenger activates RNAse L, a latent ribonuclease that cleaves ribosomal and possibly viral RNA.[87,88] Iordanov et al.[52] proposed that PKR and RNase L, through a block in translation, activate the JNK pathway to enable AP-1 activation and IFNβ transcription. If this hypothesis is correct, PKR and RNAse L are redundant and any conclusions about their role in IFN-β induction should be drawn from cells deficient in both enzymes. The display of an IFN induction phenotype by the doubly deficient cells would confirm the critical role of one of the two enzymes in IFNβ transcription.

THE AGE OF TLR3

Clearly, PKR cannot fully account for dsRNA signaling to the IFN-β promoter because priming fibroblasts with IFN restores the dsRNA pathways completely. What accounts for this effect? As more TLRs were discovered and their abilities to discern PAMPs came into focus, the notion started taking shape that TLRs may be the long-sought dsRNA sensors. Indeed, TLR3 was found to transduce a dsRNA signal.[43] Furthermore, TLR3 activation leads to IFN-β promoter induction (Figure 17.2). Since TLR3 can activate IRFs and NF-κB (see above), it certainly fits the profile of a key dsRNA sensor.

Moreover, dsRNA has long been known to not require transfection reagents for its IFN-inducing effects.[9] This fits well with the suggestion that its sensor is not cytoplasmic. By now, the signaling pathway downstream of TLR3 has been to a large extent uncovered. As noted, the crucial role in this signaling cascade belongs to TRIF (or TICAM1), the adaptor coupled to TLR3; mutations in TRIF lead to abrogation of TLR3-dependent responses.[44,46] The signal bifurcates at TRIF, as TRIF binds and activates RIP1[89] which further activates NF-κB, and TRAF3, now known to activate TBK1 and consequently IRF3 and probably IRF7.[90] Another suggestion is that TRIF can activate AP-1.[45]

Several viruses [lymphocytic choriomeningitis virus (LCMV), VSV, and reovirus] were tested on TLR3 knockouts[91] and no appreciable differences in host survival or immune responses were evident. This observation alone is not sufficient to dismiss TLR3 as an antiviral sensor. First, its effect may not be sufficiently dramatic with respect to a particular group of pathogens. Second, viruses may have evolved anti-TLR3 defenses. It has been argued that TLR3 is chiefly engaged in cross-priming of dendritic cells, thus allowing dsRNA of viruses that do not themselves infect DCs to stimulate them.[92] While the "jury is still out" on TLR3, it seems less important now by comparison with the newly discovered helicase sensors of RNA. However, the significance of TLR3 may not be fully appreciated. The age of TLR3 may have seemed short, but it is not yet over.

DICER-RELATED HELICASES: FIRST STEPS

The first clear indication of another kind of cytoplasmic dsRNA sensor came from lower organisms. In fact, PKR and TLR3 are not known to exist in invertebrates. dsRNA is recognized as a PAMP by nematode worms, fruit flies, and plants. Most eukaryotes including these organisms are endowed with a nucleic acid-based immune system called RNA interference (RNAi).[93] After dsRNA is recognized by an organism, it becomes a substrate for DICER, a double-stranded RNA nuclease that processes dsRNA into smaller (21 to 27 nucleotides) fragments.

These fragments (short interfering RNAs, or siRNAs) are then incorporated into RISC (RNA-induced silencing complex) and used as guides to direct the host RISC machinery to homologous sequences in mRNAs. It is possible that most or all mRNAs in cells are scanned by the RISC that may be associated with ribosomes.[93] RISC is a single-stranded nuclease that cleaves mRNA approximately in the middle of the 21-nt siRNA-homology region. This nucleic acid-based adaptive immune system ensures that once a particular sequence of RNA is tagged as "viral" by virtue of being double-stranded, any homologous sequences will be destroyed.

DICER is conserved in vertebrates; however, despite many efforts, no conclusive evidence suggesting its participation in a similar nucleic acid-based immune system in mammals has been found.[94] Instead, its undisputed function in mammalian cells is to process microRNAs (miRs), a family of hairpin RNA genes in which one strand is homologous to host-encoded mRNAs and serves to downregulate their expression.

DICER contains several domains, among them RNAse III (a double-stranded nuclease) and a DExH-box RNA helicase. Interestingly, the function of DICER's helicase is unclear; its ATPase activity does not seem to be required for processing of dsRNA.[95] Yet an RNA helicase encoded by a different gene has been shown to be required for RNA interference.[96] The DExH RNA helicase domain it possesses is most closely related to that of DICER. Called DRH1 (DICER-related helicase 1), it may have a very close homolog adjacent to it on chromosome IV of *C. elegans*; this putative homolog, however, seemed truncated in the strain studied by these authors.

While the Drosophila genome does not contain obvious homologs of these RNA helicases, mammalian genomes have several of them. In fact, simultaneously with their identification, one key feature of two such proteins, MDA5[97] and RIG-I[98] is apparent: their induction in response to IFN-β[97] or virus.[98] This is a key indication of their importance in antiviral immunity. Other characteristics have also come to light. Both RIG-I (whose porcine homolog was named RHIV-1 for RNA helicase induced by virus) and MDA5 contain a C terminal helicase domain and two tandem N terminal CARD domains.[97,99,100] Both are predominantly, if not exclusively, cytoplasmic[97,101] and are rather widely expressed.[97-99] MDA5 (melanoma differentiation-associated gene-5, also known as IFIH1 and Helicard[99]) was cloned by subtractive hybridization from melanoma cells undergoing differentiation in a protocol involving IFN-β.[102] Treatments with various factors have shown that type I IFN is its strongest inducer, and IFN-γ and TNF-α are also capable of upregulating MDA5 mRNA. RIG-I was discovered when it was upregulated by retinoic acid during differentiation of acute promyelocytic leukemia cells.[103] Its porcine homolog was induced by porcine reproductive and respiratory syndrome virus (PRRSV). Further study showed that RIG-I is upregulated by IFN-γ[104] and LPS.[105]

This is compelling data showing the participation of these helicases in the immune response or anti-proliferative action of IFN. RIG-I was in fact found to stimulate the expression of pro-inflammatory enzyme COX2[105] when overexpressed. However, the focus of the early research on MDA5 related to its cell growth suppression properties[97,99,106] due to its expression in melanoma differentiated with IFNβ and mezerein and the recognition of its protein–protein interaction-mediating CARD domains historically associated with apoptotic processes. Overexpression of MDA5 led to growth inhibition in the human melanoma cell line HO-1.[97] In light of our present knowledge of the role of MDA5 in inducing IFN, this result may be explained as a direct (perhaps toxic) effect of MDA5 on cells and also as a secondary consequence of IFN induction. Another interesting survey of possible connections of MDA5 with apoptotic processes, however, suggested interesting links between MDA5 and apoptosis[99] that are discussed later.

HELICASE DOMAINS

The domain structures of MDA5 and RIG-I are similar (Figure 17.3). In the mouse, MDA5 is a 1025-amino acid protein containing two CARD domains at the N terminus and a DExH-box RNA helicase domain in its C terminal part. The same domain organization is found in 926-amino acid RIG-I.

The biochemical data on RIG-I and MDA5 is still scant. Both helicases bind polyI:C.[100,107] MDA5 was found to be an RNA-dependent ATPase,[99] and RIG-I has been shown to bind single-stranded 5′ triphosphorylated RNA.[108,109] Neither has yet been shown to unwind substrates, and it is not clear what the natural substrates of these helicases are. They may be intact viral replication complexes or random pieces of viral RNA (such as dsRNA) left behind by such complexes. The possibility that the

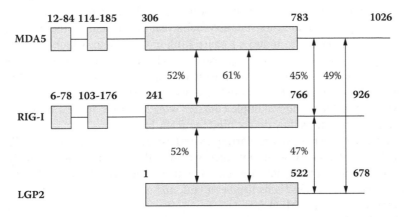

FIGURE 17.3 Domain structures of antiviral RNA helicases. Approximate domain borders of murine proteins are shown as amino acid numbers (bold), and percent homology between domains is given. Small boxes indicate CARD domains; large boxes indicate helicase domains.

substrates are host RNAs while the sensor is upstream in the pathway has not been ruled out. What is more, it is not clear that RNA is a necessary activator for either enzyme. We do know that a functional ATP-binding site is required for dsRNA- or virus-induced signaling,[100,107] but the role of ATP binding is not understood, and it is not yet possible to say that ATP is required for helicase activity.

Understanding of RNA helicases is still in its infancy. A helicase classification scheme based on primary amino acid sequence was presented in 1993.[110] A number of conserved motifs were recognized in addition to the previously known Walker A and B motifs involved in nucleotide binding.[111] A variation in the Walker B motif suggested that a superfamily of enzymes (superfamily 2) be called DExH/D helicases and set apart from other helicase families. This classification is not well correlated with function yet.

Of the long list of RNA helicase proteins, only a handful have been shown to unwind dsRNA.[112] Some (p68)[113] are processive. Some (eIF4A)[114] lose their affinity for RNA following ATP hydrolysis. As alluded to above, it is not clear whether all the proteins containing canonical helicase motifs are actually helicases. Some may strip single-stranded RNA of associated proteins; others may be merely RNA-binding proteins or simply retain the common fold.

However, progress in crystallography is advancing our understanding of RNA helicase function. Crystal structures of two DExH/D helicases in complex with a non-hydrolyzable ATP analog known as ATPPNP and a stretch of single-stranded RNA have been solved.[115–117] A number of DExH/D helicase protein structures were solved without bound RNA.[112] They demonstrated that the canonical DExH/D helicase domain folds into two RecA-like subunits joined by a flexible linker. These subunits do not seem to interact in a specific fashion in the absence of ATP and RNA. However, when co-crystallized with RNA and ATPPNP, *Drosophila vasa* and human eIF4AIII adopt strikingly similar conformations,[115–117] as ATPPNP and oligouridine RNA lock the two subunits together.

A remarkable feature of this complex is the shape assumed by the ssRNA oligonucleotide. At least four 5′-most nucleotides of the RNA conform to the features of a strand engaged in an A-form double helix that may extend in the 5′ direction. Helicase binding, however, induces a sharp kink in this strand, which is equivalent to interrupting a double helix. The suggestion which follows is that helicases work by destabilizing dsRNA duplexes through formation of an energetically favored RNA–protein structure with one RNA strand. The two proteins examined are specific for RNA over DNA due to their multiple contacts with 2′ OH groups of riboses, yet are not sequence-specific as they do not bind the RNA bases.

Unfortunately, while we are starting to understand the mechanisms of DExH/D RNA helicase interactions with substrates, we are still far from predicting likely substrates or activities of these helicases based on their sequence. However, the basis for models of helicase mechanisms has been laid, and it should aid attempts to unlock the secrets of RIG-I and MDA5.

The other domain present in the antiviral helicases is the caspase activation and recruitment domain (CARD)—a compact structure of about 75 amino acids.[118] A CARD consists of six closely packed α helices and has gained fame as a protein–protein interaction module. A well-described interaction occurs, for example, between the

CARD on Apaf-1, a mammalian sensor of cytochrome *c* leakage from mitochondria, and the CARD of procaspase-9. Apaf-1 is a multidomain protein[119] and its CARD is normally engaged in interactions with other domains of Apaf-1. Binding of cytochrome *c*, an apoptotic signal, to Apaf-1, is believed to lead to assembly of the apoptosome, a heptamer of Apaf-1 that further interacts with procaspase-9 and allosterically activates it.[120,121] This latter interaction is mediated by the CARDs of Apaf-1 and procaspase-9. The Apaf-1 CARD participates in at least three interactions: it contacts other domains of Apaf-1 in the latent state, whereas upon activation, it binds the procaspase-9 CARD while homo-oligomerizing. This kind of versatility may be at play also where the CARDs at the N terminal domains of RIG-I and MDA5 are involved.

RIG-I AND MDA5: KEY INTERFERON INDUCERS

A groundbreaking discovery by Yoneyama et al.[100] identified RIG-I as an IFN-β inducer. The authors conducted a screen for IRF promoter-inducing molecules and identified a fragment of cDNA whose expression greatly potentiated polyI:C-mediated IRF reporter and IFN-β promoter reporter activities. This fragment encoded the N terminal portion of RIG-I containing its CARD domains. While the fragment's expression elicited IFN-β transcription, the same effects by the full length RIG-I clone depended on polyI:C co-transfection. As could be expected, NF-κB-dependent transcription was also activated in RIG-I transfected cells.

Just as the CARD domain fragment of RIG-I proved constitutively active, the helicase fragment of the protein expressed alone had a dominant negative function; the same dominant negative effect was exhibited by a full-length protein carrying a point mutation in the RIG-I putative helicase ATP-binding site. These results suggested the following mechanism of RIG-I's action. The pro-IFNβ signaling by the CARD domains is normally blocked by the helicase domain; however, upon binding viral dsRNA, the helicase relieves this inhibition, and CARD domains are allowed to signal, resulting in IFNβ production.

A similar IFN-inducing function was discovered in MDA5.[122] Its significance was underscored by the fact that almost identical characteristics of MDA5-mediated IFN induction were blocked by paramyxoviral V protein that bound MDA5 and prevented its signaling. Another milestone for the two RNA helicases was reached when it was demonstrated that a dominant negative point mutation in the RIG-I gene likely accounts for the permissiveness of Huh7.5 cell line for the replication of hepatitis C virus RNA.[123] Huh7.5 cells are derived from Huh7, which are wildtype for RIG-I and do not support efficient viral RNA replication.[124] Indeed, reconstituting Huh7.5 with RIG-I led to restoration of both IFN response and suppression of viral RNA levels.[123] Based on the importance of RIG-I in hepatitis C defense, it was not surprising to find that the virus-encoded protease NS3/4A disrupted RIG-I signaling.[125,126]

A series of studies addressed the roles of RIG-I and MDA5 in inducing IFN and IFN-induced genes in response to several viruses, specifically flaviviruses, Japanese encephalitis virus (JEV), dengue,[127] the EMCV picornavirus,[128] paramyxoviruses, Sendai virus (SeV),[128] measles,[129] and orthomyxovirus influenza.[130,131] All these reports concluded that RIG-I is important for sensing all of these viruses except measles. The involvement of MDA5 was examined and found to be relevant to measles and influenza.[131] Furthermore,

VSV (a rhabdovirus) and EMCV viral yields were suppressed in a RIG-I overexpressing cell lines.[100] While supporting the general conclusion about the importance of the helicases, most of these results were based on overexpression of RIG-I or MDA5 on one hand, and on downregulating RIG-I expression by a dominant negative form of RIG-I or MDA5. These approaches, while suggestive, represented preliminary investigations and had to be followed up by RIG-I and MDA5 knockout studies.

At this point, several important findings about the RNA helicases may be summarized. First, they are capable of inducing IFN in response to dsRNA and virus infection. Second, their mRNA levels are boosted by type I IFN, demonstrating their relevance to antiviral defense. Third, a genetic lesion in an RNA helicase may lead to increased viral replication. Finally, viral proteins target them and block their IFN induction function. These compelling arguments make their importance in innate immunity nearly unquestionable. But the acknowledgement of their true role relative to other well-described dsRNA and various PAMP response mechanisms awaited knockout studies.

RNA HELICASE SIGNAL TRANSDUCTION PATHWAY

In the meantime, the signal transduction pathway utilized by the helicases started coming into view. It was likely that the CARD domains are key to RNA helicase signaling. Their target was identified by four groups simultaneously,[132–135] earning it one of the longest protein names: VISA/Cardif/MAVS/IPS-1 (virus-induced signaling adaptor/CARD adaptor inducing IFN-β/mitochondrial antiviral signaling protein/IFN-β promoter stimulator-1). Fingered mostly by homology to the RIG-I and MDA5 CARDs, this protein appears to have one N terminal CARD domain required for interaction with RIG-I and MDA5 CARDs, a proline-rich domain, and a transmembrane region.

VISA/Cardif/MAVS/IPS-1 displays mitochondrial localization,[134] which is important for its activity. Upon viral infection, no clear change in localization or obvious covalent modification is apparent. Seth et al. suggest that the protein conformation and/or interaction with other membrane components changes, based on a change in its detergent solubility.[134] VISA/Cardif/MAVS/IPS-1 overexpression stimulated IFNα/β promoters, while RNAi-mediated knockdown reduced IFN responses to viruses SeV and VSV, as well as to polyI:C.[132–135]

The mRNA is found ubiquitously, and therefore, the protein is well positioned to be a general virus alarm relay for an organism. Other interacting partners of VISA/Cardif/MAVS/IPS-1 include tumor necrosis factor receptor-associated factor 2 (TRAF2), TRAF6, and TRAF3, FADD (Fas-associated protein with a death domain), and receptor interacting protein-1 (RIP-1). FADD- and RIP1-deficient embryonic fibroblasts are significantly impaired for IFN-β induction, and their IFN-α levels are approximately at background levels.[136] Their precise relationship is still unclear, but it is likely that VISA/Cardif/MAVS/IPS-1 activates the NF-κB pathway through TRAF6, and IRF pathway, through TRAF3.[135,137]

Both TRAFs can dock to VISA/Cardif/MAVS/IPS-1 using specific, mapped TIMs (TRAF interaction motifs). The exact connection of these players to transcription factors, however, is a matter of speculation. FADD, for example, is known to activate NF-κB.[138] Rapid progress is being made and the upstream and downstream parts of the pathways will soon connect. In fact, the important kinases phosphorylating IRF3 and IRF7 were uncovered years ago.[139–142] The major kinase responsible for IRF3/7 phosphorylation is TANK-binding kinase 1 (TBK1); IκB kinase-related kinase ε (IKKε) also seems to have a role in the process.

While IRF3 and IRF7 are assumed to be involved in IFN-β and ISRE promoter induction, IRF1 knockout mice are known to be impaired under various conditions in their responses to polyI:C and EMCV but not NDV or VSV (see IRF Pathways section above). Whether this represents a specific connection of MDA5, the EMCV/polyI:C sensor, and IRF1, remains undetermined.

A broader unanswered question is how or whether MAP kinases become activated as a result of VISA/Cardif/MAVS/IPS-1 signaling. Other unanswered signaling issues concern the precise relationships of the players, a final map of the signaling pathways, and of course, sensors for the other IFN inducers such as dsDNA.

RIG-I AND MDA5: MAJOR SENSORS OF RNA VIRUSES

RIG-I and MDA5 knockout mice studies were required to elucidate the relative contributions of various PAMP sensors to viral recognition. The first surprise was almost immediate: the RIG-I disruption was mostly lethal at embryonic day 12.5 through 14. This was surprising in light of the fact that IFN-αR-deficient mice are viable[143] and we now know that the VISA/Cardif/MAVS/IPS-1 knockout is also viable. It is likely therefore that RIG-I functions beyond IFN-β and interferon-stimulated gene (ISG) induction.

One alternative explanation concerns the structure of the targeted allele. The promoter and the N terminus of the protein are intact, and the gene is disrupted at the helicase domain. This should not be a problem as a nonsense-mediated decay system generally destroys aberrant transcripts of this kind. However, it is possible that IFN or another kind of stimulus increases transcription of the truncated N terminal fragment of RIG-I, which may in turn have a constitutive IFN- and ISG-inducing activity, and thus be detrimental.

A small percentage of RIG-I knockouts survived and allowed *in vitro* experiments with DCs and derivation of embryonic fibroblasts. Negative-stranded viruses such as NDV, Sendai, influenza, VSV, and a positive-stranded flavivirus JEV elicited practically no IFN and ISG expression in RIG-I knockout cells.[144,145] VSV replicated to much higher titers in knockout cells.[144] IFN-α/β and IL-6 secretion were also abolished in bone marrow-derived conventional dendritic cells (cDCs) following NDV infection. In a remarkable show of complementarity, the MDA5-deficient cells (fibroblasts or DCs and macrophages) responded normally to all of the above viruses, but failed to secrete IFN in response to the EMCV picornavirus.[145,146] Conversely, the EMCV response was intact in the RIG-I knockout. Restoration of MDA5 to the RIG-I/MDA5 double knockout fibroblasts restored the EMCV but not SeV IFN-β promoter response; the converse was true of the RIG-I-transfected double knockout cells.

In vivo infections with JEV and EMCV demonstrated exactly the same characteristics. Lower interferon titers, higher virus titers, and a steep drop in the survival curve were observed for JEV in the RIG-I$^{-/-}$ but not MDA5$^{-/-}$ mice, and for EMCV in the MDA5$^{-/-}$ but not RIG-I$^{-/-}$ mice.[145,146]

Interesting results were also obtained with double-stranded RNA. PolyI:C-dependent IFNβ response was shown to be absolutely dependent on MDA5. The TLR3 pathway seemed to play a significant role only when polyI:C was applied extracellularly to peritoneal macrophages[146] but not bone marrow-derived DCs or macrophages.

In peritoneal macrophages, the expression of pro-inflammatory cytokines was co-regulated by MDA5 and TLR3. Similar results were obtained when injecting whole mice with polyI:C.[145,146] Pro-inflammatory cytokines depended on both MDA5 and TLR3; IFNβ was under MDA5 control. A point of difference between the two MDA5 knockout experiments was, however, the IFN response to *in vitro* synthesized dsRNA of several hundred base pairs. In one study, response to dsRNA was fully dependent on MDA5.[146] In the other, it was fully dependent on RIG-I.[145] The reason for the discrepancy is unclear. What is clear, however, is the amazing complementarity of RIG-I and MDA5 in responding to viruses, indicating some specialization we do not yet understand. In line with the grand vision of two IFN-inducing systems, one cytoplasmic and the other TLR-based, pDCs did not require RIG-I or MDA5 for NDV recognition. Instead, their ability to recognize NDV was based on MyD88, a TLR adaptor that probably transduced a single-stranded RNA, TLR7-mediated signal.[144,145]

Still uncertain are the activities of RIG-I and MDA5 in different cell types. For example, MDA5 is not easily detectable on Northern blots of the nervous system.[99] Does this mean that dsRNA-mediated IFN induction is different in the brain, at least in the absence of interferon in the milieu? Could the pattern of MDA5 (or RIG-I) expression account for viral, for example, EMCV, tropism, as IFN is known to produce?[147] Will interferon treatment force expression of otherwise undetectable sensors? The answers to these questions may vary based on the tissues involved.

Both RIG-I and MDA5 transmit their signals through VISA/Cardif/MAVS/IPS-1. Indeed, a knockout of this gene would complement our understanding of RNA helicases. Two new reports[148,149] suggest that the viral susceptibility and IFN response phenotype of VISA/Cardif/MAVS/IPS-1 is (if one discounts RIG-I lethality) similar to a sum of those of RIG-I and MDA5. The knockout does not lose responsiveness to TLR ligands or to cytoplasmic DNA.

Mitochondrial location of VISA/Cardif/MAVS/IPS-1 suggest that it may be relevant for control of apoptosis,[134] but mice lacking the protein did not display obvious apoptosis defects. In contrast, they were extremely susceptible to every tested virus including NDV, SeV, VSV (RIG-I substrates), and EMCV (MDA5 substrate). The Kumar et al. knockout was also impaired in its response to polyI:C, although in the Sun et al. knockout, polyI:C was mostly delivered by extracellular application, thus likely triggering TLR3. This may explain the mostly normal IL-6 and IFN secretion in these cells.

VIRUSES STRIKE BACK

As mentioned above, one of the most persuasive proofs for antiviral functions of proteins are countermeasures employed against them by viruses. Hepatitis C is an example of a virus that blocks RIG-I signaling. The target of HCV has been mapped: it turns out that the HCV protease NS3/4A cleaves VISA/Cardif/MAVS/IPS-1, preventing its ability to associate with RIG-I or activate the IFN promoter.[133,150–152] It has also been suggested that HCV targets more upstream components of the dsRNA signal transduction pathway[127] because, while NS3 protease inhibitor restores signaling by overexpressed VISA/Cardif/MAVS/IPS-1, the same inhibitor cannot rescue polyI:C-mediated IFN-β promoter induction. This is unrelated to the blockade imposed by HCV on TLR3 signaling, since this blockade also depends on NS3-4A that cleaves the TLR3's TRIF adaptor.[153,154]

Other viruses have already been profiled for anti-RIG-I activities. Both possible modes of such activities have been described, namely active countermeasures (such as proteolysis of host signaling components as in HCV infection) and passive avoidance of host pathway activation. MDA5 has not been carefully profiled; the full range of RIG-I activities is also unclear, although data from several reports can be summarized.

Hepatitis A virus, a picornavirus, inhibits IFN-β induction by blocking the IRF arm of the IFN-β induction pathway.[155] This affects the IFN-β upregulation imparted by RIG-I CARD fragment overexpression. Since the block is upstream of TBK1 and IKKε, it is likely to be in the VISA/Cardif/MAVS/IPS-1 or one of the interacting proteins.

Influenza virus NS1 protein has been found to bind RIG-I, but not MDA5.[109] Its co-transfection strongly decreases IFN-β promoter induction by viral RNA. The RIG-I/NS1 interaction may be direct or may be due to the RNA-binding properties of both of these proteins. In both cases, NS1 only seems to block RIG-I dependent IFN signaling caused by SeV, but not MDA5-dependent IFN signaling by EMCV. This is suggestive of an anti-RIG-I function by NS1. NS1 was previously found to block PKR-mediated translational repression and antiviral effects,[156–158] and as a dsRNA-binding protein may be well suited for interfering with many host dsRNA recognition mechanisms.

RIG-I plays an important role in the IFN response to the West Nile flavivirus (WNV),[159] although at the late stages of infection other mechanisms (possibly MDA5) compensate for it. The virus can avoid IRF3 activation at early stages of infection by an unknown mechanism even though SeV and VSV co-infections trigger IRF dimerization and nuclear translocation. This result, along with the fact that WNV grows to higher titers in the absence of RIG-I, suggests that the avoidance of RIG-I (and other sensors) is passive.

An interesting and separate issue involved the paramyxoviral V proteins that bind MDA5.[122] Amazingly, no clear role for MDA5 was found in experiments with cells from MDA5 knockout mice. Instead, RIG-I appeared to be the chief sensor of Sendai[145] (authors' unpublished data). It will nevertheless be interesting to test other paramyxoviruses and more cell types; it is likely that effects of MDA5 deficiency may be uncovered through *in vivo* experiments.

Many viruses are known to block IRF3/IRF7 activation and nuclear translocation.[160] Often, the target of the block is not known, but the helicases and especially VISA/Cardif/MAVS/IPS-1 protein upstream of the IRF3/7–NF-κB bifurcation are attractive targets because they may afford the most stealth for the virus. Furthermore, several anti-dsRNA functions found in viruses over the years were ascribed to anti-PKR and anti-RNAse L effects. They may target RIG-I and MDA5 instead of or in addition to PKR/RNase L. Clearly, we can anticipate interesting discoveries in this area.

Another point must be stressed. While it looks like only RIG-I or only MDA5 may respond to viruses, in reality we must consider the ubiquitous viral immune evasion. Therefore, it would be useful to keep in mind that a host factor that does not seem to play a role may in fact be merely rendered ineffectual by a virus. This represents a caveat for immunologists, but an opportunity for virologists hunting for new viral immune evasion strategies.

LGP2: TONING DOWN THE ALARM

The mammalian genome encodes two other members of the antiviral DExH-RNA helicase subfamily. One is the aforementioned DICER, whose helicase primary amino acid sequence is approximately 25% identical to RIG-I and MDA5. However, mammalian DICER does not appear to exert effects on the IFN-β promoter (authors' unpublished data). The other helicase is LGP2, cloned first as a gene in the Stat3/5 locus.[161] An understanding of its family connections and its interesting structure (only a helicase domain and a tail, both homologous to RIG-I and MDA5, but not containing CARDs or other identifiable signaling domains; Figure 17.3) led two groups to probe LGP2's potential role in interferon induction.[107,162]

Like RIG-I and MDA5, LGP2 is expressed ubiquitously and upregulated by type I IFNs.[107,162] Unlike RIG-I and MDA5, transfection of LGP2 did not alter yields of EMCV and VSV from a fibroblast cell line.[107] LGP2 expression downregulated IRF reporter activity in response to NDV and SeV, much like the dominant negative helicase domain-only fragments of RIG-I and MDA5.[107,162] RNAi against LGP2 led to an increase in the activity of the same promoter compared to cells expressing endogenous LGP2.[107] Since LGP2 binds dsRNA but not RIG-I or MDA5,[107,162] it was proposed that its mode of action is simply competing with other antiviral helicases for dsRNA. However, LGP2 can inhibit IFN-β signaling induced by VISA/Cardif/MAVS/IPS-1 overexpression, while not occluding the RIG-I binding site on VISA/Cardif/MAVS/IPS-1.[163]

Interestingly, LGP2 is also different from RIG-I and MDA5 because it can oligomerize; additionally, it co-immunoprecipitates with both other helicases under conditions of SeV infection. These interactions and the promoter inhibition exerted by LGP2 are not dependent on the intact Walker motif, suggesting that a helicase activity of LGP2 (if any) is not required for its signaling inhibitory function. This observation underscores the fact that we barely understand LGP2, and other potential antiviral RNA helicases.

LGP2 joins other inhibitors of the IFN-β pathways. One such long-known inhibitor is IRF2[164] that may dimerize with other IRF family members, but prevent gene activation due to the presence of a transcriptional repression motif.[165] Mice deficient for IRF2 demonstrated CD8 T cell-mediated spontaneous skin inflammation.[27] Additionally, LCMV infection, while resolved normally, resulted in death 2 to 3 weeks later.[166] Since skin disease depended on IFN-αR and IRF9 presence, and ISG's expression rather than IFN-α/β was increased, a major mode of IRF2 action was localized to attenuation of IFN-α/β signaling.

Further studies showed that IRF1-dependent gene expression also increased in IRF2 knockouts.[167] Most of these phenotypes are evident in mice that were never infected with viruses; it will be even more interesting to study LCMV and other infections to determine the mechanisms of (presumably) autoimmune diseases triggered by the virus. Finally, the full range of IRFs whose activities are attenuated by IRF2 is still unclear, and the answer to this question will help us understand how tight negative regulation must be on the transcription factor level.

Other specific inhibitors of the IFN induction pathways have been discovered. SIKE (suppressor of IKKε) interacts with IKKε and TBK1 and prevents them from interacting with TRIF, RIG-I, and IRF3, thus attenuating the IRF arm of the pathway.[168] Anti-apoptotic protein A20 has a different scope of activity.[169] It blocks RIG-I-mediated IFN-β induction upstream of TBK1 and IKKε, and prevents both IRF- and NF-κB-mediated promoter stimulation.

As we learn more about the signal transduction ladder, more inhibitors will come into view. It should not be surprising to find them at several and perhaps all rungs of the ladder. A runaway signaling cascade from any point of the ladder would risk serious consequences such as autoimmunity, so every point may need to be carefully calibrated. Precise circumstances under which such autoimmune cascades arise will be clarified by producing knockouts of inhibitor proteins and the answers should appear soon.

RECOGNITION SPECIFICITY OF RIG-I AND MDA5

We described dsRNA as a viral "Achilles' heel" at the beginning of this chapter. However, we do not yet know what the helicases actually recognize. The recognized pattern is of both theoretical and practical concern. Theoretically, we would like to identify the viral Achilles' heels because they may help us understand viral replication. (Incidentally, this represents a reversal of the usual paradigm in which the virus reveals the host's biology, as we would have the host reveal to us the viral biology). Practically, it is important to know the mechanisms of this critical point in viral recognition because it may help us design drugs directed at the invariant characteristics of the viruses.

We are faced with two questions: which features distinguish the viral structures from the host ones, and why do two separate helicases exist to recognize them? The first question may be easy. It is likely that a long (probably more than ~20 nucleotides) fragment of dsRNA can activate an antiviral RNA helicase.[170] The second question is more difficult. It may well be insufficient to recognize only dsRNA. Why do two helicases with apparently complementary recognition specificities (negative-stranded RNA viruses and flaviviruses for one and picornaviruses for the other) exist? Perhaps the recognized structures represent two different conformations of dsRNA (slightly different helix shapes) covering all likely possibilities.

Alternatively, these conformations may depend on viral replication. We know that during eukaryotic transcription in which the ends of the DNA are topologically fixed, passage of the RNA polymerase introduces positive supercoils in front of and negative supercoils behind the enzyme. If viral transcription and replication proceed from a dsRNA structure whose ends are topologically fixed, or if the virus chooses to store energy in the form of dsRNA supercoiling, the host should recognize a molecule different from relaxed dsRNA.

This may necessitate two separate RNA helicases, one of which would react to dsRNA overwound and the other to dsRNA underwound relative to a certain threshold. Unfortunately, very little is known about the dynamics of RNA replication in RNA viruses, and it is difficult to make accurate predictions and test them rigorously.

A bold hypothesis was advanced regarding RIG-I specificity. Two groups[108,109] argued, based on the fact that 5′ triphosphorylated single-stranded RNAs are known to elicit IFN responses,[170,171] that RIG-I is the receptor for these structures, and that the reason that negative-stranded viruses and JEV flavivirus were recognized by RIG-I is their triphosphorylated genomic RNA. In support of this hypothesis, RIG-I knockout fibroblasts were incapable of recognizing 5′ triphosphorylated RNAs, and RIG-I was found to bind such RNAs.

This hypothesis does not explain the apparent complementarity in the subsets of viruses recognized by the two helicases; therefore suggesting that viruses largely fall into two separate categories, dsRNA-overproducing and 5′-triphosphorylated. Besides, it would be unexpected for a 5′ triphosphate to be such a universal and invariant viral signature as suggested by the spectrum of viruses recognized by RIG-I.[145] Finally, it is unclear why a helicase domain is required if RIG-I's specificity determinant is RNA containing triphosphates. Indeed, RIG-I is known to also bind to polyI:C.

This data may be interpreted several ways, such as a possibility that RIG-I is a dual-specificity receptor for 5′ triphosphate and dsRNA; or that certain host dsRNAs are anti-determinants of RIG-I recognition (that would otherwise occur due to their triphosphates). It is important to accumulate more data using wild-type versus RIG-I and MDA5 knockout cells, reconstituted or not reconstituted with helicases, and attempt to make the RNA preparations more similar. (The triphosphorylated RNA was produced enzymatically, while 5′-OH RNA was synthesized chemically.). It would also be interesting to determine whether MDA5 has activators other than dsRNA.

POTENTIAL ROLES OF RNA HELICASES: DECISION MAKERS IN AUTOIMMUNITY AND APOPTOSIS

An interesting recent report seems to connect MDA5 to type I diabetes, an autoimmune disease involving destruction of pancreatic β-cells.[172] While the precise mutation has not been unambiguously determined, one point mutation among several, all mapping to the MDA5 locus, is likely to be a predictor of type I diabetes. It is most likely to be the only mutation in the coding region, A946T. This mutation is predicted to lead to a missense substitution of a conserved alanine residue with threonine in the C terminal portion (tail) of the protein beyond the consensus helicase domain.

As noted above, regions adjacent to helicases often play roles in determining the specificity and interactions of the enzymes. Indeed, not only is the C terminal tail conserved among MDA5 genes from different species, but a comparison of murine RIG-I and murine MDA5 tails indicates 45% homology, similar to the extent of homology within the helicase region (Figure 17.3).

What is the nature of the relationship between MDA5 and autoimmune diabetes? Autoimmune (type I) diabetes is characterized by the presence of T cells and antibodies specific for β cells of the islets of Langerhans in the pancreas, and ultimately leads to the destruction of the β cells. Type I diabetes has a strong hereditary component, although its multigenic nature has hampered an understanding of the exact contribution of genetics to the pathogenesis. However, some loci have been identified, and the major diabetes susceptibility locus is known to be MHC class II.[173] Other loci with alleles predisposing to diabetes include the insulin and CTLA4 genes.[172] Another salient characteristic is that infections of coxsackie B4 viruses are frequently associated with developing diabetes.[174] Indeed, we note here that coxsackie is a picornavirus whose chief sensor is MDA5.

How can a mutation in MDA5 translate into an autoimmune disorder? Hypotheses on this matter may be loosely grouped into three categories. The first two involve viral infections (Figure 17.4). First, MDA5 mutant may have merely lost its function of proper response to a virus, leading to alteration of viral replication (in the simplest case, an increase) to a point where the adaptive immune system begins to associate β cell antigens with the viral disease. Second, MDA5 can be at or near the point of the very decision as to what is a viral antigen and what is not, possibly in a dying cell about to be removed by APCs. Indeed, interferon production may be expected to be such a crossroads. Finally, viruses may not be involved. A946T may merely shift MDA5 into an autoimmune mode of leaky signaling through its CARD domain, which may be sufficient for activation of the adaptive immune system. In this latter case, we must explain how pancreatic β become more susceptible than other cells. However, before offering explanations, it is advisable to check other autoimmune disorders for association with MDA5 mutations.

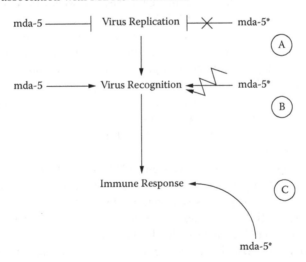

FIGURE 17.4   Models of MDA5 involvement in autoimmune disease. Model A: poor recognition of viral infection by mutant MDA5 (MDA5*) may lead to increased viral replication and dysregulation of immune response. Model B: MDA5* is the component that directs the immune response along an aberrant path. Model C: MDA5* may not require viral infection to initiate autoimmune disease.

The other developing areas involving MDA5 concern apoptosis. Kovacsovics et al.[99] presented an intriguing finding. A Western analysis of MDA5 expression across normal tissues indicated that most or all the protein detected by an anti-N terminus antibody was cleaved, most likely at the junction between the helicase and the CARDs. It is unlikely that this band arises from an alternatively spliced product because Northern hybridization detects only one band. The authors suggest caspase-mediated cleavage and demonstrate that it can happen in cultured cells upon TRAIL or FasL stimulation. However, it does not seem likely that the fragment observed on the gel would be detectable if it were limited to apoptosing cells. Therefore, it remains possible that the N terminal fragment is not a caspase cleavage product, despite the presence of functional caspase sites in the MDA5 protein. Additionally, a role of MDA5 in the induction of DNA fragmentation has been proposed on the basis of MDA5 transfection experiments; yet this suggestion is confounded by the likely presence of MDA5-included IFNβ and its secondary effects that make it very difficult to interpret experimental results. In other words, MDA5 may contribute to apoptosis via IFN; whether it has a direct pro-apoptotic effect is much less clear.

The picture is complicated by PKR- and RNAse L-induced pro-apoptotic signaling from dsRNA. If MDA5 is pro-apoptotic, it is likely that the signal it relays is from dsRNA and the pathway it will probably activate may depend on IPS-1. Two hypotheses arise. First, like many other apoptosis inducers (TNFα, PKR), MDA5 will likely initiate both pro- and anti-apoptotic signaling, the latter by virtue of NF-κB activation. Second, untangling the signaling pathways may require experiments in IFNαR–PKR–RNAse L-free background or control of variables in order to separate direct and IFN-dependent MDA5 effects.

What do we know about MDA5 and apoptosis, and what can we propose about their relationship? Initial analysis of MDA5-deficient mice argues against an essential function of MDA5 in the execution of the apoptotic program because relative numbers of T and other immune cell populations were normal and the mice appeared healthy.[145,146] However, it is unlikely that general apoptosis in mammalian development depends on dsRNA to any extent; induction of apoptosis by dsRNA is probably a specifically antiviral defense mechanism. Therefore we propose that an infected cell must make a decision. It can withstand the infection, steering toward survival and an antiviral battle, or it must die rapidly to limit viral replication. MDA5, as a key sensor of dsRNA, is a part of this decision. Other sensors like PKR may have input into the apoptotic part of the decision. MDA5 has critical input into antiviral gene expression and may be upstream of a direct pro-apoptotic pathway as well. We do not understand this process well, but its importance begs for experiments to elucidate the role of MDA5 in pro-apoptotic signaling pathways. The contribution of RIG-I is even more nebulous and should also be examined.

The importance of MDA5 and RIG-I in virus recognition places them in strategic positions. These sensors play roles in two critical decisions: self versus non-self (thus summoning the immune system), and cellular defense versus apoptosis. Understanding these decisions may afford us a clearer view of the immune system. The field is open for a variety of hypotheses and exciting experimentation. It will be interesting to discover activities of the two helicases regarding apoptosis and autoimmune reactions.

It is possible that their differences will illuminate their specificities and mechanisms of action.

SUMMARY: LOOKING TO THE FUTURE

The complexity of the dsRNA-induced signaling is evident. Numerous sensors coupled to multiple overlapping pathways, competing gene induction programs, temporal order of activity, and cell type specificities all combine to make dsRNA a challenging and exciting area of research. Taking the pathways apart and then intertwining them are important challenges for students of interferon and other transcriptional induction programs. A central piece of the puzzle recently fell into place when RIG-I and MDA5 were demonstrated to be key mammalian sensors of RNA viruses.

Based on painstaking research and a deluge of knockouts, the signaling pathways are becoming known. Does this mean that there is little point in continuing the search because twilight is descending on the era of great discoveries in innate immunity, at least related to intracellular antiviral defenses? We would like to suggest otherwise. It is likely that the evolutionary struggle of viruses with hosts is a very close contest. Removing significant contributors to the immune system will result in a critical tipping of the scales in favor of the viruses. Therefore, other contributors may be awaiting discovery. While dsRNA and RNA virus-mediated IFN-β induction appears reliant on the antiviral RNA helicases, other dsRNA-induced pathways are not, as may be expected from the amazing number of pathways turned on by dsRNA.

Regarding RNA viruses, we would like to propose that the two helicases may be threshold responders in that they sense viruses at low (albeit critical) titers or concentrations. It is possible that greater amounts of viruses will be recognized, not necessarily through dsRNA, but by different innate immune sensors, both known and unknown. Yet even a detailed understanding of these issues only scratches the surface of innate immunity, which involves induction of other cytokines, responses to diverse pathogens, specialized cells (such as macrophages and NK cells), and their interactions. We should also be concerned with the application of this knowledge to the clinic. The interesting era is just beginning.

REFERENCES

1. Isaacs, A. and J. Lindenmann, Virus interference. I. The interferon. *Proceedings of the Royal Society of London, Series B*, 1957. 147 (927): 258.
2. van Pesch, V. et al., Characterization of the murine alpha interferon gene family. *Journal of Virology*, 2004. 78 (15): 8219.
3. Der, S.D. et al., Identification of genes differentially regulated by interferon alpha, beta, or gamma using oligonucleotide arrays. *Proceedings of the National Academy of Sciences of the United States of America*, 1998. 95 (26): 15623.
4. Tough, D.F., Type I interferon as a link between innate and adaptive immunity through dendritic cell stimulation. *Leukemia & Lymphoma*, 2004. 45 (2): 257.
5. Kopp, E. and R. Medzhitov, Recognition of microbial infection by Toll-like receptors. *Current Opinion in Immunology*, 2003. 15 (4): 396.
6. Kumar, M. and G.G. Carmichael, Antisense RNA: function and fate of duplex RNA in cells of higher eukaryotes. *Microbiology & Molecular Biology Reviews*, 1998. 62 (4): 1415.

7. Tytell, A.A. et al., Inducers of interferon and host resistance, III. Double-stranded RNA from reovirus type 3 virions (Reo 3-RNA). *Proceedings of the National Academy of Sciences of the United States of America*, 1967. 58 (4): 1719.

8. Lampson, G. et al., Inducers of interferon and host resistance, I. Double-stranded RNA from extracts of *Penicillium funiculosum*. *Proceedings of the National Academy of Sciences of the United States of America*, 1967. 58 (2): 782.

9. Field, A.K. et al., Inducers of interferon and host resistance, II. Multistranded synthetic polynucleotide complexes. *Proceedings of the National Academy of Sciences of the United States of America*, 1967. 58 (3): 1004.

10. Field, A.K. et al., Inducers of interferon and host resistance, IV. Double-stranded replicative form RNA (MS2-RF-RNA) from *E. coli* infected with MS2 coliphage. *Proceedings of the National Academy of Sciences of the United States of America*, 1967. 58 (5): 2102.

11. Diebold, S.S. et al., Innate antiviral responses by means of TLR7-mediated recognition of single-stranded RNA. *Science*, 2004. 303 (5663): 1529.

12. Heil, F. et al., Species-specific recognition of single-stranded RNA via toll-like receptor 7 and 8. *Science*, 2004. 303 (5663): 1526.

13. Hemmi, H. et al., A Toll-like receptor recognizes bacterial DNA. [Erratum in *Nature* 2001 Feb 1; 409 (6820): 646]. *Nature*, 2000. 408 (6813): 740.

14. Lund, J.M. et al., Recognition of single-stranded RNA viruses by Toll-like receptor 7. *Proceedings of the National Academy of Sciences of the United States of America*, 2004. 101 (15): 5598.

15. Stetson, D.B. and R. Medzhitov, Recognition of cytosolic DNA activates an IRF3-dependent innate immune response. *Immunity*, 2006. 24 (1): 93.

16. Young, J.A.T., Virus entry and uncoating, in *Fields' Virology*, Knipe, D.M. et al., Eds. 2001, Lippincott Williams & Wilkins: Philadelphia. 87.

17. Jarrossay, D. et al., Specialization and complementarity in microbial molecule recognition by human myeloid and plasmacytoid dendritic cells. *European Journal of Immunology*, 2001. 31 (11): 3388.

18. Kadowaki, N. et al., Subsets of human dendritic cell precursors express different toll-like receptors and respond to different microbial antigens. *Journal of Experimental Medicine*, 2001. 194 (6): 863.

19. Maniatis, T., et al., Structure and function of the interferon-beta enhanceosome. *Cold Spring Harbor Symposia on Quantitative Biology*, 1998. 63: 609.

20. Peters, K.L. et al., IRF-3-dependent, NF-κB- and JNK-independent activation of the 561 and IFN-β genes in response to double-stranded RNA. *Proceedings of the National Academy of Sciences of the United States of America*, 2002. 99 (9): 6322.

21. Bocci, V., The physiological interferon response. *Immunology Today*, 1985. 6 (1): 7.

22. Gresser, I., Biologic effects of interferons. *Journal of Investigative Dermatology*, 1990. 95 (6 Suppl): 66S.

23. Taniguchi, T. and A. Takaoka, A weak signal for strong responses: interferon-alpha/beta revisited. *Nature Reviews Molecular Cell Biology*, 2001. 2 (5): 378.

24. Fujita, T. et al., Evidence for a nuclear factor(s), IRF-1, mediating induction and silencing properties to human IFN-β gene regulatory elements. *EMBO Journal*, 1988. 7 (11): 3397.

25. Miyamoto, M. et al., Regulated expression of a gene encoding a nuclear factor, IRF-1, that specifically binds to IFN-β gene regulatory elements. *Cell*, 1988. 54 (6): 903.

26. Honda, K. et al., IRF-7 is the master regulator of type-I interferon-dependent immune responses. *Nature*, 2005. 434 (7034): 772.

27. Matsuyama, T. et al., Targeted disruption of IRF-1 or IRF-2 results in abnormal type I IFN gene induction and aberrant lymphocyte development. *Cell*, 1993. 75 (1): 83.

28. Kimura, T. et al., Involvement of the IRF-1 transcription factor in antiviral responses to interferons. *Science*, 1994. 264 (5167): 1921.

29. Au, W.C. et al., Characterization of the interferon regulatory factor-7 and its potential role in the transcription activation of interferon A genes. *Journal of Biological Chemistry*, 1998. 273 (44): 29210.

30. Juang, Y.T., et al., Primary activation of interferon A and interferon B gene transcription by interferon regulatory factor 3. *Proceedings of the National Academy of Sciences of the United States of America*, 1998. 95 (17): 9837.

31. Lin, R. et al., Virus-dependent phosphorylation of the IRF-3 transcription factor regulates nuclear translocation, transactivation potential, and proteasome-mediated degradation. *Molecular & Cellular Biology*, 1998. 18 (5): 2986.

32. Sato, M. et al., Involvement of the IRF family transcription factor IRF-3 in virus-induced activation of the IFN-β gene. *FEBS Letters*, 1998. 425 (1): 112.

33. Yoneyama, M. et al., Direct triggering of the type I interferon system by virus infection: activation of a transcription factor complex containing IRF-3 and CBP/p300. *EMBO Journal*, 1998. 17 (4): 1087.

34. Marie, I., J.E. Durbin, and D.E. Levy, Differential viral induction of distinct interferon-alpha genes by positive feedback through interferon regulatory factor-7. *EMBO Journal*, 1998. 17 (22): 6660.

35. Sato, M., et al., Positive feedback regulation of type I IFN genes by the IFN-inducible transcription factor IRF-7. *FEBS Letters*, 1998. 441 (1): 106.

36. Levy, D.E., I. Marie, and A. Prakash, Ringing the interferon alarm: differential regulation of gene expression at the interface between innate and adaptive immunity. *Current Opinion in Immunology*, 2003. 15 (1): 52.

37. Sato, M. et al., Distinct and essential roles of transcription factors IRF-3 and IRF-7 in response to viruses for IFN-α/β gene induction. *Immunity*, 2000. 13 (4): 539.

38. Barchet, W. et al., Virus-induced interferon alpha production by a dendritic cell subset in the absence of feedback signaling in vivo. *Journal of Experimental Medicine*, 2002. 195 (4): 507.

39. Hata, N. et al., Constitutive IFN-α/β signal for efficient IFN-α/β gene induction by virus. *Biochemical & Biophysical Research Communications*, 2001. 285 (2): 518.

40. Prakash, A. et al., Tissue-specific positive feedback requirements for production of type I interferon following virus infection. *Journal of Biological Chemistry*, 2005. 280 (19): 18651.

41. Honda, K. et al., Role of a transductional-transcriptional processor complex involving MyD88 and IRF-7 in Toll-like receptor signaling. *Proceedings of the National Academy of Sciences of the United States of America*, 2004. 101 (43): 15416.

42. Kawai, T. et al., Interferon-alpha induction through Toll-like receptors involves a direct interaction of IRF7 with MyD88 and TRAF6. *Nature Immunology*, 2004. 5 (10): 1061.

43. Alexopoulou, L. et al., Recognition of double-stranded RNA and activation of NF-κB by Toll-like receptor 3. *Nature*, 2001. 413 (6857): 732.

44. Hoebe, K. et al., Identification of Lps2 as a key transducer of MyD88-independent TIR signalling. *Nature*, 2003. 424 (6950): 743.

45. Oshiumi, H. et al., TICAM-1, an adaptor molecule that participates in Toll-like receptor 3-mediated interferon-beta induction. *Nature Immunology*, 2003. 4 (2): 161.

46. Yamamoto, M. et al., Role of adaptor TRIF in the MyD88-independent toll-like receptor signaling pathway. *Science*, 2003. 301 (5633): 640.

47. Yamamoto, M. et al., Cutting edge: a novel Toll/IL-1 receptor domain-containing adapter that preferentially activates the IFN-β promoter in the Toll-like receptor signaling. *Journal of Immunology*, 2002. 169 (12): 6668.

48. Fitzgerald, K.A. et al., LPS-TLR4 signaling to IRF-3/7 and NF-κB involves the toll adapters TRAM and TRIF. [Erratum in *Journal of Experimental Medicine*, 2003. 198 (9): 1450]. *Journal of Experimental Medicine*, 2003. 198 (7): 1043.

49. Yamamoto, M. et al., TRAM is specifically involved in the Toll-like receptor 4-mediated MyD88-independent signaling pathway. *Nature Immunology*, 2003. 4 (11): 1144.

50. Sakaguchi, S. et al., Essential role of IRF-3 in lipopolysaccharide-induced interferon-beta gene expression and endotoxin shock. *Biochemical & Biophysical Research Communications*, 2003. 306 (4): 860.

51. Weaver, B.K., K. Kumar, and N.C. Reich, Interferon regulatory factor 3 and CREB-binding protein/p300 are subunits of double-stranded RNA-activated transcription factor DRAF1. *Molecular & Cellular Biology*, 1998. 18 (3): 1359.

52. Iordanov, M.S. et al., Activation of NF-κB by double-stranded RNA (dsRNA) in the absence of protein kinase R and RNase L demonstrates the existence of two separate dsRNA-triggered antiviral programs. *Molecular & Cellular Biology*, 2001. 21 (1): 61.

53. Thanos, D. and T. Maniatis, Identification of the rel family members required for virus induction of human beta interferon gene. *Molecular & Cellular Biology*, 1995. 15 (1): 152.

54. Chu, W.M. et al., JNK2 and IKK-β are required for activating the innate response to viral infection. *Immunity*, 1999. 11 (6): 721.

55. Kucharczak, J. et al., To be, or not to be: NF-κB is the answer: role of Rel/NF-κB in the regulation of apoptosis. [Erratum in *Oncogene*. 2004. 23 (54): 8858]. *Oncogene*, 2003. 22 (56): 8961.

56. Steer, S.A. et al., Role of MAPK in the regulation of double-stranded RNA- and encephalomyocarditis virus-induced cyclooxygenase-2 expression by macrophages. *Journal of Immunology*, 2006. 177 (5): 3413.

57. Iordanov, M.S. et al., Activation of p38 mitogen-activated protein kinase and c-Jun NH(2)-terminal kinase by double-stranded RNA and encephalomyocarditis virus: involvement of RNase L, protein kinase R, and alternative pathways. *Molecular & Cellular Biology*, 2000. 20 (2): 617.

58. Yang, Y.L. et al., Deficient signaling in mice devoid of double-stranded RNA-dependent protein kinase. *EMBO Journal*, 1995. 14 (24): 6095.

59. Marcus, I. and M.J. Sekellick, Defective interfering particles with covalently linked [+/–]RNA induce interferon. *Nature*, 1977. 266 (5605): 815.

60. Koch, F. and G. Koch, *The Molecular Biology of Poliovirus* 1985, New York: Springer.

61. Schwartz, M. et al., A positive-strand RNA virus replication complex parallels form and function of retrovirus capsids. *Molecular Cell*, 2002. 9 (3): 505.

62. Tao, Y. et al., RNA synthesis in a cage—structural studies of reovirus polymerase lambda-3. *Cell*, 2002. 111 (5): 733.

63. Wathelet, M.G. et al., Regulation of two interferon-inducible human genes by interferon, poly(rI).poly(rC) and viruses. *European Journal of Biochemistry*, 1988. 174 (2): 323.

64. Bandyopadhyay, S.K. et al., Transcriptional induction by double-stranded RNA is mediated by interferon-stimulated response elements without activation of interferon-stimulated gene factor 3. *Journal of Biological Chemistry*, 1995. 270 (33): 19624.

65. Geiss, G. et al., A comprehensive view of regulation of gene expression by double-stranded RNA-mediated cell signaling. *Journal of Biological Chemistry*, 2001. 276 (32): 30178.

66. Arnott, S., D.W. Hukins, and S.D. Dover, Optimised parameters for RNA double-helices. *Biochemical & Biophysical Research Communications*, 1972. 48 (6): 1392.

67. Ehrenfeld, E. and T. Hunt, Double-stranded poliovirus RNA inhibits initiation of protein synthesis by reticulocyte lysates. *Proceedings of the National Academy of Sciences of the United States of America*, 1971. 68 (5): 1075.

68. Kerr, I.M. and R.E. Brown, pppA2'p5'A2'p5'A: an inhibitor of protein synthesis synthesized with an enzyme fraction from interferon-treated cells. *Proceedings of the National Academy of Sciences of the United States of America*, 1978. 75 (1): 256.

69. Lebleu, B. et al., Interferon, double-stranded RNA, and protein phosphorylation. *Proceedings of the National Academy of Sciences of the United States of America*, 1976. 73 (9): 3107.

70. Williams, B.R., PKR; a sentinel kinase for cellular stress. *Oncogene*, 1999. 18 (45): 6112.

71. Dar, A.C., T.E. Dever, and F. Sicheri, Higher-order substrate recognition of eIF2-alpha by the RNA-dependent protein kinase PKR. *Cell*, 2005. 122 (6): 887.

72. Williams, B.R.G., The role of the dsRNA-activated kinase, PKR, in signal transduction. *Seminars in Virology*, 1995. 6: 191.

73. Kumar, A. et al., Double-stranded RNA-dependent protein kinase activates transcription factor NF-κ-B by phosphorylating I-κ-B. *Proceedings of the National Academy of Sciences of the United States of America*, 1994. 91 (14): 6288.

74. Maran, A. et al., Blockage of NF-κ-B signaling by selective ablation of an mRNA target by 2-5A antisense chimeras. *Science*, 1994. 265 (5173): 789.

75. Zinn, K. et al., 2-Aminopurine selectively inhibits the induction of beta-interferon, c-fos, and c-myc gene expression. *Science*, 1988. 240 (4849): 210.

76. Sugiyama, T. et al., 2-aminopurine inhibits lipopolysaccharide-induced nitric oxide production by preventing IFN-β production. *Microbiology & Immunology*, 2004. 48 (12): 957.

77. Zamanian-Daryoush, M. et al., NF-κ-B activation by double-stranded-RNA-activated protein kinase (PKR) is mediated through NF-κ-B-inducing kinase and I-κ-B kinase. *Molecular & Cellular Biology*, 2000. 20 (4): 1278.

78. Abraham, N. et al., Characterization of transgenic mice with targeted disruption of the catalytic domain of the double-stranded RNA-dependent protein kinase, PKR. *Journal of Biological Chemistry*, 1999. 274 (9): 5953.

79. Baltzis, D., S. Li, and A.E. Koromilas, Functional characterization of PKR gene products expressed in cells from mice with a targeted deletion of the N terminus or C terminus domain of PKR. *Journal of Biological Chemistry*, 2002. 277 (41): 38364.

80. Bonnet, M.C. et al., The N-terminus of PKR is responsible for the activation of the NF-[kappa]B signaling pathway by interacting with the IKK complex. *Cellular Signalling*, 2006. 18 (11): 1865.

81. Bonnet, M.C., et al., PKR stimulates NF-kappaB irrespective of its kinase function by interacting with the I-κ-B kinase complex. *Molecular & Cellular Biology*, 2000. 20 (13): 4532.

82. Gil, J., J. Alcami, and M. Esteban, Activation of NF-kappa B by the dsRNA-dependent protein kinase, PKR involves the I κ B kinase complex. *Oncogene*, 2000. 19 (11): 1369.

83. Ishii, T. et al., Activation of the I κ B alpha kinase (IKK) complex by double-stranded RNA-binding defective and catalytic inactive mutants of the interferon-inducible protein kinase PKR. *Oncogene*, 2001. 20 (15): 1900.

84. Gil, J., et al., TRAF family proteins link PKR with NF-κ B activation. *Molecular & Cellular Biology*, 2004. 24 (10): 4502.

85. Donze, O. et al., The protein kinase PKR: a molecular clock that sequentially activates survival and death programs. *EMBO Journal*, 2004. 23 (3): 564.

86. Malathi, K. et al., A transcriptional signaling pathway in the IFN system mediated by 2'-5'-oligoadenylate activation of RNase L. *Proceedings of the National Academy of Sciences of the United States of America*, 2005. 102 (41): 14533.

87. Brown, G.E. et al., Increased endonuclease activity in an extract from mouse Ehrlich ascites tumor cells which had been treated with a partially purified interferon preparation: dependence of double-stranded RNA. *Biochemical & Biophysical Research Communications*, 1976. 69 (1): 114.

88. Wreschner, D.H. et al., Ribosomal RNA cleavage, nuclease activation and 2-5A(ppp(A2′p) nA) in interferon-treated cells. *Nucleic Acids Research*, 1981. 9 (7): 1571.

89. Meylan, E. et al., RIP1 is an essential mediator of Toll-like receptor 3-induced NF-κ B activation. *Nature Immunology*, 2004. 5 (5): 503.

90. Oganesyan, G. et al., Critical role of TRAF3 in the Toll-like receptor-dependent and -independent antiviral response. *Nature*, 2006. 439 (7073): 208.

91. Edelmann, K.H. et al., Does Toll-like receptor 3 play a biological role in virus infections? *Virology*, 2004. 322 (2): 231.

92. Schulz, O. et al., Toll-like receptor 3 promotes cross-priming to virus-infected cells. *Nature*, 2005. 433 (7028): 887.

93. Meister, G. and T. Tuschl, Mechanisms of gene silencing by double-stranded RNA. *Nature*, 2004. 431 (7006): 343.

94. Cullen, B.R., Is RNA interference involved in intrinsic antiviral immunity in mammals? *Nature Immunology*, 2006. 7 (6): 563.

95. Zhang, H. et al., Human Dicer preferentially cleaves dsRNAs at their termini without a requirement for AT *EMBO Journal*, 2002. 21 (21): 5875.

96. Tabara, H. et al., The dsRNA binding protein RDE-4 interacts with RDE-1, DCR-1, and a DExH-box helicase to direct RNAi in *C. elegans*. *Cell*, 2002. 109 (7): 861.

97. Kang, D.C. et al., MDA5: An interferon-inducible putative RNA helicase with double-stranded RNA-dependent ATPase activity and melanoma growth-suppressive properties. *Proceedings of the National Academy of Sciences of the United States of America*, 2002. 99 (2): 637.

98. Zhang, X. et al., An RNA helicase, RHIV -1, induced by porcine reproductive and respiratory syndrome virus (PRRSV) is mapped on porcine chromosome 10q13. *Microbial Pathogenesis*, 2000. 28 (5): 267.

99. Kovacsovics, M. et al., Overexpression of helicard, a CARD-containing helicase cleaved during apoptosis, accelerates DNA degradation. [Erratum in *Current Biology*, 2002. 12 (18): 1633.]. *Current Biology*, 2002. 12 (10): 838.

100. Yoneyama, M. et al., The RNA helicase RIG-I has an essential function in double-stranded RNA-induced innate antiviral responses. *Nature Immunology*, 2004. 5 (7): 730.

101. Imaizumi, T. et al., Interferon-gamma induces retinoic acid-inducible gene-I in endothelial cells. *Endothelium: Journal of Endothelial Cell Research*, 2004. 11 (3-4): 169.

102. Jiang, H. and B. Fisher, Use of a sensitive and efficient subtraction hybridization protocol for the identification of genes differentially regulated during the induction of differentiation in human melanoma cells. *Molecular & Cellular Differentiation*, 1993. 1: 285.

103. Sun, Y.W., RIG-I, a human homolog gene of RNA helicase, is induced by retinoic acid during the differentiation of acute promyelocytic leukemia cell, 1997. Shanghai Second Medical University: Shanghai.

104. Cui, X.F. et al., Retinoic acid-inducible gene-I is induced by interferon-gamma and regulates the expression of interferon-gamma stimulated gene 15 in MCF-7 cells. *Biochemistry & Cell Biology*, 2004. 82 (3): 401.

105. Imaizumi, T. et al., Retinoic acid-inducible gene-I is induced in endothelial cells by LPS and regulates expression of COX-2. *Biochemical & Biophysical Research Communications*, 2002. 292 (1): 274.

106. Kang, D.C. et al., Expression analysis and genomic characterization of human mela-noma differentiation associated gene-5, MDA5: a novel type I interferon-responsive apoptosis-inducing gene. *Oncogene*, 2004. 23 (9): 1789.

107. Yoneyama, M. et al., Shared and unique functions of the DExD/H-box helicases RIG-I, MDA5, and LGP2 in antiviral innate immunity. *Journal of Immunology*, 2005. 175 (5): 2851.

108. Hornung, V. et al., 5'-Triphosphate RNA is ligand for RIG-I. *Science*, 2006. 314 (5801): 994.

109. Pichlmair, A. et al., RIG-I-mediated antiviral responses to single-stranded RNA Bear-ing 5'-phosphates. *Science*, 2006. 314 (5801): 997.

110. Gorbalenya, A.E. and E.V. Koonin, Helicases: amino acid sequence comparisons and structure-function relationships. *Current Opinion in Structural Biology*, 1993. 3 (3): 419.

111. Walker, J.E. et al., Distantly related sequences in the alpha and beta subunits of ATP synthase, myosin, kinases and other ATP-requiring enzymes and a common nucleotide binding fold. *EMBO Journal*, 1982. 1 (8): 945.

112. Cordin, O. et al., The DEAD-box protein family of RNA helicases. *Gene*, 2006. 367: 17.

113. Hirling, H. et al., RNA helicase activity associated with the human p68 protein. *Nature*, 1989. 339 (6225): 562.

114. Lorsch, J.R. and D. Herschlag, The DEAD box protein eIF4A. 1. A minimal kinetic and thermodynamic framework reveals coupled binding of RNA and nucleotide. *Biochem-istry*, 1998. 37 (8): 2180.

115. Andersen, C.B. et al., Structure of the exon junction core complex with a trapped DEAD-box ATPase bound to RNA. *Science*, 2006. 313 (5795): 1968.

116. Bono, F. et al., The crystal structure of the exon junction complex reveals how it main-tains a stable grip on mRNA. *Cell*, 2006. 126 (4): 713.

117. Sengoku, T. et al., Structural basis for RNA unwinding by the DEAD-box protein Dros-ophila Vasa. *Cell*, 2006. 125 (2): 2870.

118. Weber, C.H. and C. Vincenz, The death domain superfamily: a tale of two interfaces? *Trends in Biochemical Sciences*, 2001. 26 (8): 475.

119. Riedl, S.J. et al., Structure of the apoptotic protease-activating factor 1 bound to AD *Nature*, 2005. 434 (7035): 926.

120. Rodriguez, J. and Y. Lazebnik, Caspase-9 and APAF-1 form an active holoenzyme. *Genes & Development*, 1999. 13 (24): 3179.

121. Yu, X. et al., Structure of human apoptosome at 12.8 Å resolution provides insights into cell death platform structure, 2005. 13 (11): 17255

122. Andrejeva, J. et al., The V proteins of paramyxoviruses bind the IFN-inducible RNA helicase, MDA5, and inhibit its activation of the IFN-β promoter. *Proceedings of the National Academy of Sciences of the United States of America*, 2004. 101 (49): 17264.

123. Sumpter, R., Jr. et al., Regulating intracellular antiviral defense and permissiveness to hepatitis C virus RNA replication through a cellular RNA helicase, RIG-I. *Journal of Virology*, 2005. 79 (5): 2689.

124. Blight, K.J., J.A. McKeating, and C.M. Rice, Highly permissive cell lines for subge-nomic and genomic hepatitis C virus RNA replication. *Journal of Virology*, 2002. 76 (24): 13001.

125. Breiman, A. et al., Inhibition of RIG-I-dependent signaling to the interferon pathway during hepatitis C virus expression and restoration of signaling by IKK-ε. *Journal of Virology*, 2005. 79 (7): 3969.

126. Foy, E. et al., Control of antiviral defenses through hepatitis C virus disruption of retin-oic acid-inducible gene-I signaling. *Proceedings of the National Academy of Sciences of the United States of America*, 2005. 102 (8): 2986.

127. Chang, T.H., C.L. Liao, and Y.L. Lin, Flavivirus induces interferon-beta gene expression through a pathway involving RIG-I-dependent IRF-3 and PI3K-dependent NF-κB activation. *Microbes & Infection*, 2006. 8 (1): 157.
128. Melchjorsen, J. et al., Activation of innate defense against a paramyxovirus is mediated by RIG-I and TLR7 and TLR8 in a cell-type-specific manner. *Journal of Virology*, 2005. 79 (20): 12944.
129. Berghall, H. et al., The interferon-inducible RNA helicase, MDA5, is involved in measles virus-induced expression of antiviral cytokines. *Microbes & Infection*, 2006. 8 (8): 2138.
130. Matikainen, S. et al., Tumor necrosis factor alpha enhances influenza A virus-induced expression of antiviral cytokines by activating RIG-I gene expression. *Journal of Virology*, 2006. 80 (7): 3515.
131. Siren, J. et al., Retinoic acid inducible gene-I and MDA5 are involved in influenza A virus-induced expression of antiviral cytokines. *Microbes & Infection*, 2006. 8 (8): 2013.
132. Kawai, T. et al., IPS-1, an adaptor triggering RIG-I- and Mda5-mediated type I interferon induction. *Nature Immunology*, 2005. 6 (10): 981.
133. Meylan, E. et al., Cardif is an adaptor protein in the RIG-I antiviral pathway and is targeted by hepatitis C virus. *Nature*, 2005. 437 (7062): 1167.
134. Seth, R.B. et al., Identification and characterization of MAVS, a mitochondrial antiviral signaling protein that activates NF-κ-B and IRF 3. *Cell*, 2005. 122 (5): 669.
135. Xu, L.G. et al., VISA is an adapter protein required for virus-triggered IFN-beta signaling. *Molecular Cell*, 2005. 19 (6): 727.
136. Balachandran, S., E. Thomas, and G.N. Barber, A FADD-dependent innate immune mechanism in mammalian cells. *Nature*, 2004. 432 (7015): 401.
137. Saha, S.K. et al., Regulation of antiviral responses by a direct and specific interaction between TRAF3 and Cardif. *EMBO Journal*, 2006. 25 (14): 3257.
138. Hu, W.H., H. Johnson, and H.B. Shu, Activation of NF-κB by FADD, Casper, and caspase-8. *Journal of Biological Chemistry*, 2000. 275 (15): 10838.
139. Fitzgerald, K.A. et al., IKK-epsilon and TBK1 are essential components of the IRF3 signaling pathway. *Nature Immunology*, 2003. 4 (5): 491.
140. Hemmi, H. et al., The roles of two I κ B kinase-related kinases in lipopolysaccharide and double stranded RNA signaling and viral infection. *Journal of Experimental Medicine*, 2004. 199 (12): 1641.
141. McWhirter, S.M. et al., IFN-regulatory factor 3-dependent gene expression is defective in Tbk1-deficient mouse embryonic fibroblasts. *Proceedings of the National Academy of Sciences of the United States of America*, 2004. 101 (1): 233.
142. Sharma, S. et al., Triggering the interferon antiviral response through an IKK-related pathway. *Science*, 2003. 300 (5622): 1148.
143. Muller, U. et al., Functional role of type I and type II interferons in antiviral defense. *Science*, 1994. 264 (5167): 1918.
144. Kato, H. et al., Cell type-specific involvement of RIG-I in antiviral response. *Immunity*, 2005. 23 (1): 19.
145. Kato, H., et al., Differential roles of MDA5 and RIG-I helicases in the recognition of RNA viruses. *Nature*, 2006. 441 (7089): 101.
146. Gitlin, L. et al., Essential role of MDA5 in type I IFN responses to polyriboinosinic:polyribocytidylic acid and encephalomyocarditis picornavirus. *Proceedings of the National Academy of Sciences of the United States of America*, 2006. 103 (22): 8459.
147. Ryman, K.D. et al., Alpha/beta interferon protects adult mice from fatal Sindbis virus infection and is an important determinant of cell and tissue tropism. *Journal of Virology*, 2000. 74 (7): 3366.
148. Kumar, H. et al., Essential role of IPS-1 in innate immune responses against RNA viruses. *Journal of Experimental Medicine*, 2006. 203 (7): 1795.

149. Sun, Q. et al., The specific and essential role of MAVS in antiviral innate immune responses. *Immunity*, 2006. 24 (5): 633.
150. Cheng, G., J. Zhong, and F.V. Chisari, Inhibition of dsRNA-induced signaling in hepatitis C virus-infected cells by NS3 protease-dependent and -independent mechanisms. *Proceedings of the National Academy of Sciences of the United States of America*, 2006. 103 (22): 8499.
151. Lin, R. et al., Dissociation of a MAVS/IPS-1/VISA/Cardif-IKKε molecular complex from the mitochondrial outer membrane by hepatitis C virus NS3-4A proteolytic cleavage. *Journal of Virology*, 2006. 80 (12): 6072.
152. Loo, Y.M. et al., Viral and therapeutic control of IFN-β promoter stimulator 1 during hepatitis C virus infection. *Proceedings of the National Academy of Sciences of the United States of America*, 2006. 103 (15): 6001.
153. Ferreon, J.C. et al., Molecular determinants of TRIF proteolysis mediated by the hepatitis C virus NS3/4A protease. *Journal of Biological Chemistry*, 2005. 280 (21): 20483.
154. Li, K. et al., Immune evasion by hepatitis C virus NS3/4A protease-mediated cleavage of the Toll-like receptor 3 adaptor protein TRIF. *Proceedings of the National Academy of Sciences of the United States of America*, 2005. 102 (8): 2992.
155. Fensterl, V. et al., Hepatitis A virus suppresses RIG-I-mediated IRF-3 activation to block induction of beta interferon. *Journal of Virology*, 2005. 79 (17): 10968.
156. Bergmann, M. et al., Influenza virus NS1 protein counteracts PKR-mediated inhibition of replication. *Journal of Virology*, 2000. 74 (13): 6203.
157. Hatada, E., S. Saito, and R. Fukuda, Mutant influenza viruses with a defective NS1 protein cannot block activation of PKR in infected cells. *Journal of Virology*, 1999. 73 (3): 2425.
158. Lu, Y. et al., Binding of the influenza virus NS1 protein to double-stranded RNA inhibits the activation of the protein kinase that phosphorylates the eIF-2 translation initiation factor. *Virology*, 1995. 214 (1): 222.
159. Fredericksen, B.L. and M. Gale, Jr., West Nile virus evades activation of interferon regulatory factor 3 through RIG-I-dependent and -independent pathways without antagonizing host defense signaling. *Journal of Virology*, 2006. 80 (6): 2913.
160. Weber, F., G. Kochs, and O. Haller, Inverse interference: how viruses fight the interferon system. *Viral Immunology*, 2004. 17 (4): 498.
161. Cui, Y. et al., The Stat3/5 locus encodes novel endoplasmic reticulum and helicase-like proteins that are preferentially expressed in normal and neoplastic mammary tissue. *Genomics*, 2001. 78 (3): 129.
162. Rothenfusser, S. et al., The RNA helicase Lgp2 inhibits TLR-independent sensing of viral replication by retinoic acid-inducible gene-I. *Journal of Immunology*, 2005. 175 (8): 5260.
163. Komuro, A. and C.M. Horvath, RNA and virus-independent inhibition of antiviral signaling by RNA helicase LGP2. *Journal of Virology*, 2006.
164. Harada, H. et al., Structurally similar but functionally distinct factors, IRF-1 and IRF-2, bind to the same regulatory elements of IFN and IFN-inducible genes. *Cell*, 1989. 58 (4): 729.
165. Senger, K. et al., Gene repression by coactivator repulsion. *Molecular Cell*, 2000. 6 (4): 931.
166. Hida, S. et al., CD8(+) T cell-mediated skin disease in mice lacking IRF-2, the transcriptional attenuator of interferon-alpha/beta signaling. *Immunity*, 2000. 13 (5): 643.
167. Blanco, J.C. et al., Interferon regulatory factor (IRF)-1 and IRF-2 regulate interferon gamma-dependent cyclooxygenase 2 expression. *Journal of Experimental Medicine*, 2000. 191 (12): 2131.
168. Huang, J. et al., SIKE is an IKK ε/TBK1-associated suppressor of TLR3- and virus-triggered IRF-3 activation pathways. *EMBO Journal*, 2005. 24 (23): 4018.

169. Lin, R. et al., Negative regulation of the retinoic acid-inducible gene I-induced antiviral state by the ubiquitin-editing protein A20. *Journal of Biological Chemistry*, 2006. 281 (4): 2095.

170. Marques, J.T. et al., Structural basis for discriminating between self and nonself double-stranded RNAs in mammalian cells. *Nature Biotechnology*, 2006. 24 (5): 559.

171. Kim, D.H. et al., Interferon induction by siRNAs and ssRNAs synthesized by phage polymerase. *Nature Biotechnology*, 2004. 22 (3): 321.

172. Smyth, D.J. et al., A genome-wide association study of nonsynonymous SNPs identifies a type 1 diabetes locus in the interferon-induced helicase (IFIH1) region. *Nature Genetics*, 2006. 38 (6): 617.

173. Todd, J.A. and L.S. Wicker, Genetic protection from the inflammatory disease type 1 diabetes in humans and animal models. *Immunity*, 2001. 15 (3): 387.

174. Bach, J.F., Infections and autoimmune diseases. *Journal of Autoimmunity*, 2005. 25 Suppl: 74.

Index